T0296674

LONDON MATHEMATICAL SOCIETY LECTURE NOTE SERIES

Managing Editor: Professor M. Reid, Mathematics Institute,
University of Warwick, Coventry CV4 7AL, United Kingdom

The titles below are available from booksellers, or from Cambridge University Press at
http://www.cambridge.org/mathematics

340 Groups St Andrews 2005 II, C.M. CAMPBELL, M.R. QUICK, E.F. ROBERTSON & G.C. SMITH (eds)
341 Ranks of elliptic curves and random matrix theory, J.B. CONREY, D.W. FARMER, F. MEZZADRI & N.C. SNAITH (eds)
342 Elliptic cohomology, H.R. MILLER & D.C. RAVENEL (eds)
343 Algebraic cycles and motives I, J. NAGEL & C. PETERS (eds)
344 Algebraic cycles and motives II, J. NAGEL & C. PETERS (eds)
345 Algebraic and analytic geometry, A. NEEMAN
346 Surveys in combinatorics 2007, A. HILTON & J. TALBOT (eds)
347 Surveys in contemporary mathematics, N. YOUNG & Y. CHOI (eds)
348 Transcendental dynamics and complex analysis, P.J. RIPPON & G.M. STALLARD (eds)
349 Model theory with applications to algebra and analysis I, Z. CHATZIDAKIS, D. MACPHERSON, A. PILLAY & A. WILKIE (eds)
350 Model theory with applications to algebra and analysis II, Z. CHATZIDAKIS, D. MACPHERSON, A. PILLAY & A. WILKIE (eds)
351 Finite von Neumann algebras and masas, A.M. SINCLAIR & R.R. SMITH
352 Number theory and polynomials, J. MCKEE & C. SMYTH (eds)
353 Trends in stochastic analysis, J. BLATH, P. MÖRTERS & M. SCHEUTZOW (eds)
354 Groups and analysis, K. TENT (ed)
355 Non-equilibrium statistical mechanics and turbulence, J. CARDY, G. FALKOVICH & K. GAWEDZKI
356 Elliptic curves and big Galois representations, D. DELBOURGO
357 Algebraic theory of differential equations, M.A.H. MACCALLUM & A.V. MIKHAILOV (eds)
358 Geometric and cohomological methods in group theory, M.R. BRIDSON, P.H. KROPHOLLER & I.J. LEARY (eds)
359 Moduli spaces and vector bundles, L. BRAMBILA-PAZ, S.B. BRADLOW, O. GARCÍA-PRADA & S. RAMANAN (eds)
360 Zariski geometries, B. ZILBER
361 Words: Notes on verbal width in groups, D. SEGAL
362 Differential tensor algebras and their module categories, R. BAUTISTA, L. SALMERÓN & R. ZUAZUA
363 Foundations of computational mathematics, Hong Kong 2008, F. CUCKER, A. PINKUS & M.J. TODD (eds)
364 Partial differential equations and fluid mechanics, J.C. ROBINSON & J.L. RODRIGO (eds)
365 Surveys in combinatorics 2009, S. HUCZYNSKA, J.D. MITCHELL & C.M. RONEY-DOUGAL (eds)
366 Highly oscillatory problems, B. ENGQUIST, A. FOKAS, E. HAIRER & A. ISERLES (eds)
367 Random matrices: High dimensional phenomena, G. BLOWER
368 Geometry of Riemann surfaces, F.P. GARDINER, G. GONZÁLEZ-DIEZ & C. KOUROUNIOTIS (eds)
369 Epidemics and rumours in complex networks, M. DRAIEF & L. MASSOULIÉ
370 Theory of p-adic distributions, S. ALBEVERIO, A.YU. KHRENNIKOV & V.M. SHELKOVICH
371 Conformal fractals, F. PRZYTYCKI & M. URBAŃSKI
372 Moonshine: The first quarter century and beyond, J. LEPOWSKY, J. MCKAY & M.P. TUITE (eds)
373 Smoothness, regularity and complete intersection, J. MAJADAS & A. G. RODICIO
374 Geometric analysis of hyperbolic differential equations: An introduction, S. ALINHAC
375 Triangulated categories, T. HOLM, P. JØRGENSEN & R. ROUQUIER (eds)
376 Permutation patterns, S. LINTON, N. RUŠKUC & V. VATTER (eds)
377 An introduction to Galois cohomology and its applications, G. BERHUY
378 Probability and mathematical genetics, N. H. BINGHAM & C. M. GOLDIE (eds)
379 Finite and algorithmic model theory, J. ESPARZA, C. MICHAUX & C. STEINHORN (eds)
380 Real and complex singularities, M. MANOEL, M.C. ROMERO FUSTER & C.T.C WALL (eds)
381 Symmetries and integrability of difference equations, D. LEVI, P. OLVER, Z. THOMOVA & P. WINTERNITZ (eds)
382 Forcing with random variables and proof complexity, J. KRAJÍČEK
383 Motivic integration and its interactions with model theory and non-Archimedean geometry I, R. CLUCKERS, J. NICAISE & J. SEBAG (eds)
384 Motivic integration and its interactions with model theory and non-Archimedean geometry II, R. CLUCKERS, J. NICAISE & J. SEBAG (eds)
385 Entropy of hidden Markov processes and connections to dynamical systems, B. MARCUS, K. PETERSEN & T. WEISSMAN (eds)
386 Independence-friendly logic, A.L. MANN, G. SANDU & M. SEVENSTER
387 Groups St Andrews 2009 in Bath I, C.M. CAMPBELL *et al* (eds)
388 Groups St Andrews 2009 in Bath II, C.M. CAMPBELL *et al* (eds)
389 Random fields on the sphere, D. MARINUCCI & G. PECCATI
390 Localization in periodic potentials, D.E. PELINOVSKY

391 Fusion systems in algebra and topology, M. ASCHBACHER, R. KESSAR & B. OLIVER
392 Surveys in combinatorics 2011, R. CHAPMAN (ed)
393 Non-abelian fundamental groups and Iwasawa theory, J. COATES *et al* (eds)
394 Variational problems in differential geometry, R. BIELAWSKI, K. HOUSTON & M. SPEIGHT (eds)
395 How groups grow, A. MANN
396 Arithmetic differential operators over the p-adic integers, C.C. RALPH & S.R. SIMANCA
397 Hyperbolic geometry and applications in quantum chaos and cosmology, J. BOLTE & F. STEINER (eds)
398 Mathematical models in contact mechanics, M. SOFONEA & A. MATEI
399 Circuit double cover of graphs, C.-Q. ZHANG
400 Dense sphere packings: a blueprint for formal proofs, T. HALES
401 A double Hall algebra approach to affine quantum Schur–Weyl theory, B. DENG, J. DU & Q. FU
402 Mathematical aspects of fluid mechanics, J.C. ROBINSON, J.L. RODRIGO & W. SADOWSKI (eds)
403 Foundations of computational mathematics, Budapest 2011, F. CUCKER, T. KRICK, A. PINKUS & A. SZANTO (eds)
404 Operator methods for boundary value problems, S. HASSI, H.S.V. DE SNOO & F.H. SZAFRANIEC (eds)
405 Torsors, étale homotopy and applications to rational points, A.N. SKOROBOGATOV (ed)
406 Appalachian set theory, J. CUMMINGS & E. SCHIMMERLING (eds)
407 The maximal subgroups of the low-dimensional finite classical groups, J.N. BRAY, D.F. HOLT & C.M. RONEY-DOUGAL
408 Complexity science: the Warwick master's course, R. BALL, V. KOLOKOLTSOV & R.S. MACKAY (eds)
409 Surveys in combinatorics 2013, S.R. BLACKBURN, S. GERKE & M. WILDON (eds)
410 Representation theory and harmonic analysis of wreath products of finite groups, T. CECCHERINI-SILBERSTEIN, F. SCARABOTTI & F. TOLLI
411 Moduli spaces, L. BRAMBILA-PAZ, O. GARCÍA-PRADA, P. NEWSTEAD & R.P. THOMAS (eds)
412 Automorphisms and equivalence relations in topological dynamics, D.B. ELLIS & R. ELLIS
413 Optimal transportation, Y. OLLIVIER, H. PAJOT & C. VILLANI (eds)
414 Automorphic forms and Galois representations I, F. DIAMOND, P.L. KASSAEI & M. KIM (eds)
415 Automorphic forms and Galois representations II, F. DIAMOND, P.L. KASSAEI & M. KIM (eds)
416 Reversibility in dynamics and group theory, A.G. O'FARRELL & I. SHORT
417 Recent advances in algebraic geometry, C.D. HACON, M. MUSTAŢĂ & M. POPA (eds)
418 The Bloch–Kato conjecture for the Riemann zeta function, J. COATES, A. RAGHURAM, A. SAIKIA & R. SUJATHA (eds)
419 The Cauchy problem for non-Lipschitz semi-linear parabolic partial differential equations, J.C. MEYER & D.J. NEEDHAM
420 Arithmetic and geometry, L. DIEULEFAIT *et al* (eds)
421 O-minimality and Diophantine geometry, G.O. JONES & A.J. WILKIE (eds)
422 Groups St Andrews 2013, C.M. CAMPBELL *et al* (eds)
423 Inequalities for graph eigenvalues, Z. STANIĆ
424 Surveys in combinatorics 2015, A. CZUMAJ *et al* (eds)
425 Geometry, topology and dynamics in negative curvature, C.S. ARAVINDA, F.T. FARRELL & J.-F. LAFONT (eds)
426 Lectures on the theory of water waves, T. BRIDGES, M. GROVES & D. NICHOLLS (eds)
427 Recent advances in Hodge theory, M. KERR & G. PEARLSTEIN (eds)
428 Geometry in a Fréchet context, C. T. J. DODSON, G. GALANIS & E. VASSILIOU
429 Sheaves and functions modulo p, L. TAELMAN
430 Recent progress in the theory of the Euler and Navier–Stokes equations, J.C. ROBINSON, J.L. RODRIGO, W. SADOWSKI & A. VIDAL-LÓPEZ (eds)
431 Harmonic and subharmonic function theory on the real hyperbolic ball, M. STOLL
432 Topics in graph automorphisms and reconstruction (2nd Edition), J. LAURI & R. SCAPELLATO
433 Regular and irregular holonomic D-modules, M. KASHIWARA & P. SCHAPIRA
434 Analytic semigroups and semilinear initial boundary value problems (2nd Edition), K. TAIRA
435 Graded rings and graded Grothendieck groups, R. HAZRAT
436 Groups, graphs and random walks, T. CECCHERINI-SILBERSTEIN, M. SALVATORI & E. SAVA-HUSS (eds)
437 Dynamics and analytic number theory, D. BADZIAHIN, A. GORODNIK & N. PEYERIMHOFF (eds)
438 Random walks and heat kernels on graphs, M.T. BARLOW
439 Evolution equations, K. AMMARI & S. GERBI (eds)
440 Surveys in combinatorics 2017, A. CLAESSON *et al* (eds)
441 Polynomials and the mod 2 Steenrod algebra I, G. WALKER & R.M.W. WOOD
442 Polynomials and the mod 2 Steenrod algebra II, G. WALKER & R.M.W. WOOD
443 Asymptotic analysis in general relativity, T. DAUDÉ, D. HÄFNER & J.-P. NICOLAS (eds)
444 Geometric and cohomological group theory, P.H. KROPHOLLER, I.J. LEARY, C. MARTÍNEZ-PÉREZ & B.E.A. NUCINKIS (eds)
445 Introduction to hidden semi-Markov models, J. VAN DER HOEK & R.J. ELLIOTT
446 Advances in two-dimensional homotopy and combinatorial group theory, W. METZLER & S. ROSEBROCK (eds)
447 New directions in locally compact groups, P.-E. CAPRACE & N. MONOD (eds)
448 Synthetic differential topology, M.C. BUNGE, F. GAGO & A.M. SAN LUIS
449 Permutation groups and cartesian decompositions, C.E. PRAEGER & C. SCHNEIDER

London Mathematical Society Lecture Note Series: 449

Permutation Groups and Cartesian Decompositions

CHERYL E. PRAEGER
University of Western Australia, Perth, Australia

CSABA SCHNEIDER
Universidade Federal de Minas Gerais, Belo Horizonte, Brazil

CAMBRIDGE
UNIVERSITY PRESS

University Printing House, Cambridge CB2 8BS, United Kingdom

One Liberty Plaza, 20th Floor, New York, NY 10006, USA

477 Williamstown Road, Port Melbourne, VIC 3207, Australia

314–321, 3rd Floor, Plot 3, Splendor Forum, Jasola District Centre, New Delhi – 110025, India

79 Anson Road, #06-04/06, Singapore 079906

Cambridge University Press is part of the University of Cambridge.

It furthers the University's mission by disseminating knowledge in the pursuit of education, learning, and research at the highest international levels of excellence.

www.cambridge.org
Information on this title: www.cambridge.org/9780521675062
DOI: 10.1017/9781139194006

First published 2018

A catalogue record for this publication is available from the British Library.

ISBN 978-0-521-67506-2 Paperback

To John, James, and Tim

To Jutka, Benjámin, and Alice

Contents

Preface

This is a book about permutation groups, their fundamental theory and applications. It focuses on those permutation groups which are most useful for studying symmetric structures. We hope the book will be enjoyed and used by many students and researchers with an interest in mathematical symmetry.

We give a modern treatment not only of primitive permutation groups but also of the larger families of quasiprimitive and innately transitive groups. These groups occur naturally in applications where reductions to primitive groups are not available. For example, an appropriate quotient operation, applied to a structure with particular symmetry properties, often yields a smaller structure with the same symmetry properties and admitting a subgroup of automorphisms that in addition is quasiprimitive. The quasiprimitive group so obtained is not in general the full automorphism group. Thus knowing the possible overgroups of a given quasiprimitive group helps us understand possible additional symmetries of this smaller structure.

Preserving a cartesian decomposition of the underlying point set turns out to be an especially important property that in some cases determines the 'type' of a primitive, or quasiprimitive, or innately transitive permutation group. Thus, when describing the overgroups of a given primitive or quasiprimitive or innately transitive group, a fundamental problem we need to solve is the determination of all the cartesian decompositions it preserves. This book is the first to provide a mathematical framework to attack this problem. It treats, as far as possible, general permutation groups – not necessarily finite ones. For finite groups, it provides surprisingly complete solutions, making use of detailed information about the finite simple groups, based on the finite simple group classification.

The systematic study of the overgroups of a given finite permutation group was initiated by Cheryl's 1990 paper that characterised the overgroups of finite primitive permutation groups. For the applications of the O'Nan–Scott Theorem to the study of symmetric graphs, this theory needed to be extended to finite quasiprimitive permutation groups and Cheryl embarked on this research project from the beginning of the 1990s. The possibilities for finite quasiprimitive groups are far greater than for finite primitive groups, and it became apparent that describing the overgroups of a quasiprimitive group that are wreath products in product action was in itself a significant research project. For a quasiprimitive permutation group, the existence of such an overgroup is equivalent to preserving a homogeneous cartesian decomposition.

The origins of the research program concerning the overgroups of quasiprimitive permutation groups can be traced back to a conversation between Robert Baddeley and Cheryl during a conference at Oberwolfach in the 1990s. The project began as a collaboration between Robert and Cheryl. Both of them were profoundly influenced by discussions with L. (Laci) G. Kovács regarding the scope and objectives of their joint work. At the end of the 90s, Robert left academia and Cheryl completed several papers that presented the findings of their decade-long research into this problem. However, the work to describe the overgroups of finite quasiprimitive permutation groups that are wreath products in product action was left unfinished, and hence unpublished. Robert left Cheryl a substantial manuscript that contained an account of their joint work up to then. When Csaba started his post-doctoral research position at The University of Western Australia in September of 1999, Cheryl and he decided that further developing the ideas laid out in Robert's manuscript was an appropriate and worthwhile project. Thus the work was developed and completed by Csaba and Cheryl. The results inevitably spread across multiple journal articles and we hope that presenting them in this book will make them more accessible.

When the first draft of our book was essentially complete, we became aware that many of the questions we had addressed for finite permutation groups were important also for infinite permutation groups. In particular, discussions with Simon Smith convinced us that the book should be written, as far as possible, in the context of general groups – and that in particular the theory of cartesian decompositions should be developed for families of infinite permutation groups. The theory of cartesian decompositions is given for finite rank decompositions, and restriction to finite groups is made only when we wish to use the powerful information available from the finite simple group classification.

We are grateful to many people for helping us complete this research project and for making this book possible. In the first place, our thanks go to Robert W. Baddeley whose manuscript was an important milestone heavily influencing the early drafts of this book. We are grateful to Simon Smith for persuading us to present the material, whenever possible, without restricting to finite groups. We thank Roger Astley and Clare Dennison of Cambridge University Press for their infinite patience and encouragement.

1
Introduction

1.1 Construction and decomposition

Mathematical thinking has an interesting duality reflected in its fundamental techniques: constructing and decomposing. Whereas construction methods provide a means for producing larger objects from smaller ones, decomposition theorems enable us to identify the basic or irreducible blocks from which general mathematical structures can be obtained. A combination of these two methods usually leads to an understanding of how complicated objects are constructed from their building blocks. Perhaps the simplest example is in the category of sets. In this category we can construct a new set from given ones by taking their co-product, that is, their disjoint union. We can also easily decompose any given set as a co-product of smaller sets by identifying a partition. Similarly, in a first course on group theory, we teach the (external) direct product construction along with the criteria for identifying an (internal) direct product decomposition.

These algebraic and set theoretic examples are brought together in the study of group actions. A general construction for permutation groups takes two groups G and H acting on disjoint sets Γ and Δ and produces a permutation group isomorphic to $G \times H$ acting on the union $\Gamma \cup \Delta$. Using a more complicated method we may construct the wreath product $G \wr S_k$ acting on the disjoint union of k copies of the set Γ (see Section 5.2.1). Conversely, a first analysis of a permutation group G on a set Γ identifies the partition of Γ given by the G-orbits, and determines G as a subgroup of the direct product of the transitive permutation groups induced on these G-orbits (see Section 2.1). In this way, the class of permutation groups can be understood by focusing on the smaller class of transitive groups, and the subgroups of their direct products. In turn, many questions concerning transitive groups can be answered via a reduction to primitive groups, that is, to groups for which the point set has no non-trivial invariant co-product decomposition. Thus the co-product object of sets (that is, partitions of sets) serves as a very useful tool in the theory of permutation groups.

Equally important in the category of sets is the product object, which is the cartesian product. As above, there is a permutation group construction that, given groups G and H acting on Γ and Δ, respectively, returns a group isomorphic to $G \times H$ acting on $\Gamma \times \Delta$. Moreover, the wreath product $G \wr S_k$ also acts on Γ^k in its product action (see Section 5.2.2). This product action plays a very important role in the theory of primitive groups. For example, one interpretation of the O'Nan–Scott Theorem for finite primitive permutation groups is that there are four fundamental types of finite primitive groups and all others arise as subgroups of wreath products $G \wr S_k$ in product action where G is a fundamental primitive group. These four fundamental types also occur as maximal subgroups in finite symmetric and alternating groups; see Theorem 7.11. The O'Nan–Scott Theorem (see Chapter 7) has provided the most useful modern method for identifying the possible structures of finite primitive groups and is now used routinely for their analysis. Thus a cartesian decomposition concept, complementing the cartesian product construction, should play an important role in the study of permutation groups, especially in that of the finite primitive ones. It is therefore surprising that there is no widely used such cartesian decomposition concept. Instead, mathematicians usually work their way around introducing one, and this can lead to imprecise and inadequate treatment of groups acting in product action.

In our work we show how a properly defined cartesian decomposition concept leads to a new way of analysing (not necessarily finite) transitive permutation groups by decomposing them with respect to invariant cartesian decompositions of the point set. This leads to a new theory for the class of permutation groups with a transitive minimal normal subgroup. This theory is particularly powerful in the case when the minimal normal subgroup is of the form T^k for a simple group T and a positive integer k; in particular, this smaller class contains all finite primitive groups. The cartesian decomposition concept we use in this book first appeared in the paper by L. G. Kovács (Kovács 1989b) and was used to identify wreath decompositions of permutation groups. Kovács used the name *system of product-imprimitivity* for this concept, but we find the name *cartesian decomposition* more descriptive. Cartesian decompositions are exploited to deepen the current understanding of permutation groups contained in wreath products in product action. We demonstrate, in the finite case, that the study of cartesian decompositions combined with certain facts about finite simple groups, which depend on the finite simple group classification, leads to an unexpectedly detailed description of permutation groups that act on cartesian products. In particular, this description sheds new light on the theory of primitive permutation groups and also on the larger families of quasiprimitive and innately transitive permutation groups introduced in Section 1.3. It also applies, for example, to infinite primitive and quasiprimitive permutation groups with finite stabilisers; see Theorems 3.18 and 7.9.

1.2 Cartesian decompositions

To introduce the intuitive idea behind cartesian decompositions, consider the following example. In the three-dimensional coordinate system, the set C of points

$$\{(x_1, x_2, x_3) \mid x_1,\ x_2,\ x_3 \in \{0, 1\}\}$$

forms a cube. The cube C lives in a 3-dimensional space and it can be bisected in three different ways using planes parallel to the three fundamental planes of the coordinate system. For example, four of the 8 points lie on the plane defined by the equation $x_1 = 0$ and four lie on the plane with equation $x_1 = 1$, and this gives the first partition. Similarly, the second partition is determined by the planes with equations $x_2 = 0$ and $x_2 = 1$, and the third partition is determined by the equations $x_3 = 0$ and $x_3 = 1$. Let us denote these partitions by Γ_1, Γ_2, and Γ_3, respectively. Then

$$\begin{aligned}\Gamma_1 &= \{\{(0,0,0),(0,0,1),(0,1,0),(0,1,1)\},\\ &\quad\ \{(1,0,0),(1,0,1),(1,1,0),(1,1,1)\}\};\end{aligned}$$

$$\begin{aligned}\Gamma_2 &= \{\{(0,0,0),(0,0,1),(1,0,0),(1,0,1)\},\\ &\quad\ \{(0,1,0),(0,1,1),(1,1,0),(1,1,1)\}\};\end{aligned}$$

$$\begin{aligned}\Gamma_3 &= \{\{(0,0,0),(0,1,0),(1,0,0),(1,1,0)\},\\ &\quad\ \{(0,0,1),(0,1,1),(1,0,1),(1,1,1)\}\}.\end{aligned}$$

In other words, the partition Γ_i divides the vertices of the cube C into two blocks according to the i-th coordinate.

We note that the first coordinate of a point is determined by its block in the first partition, and similarly its second and third coordinates are determined by its blocks in the second and the third partitions, respectively. Since each point is determined by its three coordinates we see that the intersection of three blocks, one from each of the three different Γ_i, has size one. In other words,

$$|\gamma_1 \cap \gamma_2 \cap \gamma_3| = 1 \quad \text{whenever} \quad \gamma_1 \in \Gamma_1,\ \gamma_2 \in \Gamma_2,\ \gamma_3 \in \Gamma_3.$$

This motivates the definition of a cartesian decomposition.

Definition 1.1 A *cartesian decomposition*, $\mathcal{E}$, of a set Ω is a finite set of partitions, $\mathcal{E} = \{\Gamma_1, \ldots, \Gamma_\ell\}$, of Ω such that $|\Gamma_i| \geqslant 2$ for each i and

$$|\gamma_1 \cap \cdots \cap \gamma_\ell| = 1 \quad \text{for each} \quad \gamma_1 \in \Gamma_1, \ldots, \gamma_\ell \in \Gamma_\ell.$$

A cartesian decomposition is said to be *trivial* if it contains only one partition, namely the partition into singletons. A cartesian decomposition is said to be *homogeneous* if all the Γ_i have the same cardinality.

If $\{\Gamma_1, \ldots, \Gamma_\ell\}$ is a cartesian decomposition of a set Ω, then the defining property yields a well defined bijection between Ω and $\Gamma_1 \times \cdots \times \Gamma_\ell$, given by

$$\omega \mapsto (\gamma_1, \ldots, \gamma_\ell) \tag{1.1}$$

where, for $i = 1, \ldots, \ell$, the block $\gamma_i \in \Gamma_i$ is the unique block of Γ_i which contains ω. Thus the set Ω can be naturally identified with the cartesian product $\Gamma_1 \times \cdots \times \Gamma_\ell$.

Let us now turn back to the example given before Definition 1.1, and view the cube C as a graph in which two vertices are joined if and only if they only differ in one position. Let W denote the wreath product of the cyclic group of order 2 and the symmetric group of degree 3. That is, $W = C_2 \wr \mathsf{S}_3 = (C_2 \times C_2 \times C_2) \rtimes \mathsf{S}_3$ and W is a group of order 48. The group W acts on C as follows. Set $B = C_2 \times C_2 \times C_2$ and let $x = (x_1, x_2, x_3) \in C$. The first copy of C_2 in B flips the first coordinate of x (that is, interchanges 0 and 1), the second copy of C_2 flips the second coordinate, while the third copy flips the third coordinate. The group S_3 permutes the coordinates of x naturally. Easy consideration shows that each element of W is an automorphism of this 'cube graph', and simple graph theoretic consideration shows that W is the full automorphism group of C. This example illustrates the strong relationship between wreath products and cartesian decompositions. In fact, the observations we made in this simple example are generalised in Section 12.2 where we will study the general relationship between cartesian products of graphs and cartesian decompositions.

1.3 Cartesian factorisations

As mentioned in Section 1.1, modern studies of finite primitive permutation groups identified groups preserving cartesian decompositions as having fundamental significance for a theory of primitive groups, especially in the O'Nan–Scott Theorem. They are of similar importance in studying larger families of permutation groups such as quasiprimitive groups and innately transitive groups. A permutation group is said to be *quasiprimitive* if all its non-trivial normal subgroups are transitive and it is called *innately transitive* if it has a transitive minimal normal subgroup. Hence every primitive group is quasiprimitive (Corollary 2.21), and every finite quasiprimitive group is innately transitive. Moreover, for a finite group, a minimal normal subgroup is a direct product of finitely many copies of a simple group (see Lemma 3.14). We consider finite and infinite innately transitive groups with a minimal normal subgroup of this kind. The purpose of this book is to present a theory of cartesian decompositions that are invariant under the action of such a group, and to use it to present characterisations of the primitive, quasiprimitive, and innately transitive groups having a minimal normal subgroup of this form. We will also apply this theory in various group theoretic and combinatorial contexts.

A central problem we wish to solve is the following.

> *For a given innately transitive group, decide if the group action can be realised on a non-trivial cartesian product of smaller sets; that is, decide if the group leaves invariant a non-trivial cartesian decomposition.*

Our approach involves a mixture of combinatorial and group theoretic methods. The example of the cube in the previous section shows that cartesian decompositions sometimes arise naturally. However, they are not always so easy to recognise, and the following example shows that the structure of the acting group may help in finding invariant cartesian decompositions of the underlying set. We use the primitive subgroup $G = \mathsf{Aut}(\mathsf{A}_6) \cong \mathsf{P\Gamma L}_2(9)$ of S_{36} for illustration. The socle T of this group is isomorphic to A_6. The reason why G acts on a cartesian product $\Gamma \times \Gamma$, with $|\Gamma| = 6$, is that T has two subgroups A and B, both isomorphic to A_5, such that $A \cap B$ is a point stabiliser T_ω, the subgroup G_ω swaps A and B, and T can be factorised as $T = AB$. Since invariant partitions of the point set correspond to overgroups of the point stabiliser (Lemma 2.14), the socle T preserves two partitions Γ_1, Γ_2 of the point set that are orthogonal in the sense that $\gamma_1 \cap \gamma_2$ is a singleton for all $\gamma_1 \in \Gamma_1$ and $\gamma_2 \in \Gamma_2$. In other words, the underlying set of G can be naturally identified with the cartesian product $\Gamma_1 \times \Gamma_2$. Further, as G_ω swaps A and B, G_ω also swaps these partitions. Hence the group $G = TG_\omega$ can also be viewed as a permutation group acting on the cartesian product $\Gamma_1 \times \Gamma_2$.

This example suggests that we may be able to solve our original problem by studying certain partitions of the point set invariant under a suitable minimal normal subgroup, and, in turn, such partitions may be pinpointed by understanding the factorisations of this minimal normal subgroup.

In this book, after giving a thorough treatment of the fundamental theory of permutation groups (of arbitrary cardinality), we focus on permutation groups G with a minimal normal subgroup M that is transitive on the underlying point set Ω. In other words, we focus on innately transitive permutation groups. Such a subgroup M is called a plinth of G, and is characteristically simple. One of our more fundamental results shows that each partition in a G-invariant cartesian decomposition of Ω is M-invariant (Theorem 8.3). In order to find M-invariant partitions of the point set Ω that form a G-invariant cartesian decomposition, we study the overgroups of a fixed point stabiliser in M. We show in Theorem 8.2 that if $\{\Gamma_1, \ldots, \Gamma_\ell\}$ is a G-invariant cartesian decomposition of Ω, $\gamma_1 \in \Gamma_1, \ldots, \gamma_\ell \in \Gamma_\ell$, and $\{\omega\} = \gamma_1 \cap \cdots \cap \gamma_\ell$, then

$$\bigcap_{i=1}^{\ell} M_{\gamma_i} = M_\omega \text{ and } M_{\gamma_i}\left(\bigcap_{j \neq i} M_{\gamma_j}\right) = M \text{ for all } i \in \{1, \ldots, \ell\}. \quad (1.2)$$

The collection of subgroups M_{γ_j} satisfying (1.2) is called a *cartesian factorisation*† of M, and is studied in Chapter 8. The set $\{A, B\}$ of subgroups of $T = \mathsf{A}_6$ identified in the example presented above is a

† Cartesian factorisations were referred to as *cartesian systems of subgroups* in our earlier papers (Baddeley, Praeger & Schneider 2004a, Baddeley, Praeger & Schneider 2004b, Baddeley, Praeger & Schneider 2006, Baddeley, Praeger & Schneider 2007, Praeger & Schneider 2007, Baddeley, Praeger & Schneider 2008, Praeger & Schneider 2012).

cartesian factorisation for A_6. Looking at the last displayed equation, it is clear that in order to find G-invariant cartesian decompositions of Ω, we need to study factorisations of the characteristically simple group M. This study is carried out under the stronger condition that M is a direct product $M = T_1 \times \cdots \times T_k$ where the T_i are simple groups. Hence we introduce the following definition.

Definition 1.2 A group M is said to be FCR (finitely completely reducible) if M can be written as a direct product $M = T_1 \times \cdots \times T_k$ where the T_i are simple groups. We sometimes say that M is an FCR-*group*.

Note that a finite characteristically simple group is FCR (Lemma 3.14) and so is a minimal normal subgroup of an infinite quasiprimitive permutation group with finite stabilisers (Theorem 3.18). In a finite FCR-group, the factors T_i are finite simple groups, and the machinery provided by the finite simple group classification is available for our use. In particular, using the available knowledge on factorisations of finite simple groups (Liebeck, Praeger & Saxl 1990, Baddeley & Praeger 1998), the factorisations occurring in relation to cartesian decompositions of finite sets can be characterised. Moreover, using this characterisation, in the most interesting cases, the G-invariant cartesian decompositions can be described (see Theorems 8.17, 9.7, 10.13).

1.4 Primitive, quasiprimitive and innately transitive groups: 'O'Nan–Scott theories'

One of the most important outcomes of studying invariant cartesian decompositions is a better understanding of the O'Nan–Scott theory of primitive, quasiprimitive and innately transitive groups. The *primitive* permutation groups on a set Ω are those which leave invariant only the trivial partitions of Ω: the partition in which each part consists of a single point, and the partition with just one part. Since the early 1980s the study of finite primitive permutation groups has been transformed by the O'Nan–Scott Theorem which identifies several types of finite primitive groups, and asserts that each finite primitive group is of one of these types. Proofs of the O'Nan–Scott theorem for finite primitive permutation groups can be found in (Scott 1980, Aschbacher & Scott 1985, Kovács 1986, Buekenhout 1988, Liebeck, Praeger & Saxl 1988) and more detailed treatments of it in (Dixon & Mortimer 1996, Cameron 1999).

Cameron's approach (Cameron 1999) strongly influenced the exposition in this book. He divides the finite primitive groups into two families: the groups in the first family are called the *basic groups*, and the other family is formed by the *non-basic primitive groups*. In Cameron's terminology, a basic group is one that cannot be embedded into a wreath product. This definition of basic groups led to a slight inaccuracy in his treatment of primitive groups. Namely, Cameron treats almost simple groups as basic, even though $\mathsf{P\Gamma L}_2(9)$, for instance, acting primitively on 36 points, as considered in Section 1.3, preserves a non-trivial homogeneous cartesian decomposition, and hence it can be embedded into a wreath product in product action. Nevertheless, it is true, and a useful fact, that each finite primitive permutation group is either basic, or a subgroup of a wreath product $H \wr S_k$ in product action on Γ^k, where H is a basic primitive group on Γ. The latter situation is equivalent to the existence of a non-trivial homogeneous cartesian decomposition preserved by the group. In this book we focus on this situation, but we extend our scope to the class of (possibly infinite) groups that have a transitive minimal normal subgroup.

Much recent work on finite primitive permutation groups concentrated on understanding the basic groups, especially those related to non-abelian simple groups and to irreducible representations of finite groups. This information together with the wreath product construction leads to the solution of many problems in algebra, number theory and combinatorics. However, for some applications, detailed information is needed on precisely which subgroups of a wreath product $H \wr S_k$ with H primitive on Γ, are themselves primitive on Γ^k. The seminal paper of Kovács (Kovács 1989a) introduced the concept of a 'blow-up' of a primitive group and provided criteria for identifying such subgroups for almost all types of primitive groups H. Moreover, this led, in 1990, to a classification (Praeger 1990) of all embeddings of finite primitive groups into wreath products in product action; that is, a classification of all homogeneous cartesian decompositions invariant under primitive groups. In his study of the finite lattice representation problem (see (Pálfy & Pudlák 1980)), Aschbacher (Aschbacher 2009a, Aschbacher 2009b) addressed the same questions and obtained a similar solution.

We extend Kovács's blow-up concept to a larger class of permutation groups in Section 11.1.

The product action of a wreath product was also a pivotal concept in describing finite quasiprimitive groups, a strictly larger class of groups than that of the primitive groups, and one which arises naturally in many combinatorial applications. The term was coined in the 1970s by

Wielandt (private communication from W. Knapp to the first author in January 1994) and first appeared in print in works of Knapp (for example in (Knapp 1973)). In fact Peter Neumann's critical analysis of the 2eme Mémoire of Évariste Galois, in (Neumann 2006), suggests that the 'primitive' permutation groups Galois studied in the early 19th century were in fact the quasiprimitive groups.

To date a great many applications of group theory in combinatorics and other subjects have depended upon a reduction to a case involving a primitive permutation group. However such a reduction is not always possible. The first author consequently initiated an investigation into the suitability of quasiprimitivity, rather than primitivity, as a reduction tool in applications of group theory. This investigation was two-pronged; it involved, on the one hand, an attempt to understand quasiprimitive permutation groups in a purely group-theoretic setting, and on the other, it also involved several applications of quasiprimitivity to classification problems for combinatorial structures, such as incidence geometries (Cara, Devillers, Giudici & Praeger 2012), line-transitive linear spaces (Camina & Praeger 2001), k-arc transitive graphs (Ivanov & Praeger 1993, Baddeley 1993b, Fang 1995, Li 2001, Hassani, Nochefranca & Praeger 1999, Praeger 1993), k-arc-transitive Cayley graphs (Li 2005), locally primitive graphs (Praeger, Pyber, Spiga & Szabó 2012), locally quasiprimitive graphs (Li, Praeger, Venkatesh & Zhou 2002), and strongly regular edge-transitive graphs (Morris, Praeger & Spiga 2009).

Understanding finite quasiprimitive groups in sufficient detail for these applications (for example, to identify the full automorphism groups of these combinatorial structures) requires thorough knowledge of the set of primitive and quasiprimitive overgroups of a given quasiprimitive subgroup of the full symmetric group, a problem addressed in work of the first author and R. W. Baddeley (Baddeley & Praeger 2003). If such a subgroup preserves a non-trivial cartesian decomposition, then the full stabiliser of this decomposition, which is a wreath product in product action, is a natural overgroup. It turns out that the best way to find such overgroups is to locate the corresponding invariant cartesian decompositions.

Quasiprimitivity may be equivalently defined for finite groups as the requirement that all minimal normal subgroups are transitive. Weakening this to the requirement that at least one minimal normal subgroup is transitive gives the larger family of innately transitive groups. An 'O'Nan–Scott theory' of finite innately transitive permutation groups was developed by Bamberg and the first author (Bamberg 2003, Bamberg & Praeger 2004). Applications of this theory also require detailed knowledge of homogeneous cartesian decompositions left invariant by these groups.

Our current knowledge of finite primitive permutation groups is very strong, because not only do we have the O'Nan–Scott Theorem, which divides the primitive permutation groups into families according to the action and the structure of the socle, but we also have available a description which divides the primitive permutation groups into families according to the nature of their primitive overgroups, and most significantly we know that these two descriptions are essentially the same (Praeger 1990). Information concerning both the socle and all overgroups is central to our understanding of primitive permutation groups. This book attempts to advance our understanding of innately transitive and quasiprimitive permutation groups in a similar fashion by analysing such groups from both viewpoints and we do this for the family of (possibly infinite) groups with an FCR-plinth. Indeed, the quasiprimitive version of the O'Nan–Scott Theorem (Praeger 1993) reaches its weakest conclusions when considering finite quasiprimitive permutation groups G whose socle M is a minimal normal subgroup of G and is such that a point stabiliser M_α in M is neither trivial nor projects onto each minimal normal subgroup of M (see Section 7.5). (This is case III(b)(i) in the terminology of (Praeger 1993), and case PA in (Baddeley & Praeger 2003).) For primitive groups the analogous case corresponds essentially to those primitive permutation groups that arise as subgroups of wreath products in product action. It is to be hoped that the overgroup viewpoint can shed more light on this problematic case.

The theory of invariant cartesian decompositions is also useful for studying certain problems in geometry and combinatorics. For example, Baumeister (Baumeister 1997) determined the class of finite two-dimensional grids in which the stabiliser of one of the parallel classes of lines acts primitively on this class. Applying the theory of cartesian decompositions can give a new proof of Baumeister's results, and can extend them to results on flag-transitive multi-dimensional grids with an automorphism group innately transitive on the vertices.

Wreath products and cartesian decompositions also play a role in the reduction theorem for finite primitive distance transitive graphs in (Praeger, Saxl & Yokoyama 1987). The automorphism group of such a graph is either almost simple, or affine, or a wreath product in product action preserving a cartesian decomposition (as, for instance, with the

Hamming graphs). Further applications similar to these are developed in Chapter 12.

1.5 The structure of this book

The book is divided into three parts. Part I presents a general theory of group actions and permutation groups. In Chapter 2, we begin with the definitions of basic concepts, such as orbits, stabilisers, and coset actions. We study block systems and invariant partitions in Sections 2.4–2.5. In Section 2.6, we present a result of Manning that describes the structure of the fixed-point space of a subgroup of a transitive group. Most of Part I is focused on the theory of innately transitive permutation groups, including in particular primitive and quasiprimitive permutation groups. As defined above, a transitive minimal normal subgroup of such a group is called a plinth. A study of such minimal normal subgroups is carried out in Chapter 3. After addressing various topics for not necessarily finite groups, the chapter culminates in two important results; the first of these is Burnside's Theorem that characterises finite 2-transitive permutation groups as either almost simple or affine groups (Theorem 3.21), while the second concerns finite primitive permutation groups with a regular normal subgroup (Theorem 3.22). Chapter 4 is devoted to direct products of groups. In several classes of innately transitive permutation groups, the point stabiliser in a minimal normal FCR-subgroup is *subdirect*, that is to say, it projects onto each of the simple direct factors. In order to investigate this situation, we introduce the concept of strips in Section 4.4. Then in Section 4.6 we prove a general version of Scott's Lemma that a subdirect subgroup in an FCR-group (Definition 1.2) is a direct product of full diagonal subgroups (Theorem 4.16).

In our treatment we place special emphasis on innately transitive groups (both finite and infinite) which leave invariant a non-trivial cartesian decomposition. As the full stabiliser of such a non-trivial homogeneous cartesian decomposition is a wreath product in product action, we present a detailed treatment of wreath products and twisted wreath products in Chapters 5 and 6. In particular we study when such wreath products are primitive, quasiprimitive or innately transitive. Furthermore we prove a Wreath Embedding Theorem (Theorem 5.13) for groups preserving a non-trivial homogeneous cartesian decomposition, which identifies a wreath product in product action into which such a group can be embedded. We end Part I with Chapter 7 where we give a summary of the O'Nan–Scott Theory for (possibly infinite) primitive, quasiprimitive and innately transitive groups that have a transitive minimal normal FCR-subgroup. At the end of Chapter 7, we review the classification of the maximal subgroups of finite symmetric and alternating groups.

Based on Part I, Part II presents the theory of cartesian decompositions preserved by innately transitive groups. An innately transitive group with abelian plinth preserves a cartesian decomposition if and only if a point stabiliser preserves a direct sum decomposition of the plinth, which, in this case, is an elementary abelian p-group; see Theorem 8.4. Hence the problem of finding invariant cartesian decompositions for such groups becomes a problem in representation theory; see Section 8.3 for the details. Thus we will mostly concentrate on innately transitive groups with non-abelian plinth. Our treatment is an interplay between algebraic and combinatorial techniques. The problem that provided the original motivation for this work is essentially group theoretic.

Decide if a given innately transitive permutation group G can be embedded into a wreath product in product action.

As explained above, this is translated into the combinatorial problem of finding G-invariant cartesian decompositions. This combinatorial problem is attacked using Theorem 8.2 by searching for cartesian factorisations of the plinth M, as in (1.2), which, again, is essentially a group theoretic problem.

In Chapter 8 we describe, for innately transitive groups G with plinth M, the precise connection between the G-invariant cartesian decompositions of the point set and the factorisations of M that satisfy (1.2). The importance of factorisations in the inclusion problem for permutation groups has long been recognised. If G is a transitive subgroup of a permutation group H, then the factorisation $H = GH_\alpha$ holds for any point stabiliser H_α. If H is primitive and it is a minimal overgroup of G, then both G and H_α are maximal in H and the resulting factorisation is said to be a *maximal factorisation* of H. Using the O'Nan–Scott Theorem, the inclusion problem for finite primitive groups can be essentially reduced to the case when H is almost simple, and this problem can be solved by studying the maximal factorisations of almost simple groups, as described in (Liebeck *et al.* 1990).

The reason why a tight theory of invariant cartesian decompositions for finite innately transitive groups can be given is that the factorisations of the plinth required by (1.2) can be characterised. The first clear step in the process of describing these factorisations is to consider the case when M is simple. As proved in (Baddeley & Praeger 1998), finite simple groups rarely admit such factorisations, and the interesting cartesian factorisations of finite simple and almost simple groups can be fully classified. In fact, a cartesian factorisation of a finite simple group can have at most 3 members. Further, if a cartesian factorisation in a finite simple group has subgroups of equal order, then it has precisely two subgroups and these subgroups are conjugate under the automorphism group. Using these results, we classify the cartesian decompositions preserved by finite innately transitive groups with simple plinth in Chapter 8.

Interestingly, the classification of factorisations that we call cartesian factorisations in finite almost simple groups is the essential tool underpinning Baumeister's work (Baumeister 1997) on 2-dimensional primitive grids. Such factorisations also appear in the work by Pálfy and Saxl (Pálfy & Saxl 1990) on congruence lattices of finite algebras.

When passing from non-abelian simple groups to non-abelian characteristically simple groups M, we need to consider the possibility that a member of a cartesian factorisation may contain a diagonal subgroup that covers some simple direct factors of M. We call such a diagonal subgroup a *strip*; see the definition in Section 4.4. For finite groups our description of cartesian factorisations relies on a theorem by the first author and Baddeley (Baddeley & Praeger 2003, Lemma 2.2) that a finite non-abelian characteristically simple group M cannot be factorised as $M = AB$ where A and B are subdirect subgroups of M neither containing a simple direct factor of M. We extended this theorem for direct powers of arbitrary, possibly infinite, groups that do not admit uniform automorphisms in Theorem 4.30. Thus the possible structures of cartesian factorisations in non-abelian characteristically simple groups are analysed in Chapters 9–10.

Using the results of Chapter 9 about diagonal subgroups and about factorisations of simple groups, we are able to give in Chapters 9–10 a detailed treatment of cartesian decompositions preserved by finite innately transitive groups G with non-abelian plinth. One of our strongest results in this book is that if G is a finite innately transitive group with a non-abelian plinth preserving a cartesian decomposition $\mathcal{E}$, then G has at most 3 orbits on $\mathcal{E}$. If $\mathcal{E}$ is homogeneous, then the number of orbits is at most 2 (Theorems 10.7 and 10.11). In the case of transitive cartesian decompositions, the study of factorisations of finite simple groups and of the possible arrangements for the diagonal subgroups in the members of a cartesian factorisation enables us to identify that there are precisely 6 types of cartesian decompositions that can be transitively acted upon by a finite innately transitive group with a non-abelian plinth. A detailed description is given for all the 6 types in Chapter 9. In the same chapter, we give a similar characterisation of cartesian factorisations of FCR-plinths for infinite innately transitive groups.

Group theoretic and combinatorial applications of the theory of cartesian decompositions will be presented in Part III. In Chapter 11, we extend the blow-up concept introduced by Kovács for primitive permutation groups in (Kovács 1989a). Using this blow-up concept, we characterise the members of several classes of the O'Nan–Scott Theorem for quasiprimitive groups as blow-ups of smaller groups (Theorem 11.13). Then in Section 11.4 we present a detailed treatment of the inclusion problem of finite innately transitive permutation groups into wreath products in product action.

In Chapter 12, we consider several graph theoretic applications of the theory developed in this book. First we define a graph that is associated with a cartesian decomposition and identify this graph as the cartesian product of complete graphs. Then we prove that the full automorphism group of such a graph coincides with the stabiliser of the cartesian decomposition (Theorem 12.3). Since a Hamming graph is the cartesian product of finite complete graphs of constant size, it is natural that our theory has consequences for Hamming graphs. Indeed Theorem 12.10 shows that arc-transitive subgroups of the automorphism groups of finite Hamming graphs can be described in terms of the blow-up construction introduced in Section 11.1. In the final Sections 12.6–12.7 we investigate arc-transitive, and 2-arc-transitive graphs whose vertex sets are cartesian products and whose automorphism groups preserve the cartesian decomposition. Such finite 2-arc-transitive graphs are analysed using the 6-Class Theorem (Theorem 12.22).

Part I

Permutation groups: Fundamentals

2
Group actions and permutation groups

In this chapter we give a brief summary of the basic theory of permutation groups. Some standard notions of permutation group theory, such as permutational isomorphism, are extended to incorporate cartesian decompositions; see Section 2.2. In Section 2.7, we introduce the concepts of orbital graphs and orbital digraphs. *The results of this chapter usually hold for general (not necessarily finite) groups, and if the group needs to be finite, this will be stated explicitly.*

2.1 Group actions, stabilisers, orbits

The group of all permutations of a set Ω is the *symmetric group* on Ω and is denoted by $\mathsf{Sym}\,\Omega$. If Ω is equal to the set $\{1,\ldots,n\}$, which we will usually denote by the symbol $\underline{n}$, then $\mathsf{Sym}\,\Omega$ is written more briefly as S_n. A *permutation group* on a set Ω is a subgroup of $\mathsf{Sym}\,\Omega$. If $g \in \mathsf{Sym}\,\Omega$ and $\omega \in \Omega$ then the image of ω under the action of g is written as ωg.

If G is a group and Ω is a set then an *action* of G on Ω is a homomorphism $\alpha\colon G \to \mathsf{Sym}\,\Omega$. If there is no danger of confusion, then, for $\omega \in \Omega$ and $g \in G$, the image $\omega(g\alpha)$ is usually written as ωg. An action α is said to be *faithful*, if it is a faithful homomorphism; that is, the kernel $\ker\alpha$ is the trivial subgroup. The *degree* of such an action is $|\Omega|$. If G is a permutation group on Ω then G acts on Ω via the identity homomorphism and this is a faithful action. Further, if G acts on Ω faithfully, then we often identify G with the permutation group $G\alpha$.

Let G be a group acting on Ω. If $\omega \in \Omega$, then the *point stabiliser*

G_ω in G of ω is defined by

$$G_\omega = \{g \in G \mid \omega g = \omega\}.$$

It is straightforward to show that G_ω is a subgroup of G. The proof of the following lemma is an easy computation and so is left to the reader.

Lemma 2.1 *If $\omega_1,\ \omega_2 \in \Omega$ and $g \in G$ such that $\omega_1 g = \omega_2$ then $(G_{\omega_1})^g = G_{\omega_2}$.*

If Ω is a set and $k \geqslant 1$ then let $\mathcal{P}^k(\Omega)$ denote the k-th power set of Ω. That is, $\mathcal{P}^0(\Omega) = \Omega$, and, for $k \geqslant 1$, $\mathcal{P}^k(\Omega)$ is the set of subsets of $\mathcal{P}^{k-1}(\Omega)$. If G acts on a set Ω then, for each $k \geqslant 1$, there is a natural G-action on $\mathcal{P}^k(\Omega)$. If $X \in \mathcal{P}^k(\Omega)$ for some positive integer k, then the image Xg of X under the element $g \in G$ is given by iteratively applying the formula

$$Xg = \{xg \mid x \in X\}.$$

Further, G_X and $G_{(X)}$ are respectively the *setwise* and the *elementwise stabilisers* in G of X as given by

$$\begin{aligned} G_X &= \{g \in G \mid Xg = X\}, \\ G_{(X)} &= \{g \in G \mid xg = x \text{ for all } x \in X\}. \end{aligned}$$

The quotient $G_X/G_{(X)}$ is denoted by G^X and is viewed as a permutation group on X in the natural way; if $G_X = G$, that is $Xg = X$ for all $g \in G$, then X is said to be G-*invariant.* The group G^X is referred to as the *permutation group on X induced by G_X*.

In particular, a cartesian decomposition $\mathcal{E} = \{\Gamma_1, \ldots, \Gamma_\ell\}$ of Ω, as defined in Section 1.2, is a member of $\mathcal{P}^3(\Omega)$ with each of the Γ_i a partition of Ω. For $G \leqslant \mathsf{Sym}\,\Omega$, $\mathcal{E}$ is G-invariant if $\mathcal{E}g = \mathcal{E}$ for all $g \in G$, that is to say, $\Gamma_i g \in \mathcal{E}$ for all $\Gamma_i \in \mathcal{E},\ g \in G$.

If G acts on a set Ω and $\omega \in \Omega$ then the set

$$\omega G = \{\omega g \mid g \in G\}$$

is called a G-*orbit.*

Lemma 2.2 *If G is a group acting on a set Ω, then the G-orbits form a partition of Ω.*

Proof Define a relation on Ω by $\omega_1 \equiv \omega_2$ if and only if $\omega_1 g = \omega_2$ for some $g \in G$. It is straightforward to verify that $\equiv$ is an equivalence relation. The G-orbits are the equivalence classes of the relation $\equiv$ and hence they partition Ω. □

The G-action is said to be *transitive* if Ω is a G-orbit; that is, for all ω_1, $\omega_2 \in \Omega$ there is an element $g \in G$ such that $\omega_1 g = \omega_2$. If G is not transitive, then it is called *intransitive.* Lemma 2.1 implies that the point stabilisers in a transitive group form a conjugacy class of subgroups. If Δ is an orbit for an intransitive permutation group G, then the induced permutation group G^Δ is a transitive group of smaller degree. The group G^Δ is called the *transitive constituent* of G on Δ and is, by definition, a subgroup of $\mathsf{Sym}\,\Delta$. Hence the investigation of an intransitive group can often be reduced to the study of transitive groups.

The *core* $\mathsf{Core}_G\,H$ of a subgroup H of a group G is defined as the intersection of its conjugates: $\mathsf{Core}_G H = \bigcap_{g \in G} H^g$. The subgroup H is said to be *core-free* if $\mathsf{Core}_G H = 1$.

Lemma 2.3 *If G is a group and α is a transitive G-action on a set Ω, then, for $\omega \in \Omega$, we have that $\ker\alpha = \mathsf{Core}_G G_\omega$. Consequently, α is faithful if and only if G_ω is core-free.*

Proof The kernel of α is the intersection of the point stabilisers. If G is transitive, then Lemma 2.1 shows that the set of stabilisers coincides with the set of G-conjugates of a stabiliser G_ω. Hence $\ker\alpha = \mathsf{Core}_G G_\omega$. The second assertion of the lemma follows from the definition of a core-free subgroup. □

A permutation group is said to be *semiregular* if all its point stabilisers are trivial. A permutation group is *regular* if it is transitive and semiregular.

Lemma 2.4 *A transitive abelian permutation group is regular.*

Proof Let G be an abelian group acting transitively on a set Ω, and let $\omega \in \Omega$. Lemma 2.1 implies that G_ω fixes every point of Ω, and so, as G is faithful on Ω, $G_\omega = 1$. Thus G is semiregular, and, since G is transitive, it is regular. □

2.2 Isomorphic actions

Let G and H be groups acting on the sets Ω and Δ, respectively. The two actions are said to be *permutationally isomorphic* if there exist a bijection $\vartheta\colon \Omega \to \Delta$ and an isomorphism $\chi\colon G \to H$ such that

$$\omega g\vartheta = \omega\vartheta(g\chi) \quad \text{for all} \quad \omega \in \Omega,\ g \in G.$$

If such conditions hold, the pair (ϑ, χ) is said to be a *permutational isomorphism*. Similarly, the pair (ϑ, χ) is a *permutational embedding* of the permutation group G on Ω into the permutation group H on Δ, if $\chi\colon G \to H$ is a monomorphism and $(\vartheta, \hat{\chi})$ is a permutational isomorphism, where $\hat{\chi}\colon G \to G\chi$ is obtained from χ by simply restricting the range of χ.

For a set X, the map id_X is the *identity map* on X defined as $\mathrm{id}_X\colon x \mapsto x$ for all $x \in X$. We write only id when X is clear from the context. A G-action on a set Ω can be carried over to another set Δ if there is a bijection between Ω and Δ.

Lemma 2.5 *Let G act on a set Ω. Assume further that Δ is a set and $\vartheta\colon \Omega \to \Delta$ is a bijection. Define a G-action on Δ as follows. For $\delta \in \Delta$ and $g \in G$ let δg be the element δ_1 of Δ that satisfies $\delta_1\vartheta^{-1} = \delta\vartheta^{-1}g$. Then $(\vartheta, \mathrm{id}_G)$ is a permutational isomorphism from the G-action on Ω to the G-action on Δ.*

Proof Note, for $\omega_1,\ \omega_2 \in \Omega$ and $g \in G$ that $\omega_1 g = \omega_2$ if and only if $\omega_1\vartheta g = \omega_2\vartheta$. This implies that the G-action on Δ is well defined and is permutationally isomorphic to the G-action on Ω. The details are left to the reader. □

If G_1 and G_2 act transitively on (not necessarily distinct) sets then we can decide whether the actions are permutationally isomorphic using the following lemma. Recall that the *inner automorphisms* of a group G are the automorphisms of the form $x \mapsto x^g$ where $x^g = g^{-1}xg$ for some $g \in G$. The set of inner automorphisms of a group G forms a normal subgroup of the group $\mathsf{Aut}(G)$ of automorphisms of G and is denoted by $\mathrm{Inn}(G)$.

Lemma 2.6 *Let G_1 and G_2 be groups acting transitively on Ω_1 and Ω_2, respectively. Then the following are equivalent:*

(i) *the actions of G_1 and G_2 on Ω_1 and Ω_2, respectively, are permutationally isomorphic;*

(ii) *there exist* $\omega_1 \in \Omega_1$ *and* $\omega_2 \in \Omega_2$ *and an isomorphism* $\alpha\colon G_1 \to G_2$ *such that* $(G_1)_{\omega_1}\alpha = (G_2)_{\omega_2}$;

(iii) *for all* $\omega_1 \in \Omega_1$ *and* $\omega_2 \in \Omega_2$ *there exists an isomorphism* $\alpha\colon G_1 \to G_2$ *such that* $(G_1)_{\omega_1}\alpha = (G_2)_{\omega_2}$.

Proof First we prove that (i) implies (ii). Let (γ, α) be a permutational isomorphism such that $\alpha\colon G_1 \to G_2$ and $\gamma\colon \Omega_1 \to \Omega_2$. Let $\omega_1 \in \Omega_1$ and $g \in (G_1)_{\omega_1}$. Then $\omega_1\gamma(g\alpha) = \omega_1 g\gamma = \omega_1\gamma$, and so $g\alpha \in (G_2)_{\omega_1\gamma}$. Therefore $(G_1)_{\omega_1}\alpha \leqslant (G_2)_{\omega_1\gamma}$. On the other hand if $g_2 \in (G_2)_{\omega_1\gamma}$ then there is some $g_1 \in G_1$ such that $g_1\alpha = g_2$. Then $\omega_1\gamma = \omega_1\gamma g_2 = \omega_1\gamma(g_1\alpha) = \omega_1 g_1\gamma$. Applying γ^{-1} we obtain $\omega_1 = \omega_1 g_1$. Therefore $g_1 \in (G_1)_{\omega_1}$, and hence $g_2 = g_1\alpha \in (G_1)_{\omega_1}\alpha$. This shows that $(G_1)_{\omega_1}\alpha = (G_2)_{\omega_1\gamma}$, and assertion (ii) is valid.

Next we show that (ii) implies (iii). Since both G_1 and G_2 are transitive, $\{(G_1)_{\omega_1'} \mid \omega_1' \in \Omega_1\}$ and $\{(G_2)_{\omega_2'} \mid \omega_2' \in \Omega_2\}$ are conjugacy classes in G_1 and G_2, respectively, by Lemma 2.1. Let $\alpha\colon G_1 \to G_2$ be an isomorphism, and $\omega_1 \in \Omega_1$, $\omega_2 \in \Omega_2$ such that $(G_1)_{\omega_1}\alpha = (G_2)_{\omega_2}$, and let $\omega_1' \in \Omega_1$ and $\omega_2' \in \Omega_2$. Then there are $\sigma_1 \in \mathsf{Inn}(G_1)$ and $\sigma_2 \in \mathsf{Inn}(G_2)$ such that $(G_1)_{\omega_1'}\sigma_1 = (G_1)_{\omega_1}$ and $(G_2)_{\omega_2}\sigma_2 = (G_2)_{\omega_2'}$. If $\varphi = \sigma_1\alpha\sigma_2$ then clearly $(G_1)_{\omega_1'}\varphi = (G_2)_{\omega_2'}$.

Finally we show that (iii) implies (i). Let $\omega_1 \in \Omega_1$, $\omega_2 \in \Omega_2$, and $\alpha\colon G_1 \to G_2$ be an isomorphism such that $(G_1)_{\omega_1}\alpha = (G_2)_{\omega_2}$. Then define $\gamma\colon \Omega_1 \to \Omega_2$ by $\omega_1 g\gamma = \omega_2(g\alpha)$ for $g \in G$. Let us show that γ is well defined. Since G_1 is transitive, $\omega_1 g$ runs through all elements of Ω_1 as g runs through the elements of G_1. If $\omega_1 g_1 = \omega_1 g_2$ for some $g_1, g_2 \in G_1$, then, $g_1 g_2^{-1} \in (G_1)_{\omega_1}$, and so $(g_1 g_2^{-1})\alpha = (g_1\alpha)(g_2\alpha)^{-1} \in (G_2)_{\omega_2}$. Hence $\omega_2 g_1\alpha = \omega_2 g_2\alpha$, and γ is well defined. Since $G_2 = G_1\alpha$ is transitive, γ is onto. If $\omega_1 g_1\gamma = \omega_1 g_2\gamma$ for some $g_1, g_2 \in G_1$, then, by the definition of γ, we have $\omega_2(g_1\alpha) = \omega_2(g_2\alpha)$ and so $(g_1\alpha)(g_2\alpha)^{-1} = (g_1 g_2^{-1})\alpha \in (G_2)_{\omega_2}$. Thus $g_1 g_2^{-1} \in (G_1)_{\omega_1}$, and so $\omega_1 g_1 = \omega_1 g_2$. Therefore γ is one-to-one, and hence γ is a bijection. If $\omega \in \Omega_1$ and $g \in G_1$ then there is a $g_1 \in G_1$ such that $\omega = \omega_1 g_1$. Then

$$\omega g\gamma = (\omega_1 g_1 g)\gamma = \omega_2((g_1 g)\alpha) = \omega_2(g_1\alpha)(g\alpha) = \omega_1 g_1\gamma(g\alpha) = \omega\gamma(g\alpha).$$

Thus (γ, α) is a permutational isomorphism. □

Permutationally isomorphic groups that act on the same set can be characterised as follows.

Lemma 2.7 *Let* Ω *be a set and let* $G_1, G_2 \leqslant \mathsf{Sym}\,\Omega$. *Then* G_1 *and* G_2

are permutationally isomorphic if and only if G_1 *and* G_2 *are conjugate subgroups of* $\mathsf{Sym}\,\Omega$*. Further, if* (γ, α) *is a permutational isomorphism, then* $\gamma \in \mathsf{Sym}\,\Omega$ *and* $g\alpha = \gamma^{-1} g\gamma$*, for all* $g \in G_1$*.*

Proof Assume first that G_1 and G_2 are permutationally isomorphic, and let (γ, α) be a permutational isomorphism. Then, for all $g \in G_1$ and $\omega \in \Omega$, we have $\omega g\gamma = \omega\gamma(g\alpha)$; that is $\omega g = \omega\gamma(g\alpha)\gamma^{-1}$. This shows that $\gamma^{-1} g\gamma = g\alpha$, and so G_1 is conjugate to $G_1\alpha = G_2$.

Suppose conversely that $\gamma \in \mathsf{Sym}\,\Omega$ such that $\gamma^{-1} G_1 \gamma = G_2$ and let $\alpha\colon G_1 \to G_2$ denote the isomorphism induced by conjugating by γ; that is, for $g \in G_1$, we define $g\alpha = \gamma^{-1} g\gamma$. Then, for $g \in G$ and $\omega \in \Omega$, $\omega\gamma(g\alpha) = \omega\gamma\gamma^{-1} g\gamma = \omega g\gamma$. Hence (γ, α) is a permutational isomorphism. □

It is useful to extend the notion of permutational isomorphism and embedding to take into account invariant cartesian decompositions. Let G and H be groups acting on the sets Ω and Δ, and leaving invariant the cartesian decompositions $\mathcal{E} = \{\Gamma_1, \ldots, \Gamma_\ell\}$ and $\mathcal{E}' = \{\Gamma_1', \ldots, \Gamma_\ell'\}$ of Ω and Δ, respectively. We say that (ϑ, χ) is a *permutational isomorphism from* $(G, \Omega, \mathcal{E})$ *to* $(H, \Delta, \mathcal{E}')$ if (ϑ, χ) is a permutational isomorphism from G on Ω to H on Δ such that $\mathcal{E}\vartheta = \mathcal{E}'$, that is to say, relabelling the Γ_i' if necessary, $\Gamma_i\vartheta = \Gamma_i'$ for each i. We write $(G, \Omega, \mathcal{E}) \cong (H, \Delta, \mathcal{E}')$. Similarly, we call (ϑ, χ) a *permutational embedding from* $(G, \Omega, \mathcal{E})$ *to* $(H, \Delta, \mathcal{E}')$ if (ϑ, χ) is a permutational embedding from G on Ω to H on Δ such that $\mathcal{E}\vartheta = \mathcal{E}'$.

2.3 Coset actions

An important class of permutation actions is formed by the coset actions. If G is a group and H is a subgroup then we can define the *right coset action* ϱ_H^G of G on the set $[G\colon H]$ of right cosets of H as follows: if $g,\ h \in G$ then

$$(Hg)(h\varrho_H^G) = Hgh. \tag{2.1}$$

It is easy to see that this is indeed an action of G on $[G\colon H]$ and that G is transitive on $[G\colon H]$ with point stabiliser H. If there is no danger of confusion, then the action ϱ_H^G is simply denoted by ϱ_H. The action ϱ_1^G is usually referred to as the *right regular action* of G, and is often denoted simply by ϱ. The set on which ϱ_1^G acts is, by definition, the set of right cosets of the trivial subgroup of G, which we will simply identify with G. Hence we write, for $g,\ h \in G$, that

$$g(h\varrho_1^G) = gh. \tag{2.2}$$

As shown by the following lemma, every transitive action is permutationally isomorphic to a coset action.

Lemma 2.8 *Suppose that G acts on a set Ω transitively and let $\omega \in \Omega$. Define the map $\gamma\colon \Omega \to [G\colon G_\omega]$ by the rule $\omega' \mapsto G_\omega g$ where $g \in G$ is chosen such that $\omega g = \omega'$. Then γ is a well defined bijection, and (γ, id_G) is a permutational isomorphism.*

Proof If $g_1,\ g_2 \in G$ such that $\omega g_1 = \omega g_2$ then $g_1 g_2^{-1} \in G_\omega$ and so $G_\omega g_1 = G_\omega g_2$. Thus γ is well defined. By definition, γ is onto. Suppose that $\omega_1\gamma = \omega_2\gamma$. Then $G_\omega g_1 = G_\omega g_2$ where $\omega g_1 = \omega_1$ and $\omega g_2 = \omega_2$. Hence $g_1 g_2^{-1} \in G_\omega$ and so

$$\omega_1 = \omega g_1 = \omega g_1 g_2^{-1} g_2 = \omega g_2 = \omega_2.$$

Therefore γ is one-to-one, and so γ is a bijection. Finally we prove that (γ, id_G) is a permutational isomorphism. Let $\omega' \in \Omega$ and $g \in G$. As G is transitive, there is some $g_1 \in G$ such that $\omega g_1 = \omega'$. Then

$$\omega' g\gamma = \omega g_1 g\gamma = G_\omega g_1 g = \omega g_1 \gamma g = \omega' \gamma g.$$

This shows that (γ, id_G) is a permutational isomorphism. □

Lemma 2.8 allows us to identify the underlying set of a transitive group action with the set of right cosets of the stabiliser of a designated point.

Corollary 2.9 (Orbit-Stabiliser Theorem) *If G is a group acting transitively on a set Ω and $\omega \in \Omega$ then $|\Omega| = |G\colon G_\omega|$.*

Proof By Lemma 2.8, the action of G on Ω is equivalent to its coset action on the set of right cosets $[G\colon G_\omega]$. Therefore $|\Omega| = |[G\colon G_\omega]| = |G\colon G_\omega|$. □

Corollary 2.10 *Suppose that G is a semiregular subgroup of $\mathsf{Sym}\,\Omega$.*

(i) *If Δ is a G-orbit, then G^Δ is regular and $|\Delta| = |G|$. In particular, if Ω is finite, then G is regular if and only if $|G| = |\Omega|$.*
(ii) *If G is regular and $H \leqslant G$, then H is semiregular and the number of H-orbits in Ω is equal to $|G\colon H|$. In particular, if $H < G$, then H is intransitive.*

Proof (i) Suppose that $\delta \in \Delta$ and $\overline{g} \in G^{\Delta}$ such that $\delta\overline{g} = \delta$. Then there exists some $g \in G$ such that $\overline{g}$ is induced by g and $\delta g = \delta$. Since G is semiregular, $g = 1$, and hence $\overline{g} = 1$. Thus G^{Δ} is also semiregular. Since Δ is a G-orbit, G^{Δ} is transitive on Δ, and hence G^{Δ} is regular. By the Orbit-Stabiliser Theorem (Corollary 2.9), $|\Delta| = |G|/|G_{\delta}| = |G|$. If Ω is finite, then G is regular if and only if Ω is a G-orbit, which is equivalent to $|G| = |\Omega|$.

(ii) Since a subgroup of a semiregular group is semiregular, H is semiregular. To show that the number of H-orbits is equal to $|G\colon H|$, assume without loss of generality that G acts by the right coset action on the cosets of the trivial subgroup of G. Since the cosets of the trivial subgroup correspond to the elements of G, we may assume without loss of generality that $\Omega = G$ and G acts by right multiplication. Then every left coset gH is an H-orbit. Since these left cosets partition Ω, they must be the only H-orbits and their number is $|G\colon H|$. □

The following lemma is a useful characterisation of the transitive subgroups of a group G.

Lemma 2.11 *Suppose that G is a group acting transitively on a set Ω, and let $\omega \in \Omega$. A subgroup $H \leqslant G$ is transitive if and only if $G_{\omega}H = G$. Further, if H is transitive and $H_{\omega} = G_{\omega}$, then $H = G$.*

Proof By Lemma 2.8, we may assume without loss of generality that $\Omega = [G\colon G_{\omega}]$ and the action of G is the right coset action. Suppose that H is transitive and $g \in G$. Then there is some $h \in H$ such that $G_{\omega}g = G_{\omega}h$, and so $gh^{-1} = g_1$ for some $g_1 \in G_{\omega}$. Therefore $g = g_1 h$, as claimed. Reversing this argument gives the converse statement.

If H is transitive and $H_{\omega} = G_{\omega}$, then $G = G_{\omega}H = H_{\omega}H = H$, which proves the second statement. □

2.4 Primitive and imprimitive groups

Suppose that G is a group acting transitively on a set Ω. A non-empty subset $\Delta \subseteq \Omega$ is said to be a *block of imprimitivity* or simply a *block* if, for all $g \in G$, either $\Delta g = \Delta$ or $\Delta g \cap \Delta = \emptyset$.

Lemma 2.12 *Suppose that G is a group acting transitively on a set Ω and let Δ be a block of imprimitivity. Then the following hold.*

(i) *G_{Δ} is transitive on Δ.*

(ii) *For each* $g \in G$, Δg *is a block of imprimitivity and* $|\Delta g| = |\Delta|$.

Moreover, a subset Δ *of* Ω *is a block of imprimitivity if and only if* $\{\Delta g \mid g \in G\}$ *is a partition of* Ω.

Proof (i) Suppose that δ_1, $\delta_2 \in \Delta$. As G is transitive, there is some element $g \in G$ such that $\delta_1 g = \delta_2$. Now $\delta_2 \in \Delta \cap \Delta g$, and so the definition of a block implies that $\Delta g = \Delta$. This shows that $g \in G_\Delta$, and we obtain that G_Δ is transitive on Δ.

(ii) Let $h \in G$ be such that $\Delta g \cap (\Delta g)h \neq \emptyset$. Then $\Delta \cap (\Delta g h g^{-1}) = (\Delta g \cap \Delta g h) g^{-1} \neq \emptyset$. Since Δ is a block of imprimitivity it follows that $\Delta = \Delta g h g^{-1}$, and hence that $\Delta g = \Delta g h$. Thus Δg is a block of imprimitivity. As g is a bijection, $|\Delta g| = |\Delta|$.

Finally, to prove the last statement, assume that Δ is a block of imprimitivity and set $\mathcal{P} = \{\Delta g \mid g \in G\}$. We only need to show that $\mathcal{P}$ is a partition of Ω. If $\omega \in \Omega$ then, as G is transitive and $\Delta \neq \emptyset$, there is some $g \in G$ such that $\omega g \in \Delta$; that is $\omega \in \Delta g^{-1}$ which shows that $\bigcup_{g \in G} \Delta g = \Omega$. We show that distinct elements of $\mathcal{P}$ are disjoint. Suppose that $\Delta g_1, \Delta g_2 \in \mathcal{P}$, and $\Delta g_1 \cap \Delta g_2 \neq \emptyset$. Then $\Delta g_1 \cap ((\Delta g_1) g_1^{-1} g_2) \neq \emptyset$. Since Δg_1 is a block of imprimitivity, by part (ii), it follows that $\Delta g_1 = (\Delta g_1) g_1^{-1} g_2 = \Delta g_2$. Thus distinct elements of $\mathcal{P}$ are disjoint, so $\mathcal{P}$ is a partition of Ω. To prove the converse, observe that if the Δg form a partition of Ω, then $\Delta g = \Delta$ or $\Delta g \cap \Delta = \emptyset$ for all $g \in G$. □

Suppose that G is a group acting on Ω. Then a partition $\mathcal{P}$ of Ω is said to be G-*invariant* if $\Delta g \in \mathcal{P}$ for all $\Delta \in \mathcal{P}$ and $g \in G$. By Lemma 2.12, if Δ is a block for a transitive permutation group, then $\mathcal{P} = \{\Delta g \mid g \in G\}$ is a G-invariant partition. By the following easy result, the converse of this statement also holds.

Lemma 2.13 *Let* G *be a transitive group acting on* Ω *and let* $\mathcal{P}$ *be a* G*-invariant partition of* Ω. *Then each* $\Delta \in \mathcal{P}$ *is a block of imprimitivity for* G.

Proof Let $\Delta \in \mathcal{P}$ and $g \in G$. Since $\mathcal{P}$ is G-invariant, $\Delta g \in \mathcal{P}$. Since $\mathcal{P}$ is a partition of Ω, either $\Delta g = \Delta$ or $\Delta \cap \Delta g = \emptyset$. Hence Δ is a block for G. □

When G is transitive, a G-invariant partition of Ω is called a *system of imprimitivity* or a *block system* for G. The parts in a system of imprimitivity are blocks, and the previous lemma shows that if Δ is a block then the translates Δg of Δ, for $g \in G$, form a system of imprimitivity. The set Ω and, for $\omega \in \Omega$, the singleton $\{\omega\}$ are always blocks, and these are called *trivial blocks*. Similarly, the partitions $\{\{\omega\} \mid \omega \in \Omega\}$ and $\{\Omega\}$ are always G-invariant partitions, and they are called *trivial partitions* or *trivial systems of imprimitivity*. The group G is said to be *primitive* if its only blocks of imprimitivity in Ω are the trivial blocks of imprimitivity. Otherwise its action is *imprimitive*.

Lemma 2.12 can be used to construct, for a given imprimitive group G and a given non-trivial block Δ, two groups with smaller degrees, namely G^{Δ} (the permutation group induced on Δ by G_{Δ}) and $G^{\mathcal{P}}$, where $\mathcal{P}$ is the system of imprimitivity $\{\Delta g \mid g \in G\}$. In Section 5.2.1, we will show that G is permutationally isomorphic to a subgroup of $G^{\Delta} \wr G^{\mathcal{P}}$. If one of the groups G^{Δ} or $G^{\mathcal{P}}$ is still imprimitive, then the construction can be repeated, and in this way the study of imprimitive groups is often reduced to the study of primitive groups.

Systems of imprimitivity can be characterised using the subgroup structure of the acting group.

Lemma 2.14 *Let G be a transitive group acting on a set Ω and let ω be a fixed element of Ω. If H is a subgroup of G containing G_{ω}, then the orbit ωH is a block which contains ω. Further, the correspondence $H \mapsto \omega H$ is an inclusion preserving bijection between the set of subgroups containing G_{ω} and the set of blocks containing ω.*

Proof First we show that ωH is a block. Let $g \in G$ and assume that $\omega H \cap \omega H g \neq \emptyset$. Let ω' be a point in this intersection. Then there are $h_1,\ h_2 \in H$ such that $\omega' = \omega h_1 = \omega h_2 g$. Hence $h_1 g^{-1} h_2^{-1} \in G_{\omega}$. Since $G_{\omega} \leqslant H$, we obtain that $h_1 g^{-1} h_2 \in H$, and hence $g \in H$. Therefore $Hg = H$ and so $\omega H = \omega H g$. Thus ωH is a block which contains ω.

In the previous paragraph we established a map $H \mapsto \omega H$ between the set of subgroups of G that contain G_{ω} and the set of blocks that contain ω. Let us show that this map is onto. Assume that Δ is a block containing ω and let H denote the setwise stabiliser of Δ. By Lemma 2.12(i), $\Delta = \omega H$. We must show that $G_{\omega} \leqslant H$. Choose an element g from G_{ω}. Then $\omega \in \Delta \cap \Delta g$, and so the definition of a block implies that $\Delta = \Delta g$ and $g \in H$. This shows that the map is onto.

If $H_1 \leqslant H_2$ then clearly $\omega H_1 \subseteq \omega H_2$, therefore this map preserves inclusion. Assume now that $H_1,\ H_2$ are two subgroups of G such that

$G_\omega \leqslant H_1,\ H_2$ and that $\omega H_1 \subseteq \omega H_2$. We assert that $H_1 \leqslant H_2$. Let $h_1 \in H_1$ and consider the element ωh_1 of ωH_1. Then $\omega h_1 \in \omega H_2$ and so there is some $h_2 \in H_2$ such that $\omega h_1 = \omega h_2$. Therefore $h_1 h_2^{-1} \in G_\omega$. As $G_\omega \leqslant H_2$ and $h_2 \in H_2$, we obtain that $h_1 \in H_2$, and hence the assertion $H_1 \leqslant H_2$ is proved. The argument above implies, for $H_1,\ H_2 \leqslant G$ such that $G_\omega \leqslant H_1 \cap H_2$, that the orbits ωH_1 and ωH_2 coincide if and only if $H_1 = H_2$. Therefore the map is an inclusion preserving bijection between the set of subgroups that contain G_ω and the set of blocks that contain ω. □

In the proof of Lemma 2.14 we showed that the map $H \mapsto \omega H$ is an inclusion preserving bijection between the set of subgroups H that contain G_ω and the set of blocks containing ω. Since both of these sets can be considered as lattices with respect to the set-theoretic inclusion, we conclude that this map is a lattice isomorphism. In a group G, a subgroup H is said to be a *maximal subgroup* if H is a proper subgroup of G, and the only other proper subgroup of G containing H is G itself.

Corollary 2.15 *Let G be a group acting transitively on a set Ω of size at least 2, and let $\omega \in \Omega$. Then the action of G is primitive if and only if G_ω is a maximal subgroup of G.*

Proof For a fixed $\omega \in \Omega$, the group G is primitive if and only if there are precisely two blocks that contain ω; namely, $\{\omega\}$ and Ω. By Lemma 2.14, this amounts to saying that there are precisely two subgroups of G that contain G_ω and these subgroups must be G_ω and G, or equivalently, G_ω is a maximal subgroup of G. □

The following result shows how block systems can be obtained by taking intersections. Suppose that $\mathcal{P}_1, \ldots, \mathcal{P}_\ell$ are partitions of a set Ω. Set

$$\mathcal{P} = \{\gamma_1 \cap \cdots \cap \gamma_\ell \mid \gamma_i \in \mathcal{P}_i \text{ for } i \in \underline{\ell}\} \setminus \{\emptyset\}. \tag{2.3}$$

Let $G \leqslant \mathsf{Sym}\,\Omega$ be a transitive group. If each of the $\mathcal{P}_i$ is a block system for G, then so is $\mathcal{P}$. A more general result holds if the set $\{\mathcal{P}_1, \ldots, \mathcal{P}_\ell\}$ is G-invariant, but the individual members $\mathcal{P}_i$ are not necessarily G-invariant.

Lemma 2.16 *Suppose that G is transitive on Ω and let $\mathcal{P}_1, \ldots, \mathcal{P}_\ell$ be partitions of Ω. Define $\mathcal{P}$ as in (2.3). If the set $\{\mathcal{P}_1, \ldots, \mathcal{P}_\ell\}$ is G-invariant, then $\mathcal{P}$ is a block system for G. In particular, if each $\mathcal{P}_i$ is a block system, then so is $\mathcal{P}$.*

Proof If $\omega \in \Omega$, then, for all i, there is $\gamma_i \in \mathcal{P}_i$, such that $\omega \in \gamma_i$, and so $\omega \in \gamma_1 \cap \cdots \cap \gamma_\ell$. Further, if two elements of $\mathcal{P}$, $\gamma_1 \cap \cdots \cap \gamma_\ell$ and $\gamma_1' \cap \cdots \cap \gamma_\ell'$, say, are distinct, then there is some i such that $\gamma_i \neq \gamma_i'$ and hence $\gamma_i \cap \gamma_i' = \emptyset$. This shows that $(\gamma_1 \cap \cdots \cap \gamma_\ell) \cap (\gamma_1' \cap \cdots \cap \gamma_\ell') = \emptyset$, and so $\mathcal{P}$ is a partition of Ω. Suppose that $\gamma_i \in \mathcal{P}_i$ for all $i \in \underline{\ell}$ and $g \in G$. As g permutes the partitions $\mathcal{P}_i$, there is a permutation $\sigma \in \mathsf{S}_\ell$ such that $\gamma_i g \in \mathcal{P}_{i\sigma}$ for each i, which shows that $(\gamma_1 \cap \cdots \cap \gamma_\ell)g = \gamma_1 g \cap \cdots \cap \gamma_\ell g \in \mathcal{P}$. Therefore $\mathcal{P}$ is G-invariant, and hence $\mathcal{P}$ is a block system.

The last assertion of the lemma follows trivially from the first. □

We obtain an important consequence for cartesian decompositions (see Definition 1.1) preserved by primitive groups.

Corollary 2.17 *Suppose that G is a primitive group acting on Ω and that $\mathcal{E}$ is a G-invariant cartesian decomposition of Ω. Then G is transitive on $\mathcal{E}$.*

Proof Assume that G is intransitive on $\mathcal{E}$, that $\Delta = \{\Gamma_1, \ldots, \Gamma_m\}$ is a G-orbit in $\mathcal{E}$, and set

$$\Gamma = \{\gamma_1 \cap \cdots \cap \gamma_m \mid \gamma_i \in \Gamma_i \text{ for } i = 1, \ldots, m\} \setminus \{\emptyset\}.$$

By Lemma 2.16, Γ is a block system for G in Ω. If $\gamma \in \Gamma$, then $\gamma \neq \Omega$ as, by Definition 1.1, whenever $\gamma_i \in \Gamma_i$, $\gamma_i \neq \Omega$. Further, if $\Gamma_s \in \mathcal{E} \setminus \Delta$, $\gamma \in \Gamma$, and γ_s, γ_s' are distinct elements of Γ_s, then $\gamma_s \cap \gamma_s' = \emptyset$, but, by the intersection property in Definition 1.1, $\gamma \cap \gamma_s \neq \emptyset$ and $\gamma \cap \gamma_s' \neq \emptyset$. Hence $|\gamma| \geqslant 2$ and γ is a non-trivial block for G, which is a contradiction. □

2.5 Examples of blocks

Deciding whether a permutation group is primitive is not always easy. In this section we show that block systems can sometimes be found by looking at certain natural candidates.

Lemma 2.18 *Let G be a group acting in its right-regular representation ϱ on G as defined in (2.2). If Δ is a block for G then Δ is a right coset of a subgroup of G. Conversely, every such right coset is a block for G.*

Proof Suppose that Δ is a block for G. As each of the translates of Δ is also a block for G (Lemma 2.12), we may assume without loss of generality that $1 \in \Delta$. In this case we are required to show that Δ is a subgroup; that is, $g_1 g_2^{-1} \in \Delta$, for all g_1, $g_2 \in \Delta$. Indeed, as $g_2 g_2^{-1} = 1$, we obtain that $1 \in \Delta g_2^{-1} \cap \Delta$, and so the definition of a block gives that $\Delta g_2^{-1} = \Delta$. Therefore $g_1 g_2^{-1} \in \Delta$, and this is precisely what we needed. On the other hand, if Δ is a subgroup, then the translates $\{\Delta g \mid g \in G\}$ partition G into the cosets of Δ, and so $\{\Delta g \mid g \in G\}$ is a G-invariant partition. Thus Δ is a block by Lemma 2.13.

The second assertion of the lemma follows trivially from the first. □

For a group G acting on a set Ω, let $\mathsf{Fix}_\Omega(G)$ denote the set of fixed points of G in Ω; that is,

$$\mathsf{Fix}_\Omega(G) = \{\omega \in \Omega \mid \omega g = \omega \text{ for all } g \in G\}.$$

The *normaliser* $\mathsf{N}_G(H)$ of a subgroup H of G is

$$\mathsf{N}_G(H) = \{g \in G \mid H^g = H\}.$$

Lemma 2.19 *If G is transitive on Ω, $\omega \in \Omega$, and N is a normal subgroup of G, then $\mathsf{Fix}_\Omega(N_\omega) = \omega \mathsf{N}_G(N_\omega)$. In particular, $G_\omega \leqslant \mathsf{N}_G(N_\omega)$, $|\mathsf{Fix}_\Omega(N_\omega)| = |\mathsf{N}_G(N_\omega) : G_\omega|$, and $\mathsf{Fix}_\Omega(N_\omega)$ is a block for G in Ω.*

Proof If $g \in \mathsf{N}_G(N_\omega)$ and $n \in N_\omega$ then, as $n^{g^{-1}} \in N_\omega$,

$$\omega g n = \omega n^{g^{-1}} g = \omega g.$$

Hence $\omega g \in \mathsf{Fix}_\Omega(N_\omega)$, and so $\omega \mathsf{N}_G(N_\omega) \subseteq \mathsf{Fix}_\Omega(N_\omega)$. On the other hand, if $\omega_1 \in \mathsf{Fix}_\Omega(N_\omega)$, then, as G is transitive, there is some $g \in G$ such that $\omega_1 = \omega g$. Now, as $\omega_1 \in \mathsf{Fix}_\Omega(N_\omega)$ we have, for all $n \in N_\omega$, that $\omega g n = \omega g$, and as N is normal in G, therefore $g n g^{-1} \in G_\omega \cap N = N_\omega$. As this holds for all $n \in N_\omega$, we find that $g \in \mathsf{N}_G(N_\omega)$, and hence $\omega_1 = \omega g \in \omega \mathsf{N}_G(N_\omega)$. Therefore $\mathsf{Fix}_\Omega(N_\omega) = \omega \mathsf{N}_G(N_\omega)$, as claimed.

As N is a normal subgroup of G, $N_\omega = N \cap G_\omega$ is normal in G_ω, and so $G_\omega \leqslant \mathsf{N}_G(N_\omega)$. Hence $\mathsf{Fix}_\Omega(N_\omega)$ is a block for G in Ω by Lemma 2.14. We showed in the previous paragraph that $\mathsf{Fix}_\Omega(N_\omega)$ is an $\mathsf{N}_G(N_\omega)$-orbit. The stabiliser of ω in $\mathsf{N}_G(N_\omega)$ is G_ω, and so the Orbit-Stabiliser Theorem (Corollary 2.9) shows that $|\mathsf{Fix}_\Omega(N_\omega)| = |\mathsf{N}_G(N_\omega) : G_\omega|$. □

In particular, if G is transitive on Ω and $\omega \in \Omega$, then Lemma 2.19 shows that $\mathsf{Fix}_\Omega(G_\omega)$ is a block for G.

Lemma 2.20 *Let G be a group acting transitively on a set Ω, let N be a normal subgroup of G, and let Δ be an N-orbit. Then Δ is a block for the action of G on Ω.*

Proof Let us first show that Δ is a block. If $g \in G$ and $n \in N$, then

$$\Delta gn = \Delta n^{g^{-1}} g = \Delta g,$$

which shows that Δg is N-invariant, and, for $\delta g \in \Delta g$,

$$(\delta g)N = \{\delta gn \mid n \in N\} = \{\delta n^{g^{-1}} g \mid n \in N\} = \{\delta ng \mid n \in N\} = \Delta g.$$

Thus Δg is an N-orbit. By Lemma 2.2, either $\Delta = \Delta g$ or $\Delta \cap \Delta g = \emptyset$, and hence Δ is a block for G. □

Corollary 2.21 *Let G be a group acting primitively on a set Ω and let N be a normal subgroup of G which is not contained in the kernel of the action. Then N is transitive, and either N induces a regular permutation group on Ω, or $\mathsf{Fix}_\Omega(N_\omega) = \{\omega\}$ for each $\omega \in \Omega$. In particular, a non-trivial normal subgroup of a primitive permutation group is transitive.*

Proof As N is normal and is not contained in the kernel of the G-action on Ω, N is not contained in a point stabiliser. Thus an N-orbit Δ has size at least 2. As G is primitive and, by Lemma 2.20, Δ is a block, $\Delta = \Omega$ must hold. Therefore N is transitive. Let $\omega \in \Omega$ so that, by Lemma 2.19, $\mathsf{Fix}_\Omega(N_\omega)$ is a block for G containing ω. Since G acts primitively either $\mathsf{Fix}_\Omega(N_\omega) = \{\omega\}$ or $\mathsf{Fix}_\Omega(N_\omega) = \Omega$, and in the latter case N acts regularly on Ω. If G is a permutation group, then the kernel of the action is the trivial subgroup, and so no non-trivial normal subgroup can be contained in this kernel. This implies the last assertion of the corollary. □

Corollary 2.22 *If G is a primitive permutation group with a regular normal subgroup N, then N is a minimal normal subgroup.*

Proof Suppose that M is a normal subgroup of G and M is a proper subgroup of N. By Corollary 2.10(ii), M is intransitive (since N is regular). Since G is primitive, it follows from Corollary 2.21, that $M = 1$. Thus N is a minimal normal subgroup. □

Corollary 2.21 leads to two generalisations of the concept of primitivity. As defined in the Introduction, a permutation group is called *quasiprimitive* if each of its non-trivial normal subgroups is transitive. In the case of finite permutation groups, this is equivalent to the condition that every minimal normal subgroup is transitive. (See Chapter 3 for a detailed treatment of minimal normal subgroups of permutation groups.) Further, a permutation group is called *innately transitive* if it has a transitive minimal normal subgroup. Such a minimal normal subgroup is called a *plinth* of the innately transitive group. Thus a finite quasiprimitive group is innately transitive, and Corollary 2.21 implies that a primitive group is quasiprimitive.

2.6 The fixed points of subgroups of a point stabiliser

Lemma 2.19 shows us that, for $H \leqslant G_\omega$, the normaliser $\mathsf{N}_G(H)$ is transitive on the set $\mathsf{Fix}_\Omega(H)$ when $H = N_\omega$, for some normal subgroup N of G. We might ask whether the same is true for other subgroups H of G_ω. The following result of Manning (Manning 1918, Theorem XIV) dating from 1918 gives useful information about this in terms of the G-conjugacy class of H. We give a version for general (not necessarily finite) transitive groups.

Lemma 2.23 *Let $G \leqslant \mathsf{Sym}(\Omega)$, with G transitive, let $\omega \in \Omega$, $H \leqslant G_\omega$, and let $\mathcal{C} := H^G$ be the G-conjugacy class of H. Then the subset $\mathcal{C}'$ of those subgroups in $\mathcal{C}$ which lie in G_ω is a union of G_ω-classes, say $\mathcal{C}_i$, for $i \in I$. Then the following hold.*

(i) *For each $i \in I$, let $H_i \in \mathcal{C}_i$, with $H_1 = H$, let $g_i \in G$ such that $H_i = H^{g_i}$, and set $\omega_i := \omega g_i^{-1}$. Then $\omega_i \in \mathsf{Fix}_\Omega(H)$, and the action of $\mathsf{N}_G(H)$ on the orbit $\omega_i\mathsf{N}_G(H)$ is equivalent to the right coset action of $\mathsf{N}_G(H_i)$ on the coset space $[\mathsf{N}_G(H_i) : \mathsf{N}_{G_\omega}(H_i)]$. In particular $|\omega_i\mathsf{N}_G(H)| = |\mathsf{N}_G(H_i) : \mathsf{N}_{G_\omega}(H_i)|$.*

(ii) *The map $i \mapsto \omega_i\mathsf{N}_G(H)$ defines a one-to-one correspondence between I and the set of $\mathsf{N}_G(H)$-orbits in $\mathsf{Fix}_\Omega(H)$.*

Proof (i) Set $F = \mathsf{Fix}_\Omega(H)$. Since $H_i = H^{g_i} \leqslant G_\omega$, Lemma 2.1 implies that $H \leqslant (G_\omega)^{g_i^{-1}} = G_{\omega_i}$. Hence $\omega_i \in F$. The stabiliser of ω_i under the action of $\mathsf{N}_G(H)$ is $\mathsf{N}_{G_{\omega_i}}(H)$, and so, by Lemma 2.8, the $\mathsf{N}_G(H)$-action on the orbit $\omega_i\mathsf{N}_G(H)$ is equivalent to the coset action of

$\mathsf{N}_G(H)$ on $[\mathsf{N}_G(H) : \mathsf{N}_{G_{\omega_i}}(H)]$. Conjugation by g_i determines a permutational isomorphism from this action to the coset action of $\mathsf{N}_G(H_i)$ on $[\mathsf{N}_G(H_i) : \mathsf{N}_{G_\omega}(H_i)]$. In particular the cardinality of the orbit is as stated.

(ii) Consider an arbitrary point $\omega' \in F$. Then $H \leqslant G_{\omega'}$. If $h \in H$ and $n \in \mathsf{N}_G(H)$, then

$$\omega' nh = \omega' h^{n^{-1}} n = \omega' n,$$

and hence $\omega' n \in \mathsf{Fix}_\Omega(H)$. Thus F is invariant under the action of $\mathsf{N}_G(H)$ and hence F is a union of $\mathsf{N}_G(H)$-orbits. Let $g \in G$ such that $\omega' g = \omega$. Then $H^g \leqslant G_\omega$, and so $H^g \in \mathcal{C}_i$ for some i. Hence there exists $x \in G_\omega$ such that $H^g = H_i^x = H^{g_i x}$. Thus $g_i x g^{-1} \in \mathsf{N}_G(H)$, and so the $\mathsf{N}_G(H)$-orbit containing ω_i contains the point $\omega_i g_i x g^{-1} = \omega x g^{-1} = \omega g^{-1} = \omega'$. Thus each $\mathsf{N}_G(H)$-orbit in F contains at least one of the ω_i.

We are required to show that each $\mathsf{N}_G(H)$-orbit contained in F contains a unique ω_i with $i \in I$. Suppose that ω_i and ω_j lie in the same $\mathsf{N}_G(H)$-orbit; that is, $\omega_i = \omega_j x$ for some $x \in \mathsf{N}_G(H)$. Then $\omega g_i^{-1} = \omega_i = \omega_j x = \omega g_j^{-1} x$, and so $g_j^{-1} x g_i \in G_\omega$. Hence the class $\mathcal{C}_j$ contains $H_j^{g_j^{-1} x g_i} = H^{x g_i} = H^{g_i} = H_i$, and hence $i = j$. Hence for distinct i, j, the points ω_i, ω_j lie in distinct $\mathsf{N}_G(H)$-orbits in F. □

Corollary 2.24 *Let $G \leqslant \mathsf{Sym}(\Omega)$, with G transitive, let $\omega \in \Omega$, $H \leqslant G_\omega$. Then the group $\mathsf{N}_G(H)$ is transitive on $\mathsf{Fix}_\Omega(H)$ if and only if every G-conjugate of H lying in G_ω is in fact a G_ω-conjugate. In particular if G is finite and H is a Sylow subgroup of G_ω then $\mathsf{N}_G(H)$ is transitive on $\mathsf{Fix}_\Omega(H)$.*

The results in this section prove useful in various contexts. For instance, as shown in Example 2.25, they can be used to determine the orbit sizes of point stabilisers for a family of transitive group actions.

Example 2.25 Consider the group $G = \mathsf{PSL}_2(q)$ with $q \equiv 1 \pmod 4$ and let H be a maximal subgroup of G isomorphic to the dihedral group $D_{(q+1)/2}$ of order $q+1$. Consider the right coset action of G on $\Omega = [G\colon H]$ and let ω denote the trivial coset H in Ω. We claim that $H = G_\omega$ has one orbit of length 1, $(q-3)/2$ orbits of length $(q+1)/2$, and $(q-1)/4$ orbits of length $q+1$.

Note that $G_\omega \cong D_{(q+1)/2}$, and so $|G_\omega| = q+1 \equiv 2 \pmod 4$. Also since G_ω is maximal in G, G is primitive and G_ω fixes only the point

ω (Corollary 2.21). First set $K = \langle x \rangle < G_\omega$, with $|x| > 2$. Then $\mathsf{N}_G(K) = G_\omega$ and also K is the only cyclic subgroup of G_ω of order $|K|$. Hence Corollary 2.24 gives that $\mathsf{N}_G(K) = G_\omega$ is transitive on $\mathsf{Fix}_\Omega(K)$, and so $\mathsf{Fix}_\Omega(K) = \{\omega\}$. Thus for any $\omega' \neq \omega$, the two-point stabiliser $G_{\omega,\omega'}$ has order at most 2.

Next take $K = \langle x \rangle = C_2$ in G_ω. Since $|G_\omega|/2$ is odd, all involutions in G_ω are conjugate, so by Corollary 2.24 $\mathsf{N}_G(K)$ is transitive on $\mathsf{Fix}_\Omega(K)$, and by Lemma 2.23 $|\mathsf{Fix}_\Omega(K)| = |\mathsf{N}_G(K) : \mathsf{N}_{G_\omega}(K)|$.

Since $\mathsf{N}_G(K) \cong D_{(q-1)/2}$ and $\mathsf{N}_{G_\omega}(K) = K$, we have $|\mathsf{Fix}_\Omega(K)| = (q-1)/2$. For each point ω' in $\mathsf{Fix}_\Omega(K)$ with $\omega' \neq \omega$, the G_ω-orbit containing ω' has length $(q+1)/2$, and the permutation group induced by G_ω on this orbit is the natural action of $D_{(q+1)/2}$ with a stabiliser fixing just one point of the orbit. Hence there are exactly $(q-1)/2-1 = (q-3)/2$ orbits of G_ω of length $(q+1)/2$ with exactly one fixed point of K in each. Note that each involution in G_ω fixes one point in each of these orbits, since all involutions of G_ω are conjugate.

The remaining points all lie in G_ω-orbits of length $q+1$: there are $|G : G_\omega| - 1 - ((q-3)/2)((q+1)/2) = (q^2-1)/4$ remaining points, and hence $(q-1)/4$ orbits of length $q+1$.

2.7 Orbitals and orbital graphs

In this section we introduce a family of graphs and digraphs associated with a transitive permutation group. Such graphs and digraphs were first defined by D. G. Higman in the 1960s and they can be used to characterise several classes of permutation groups; in particular, they give beautiful primitivity criteria (see Propositions 2.38 and 2.40).

For a set Ω, let us define

$$\begin{aligned} \Omega^2 &= \{(\alpha,\beta) \mid \alpha,\ \beta \in \Omega\}; \\ \Omega^{(2)} &= \{(\alpha,\beta) \mid \alpha,\ \beta \in \Omega,\ \alpha \neq \beta\}; \\ \Omega^{\{2\}} &= \{\{\alpha,\beta\} \mid \alpha,\ \beta \in \Omega,\ \alpha \neq \beta\}. \end{aligned}$$

If G is a permutation group acting on Ω, then the G-action on Ω can be extended to Ω^2, $\Omega^{(2)}$, and to $\Omega^{\{2\}}$ by defining $(\alpha,\beta)g = (\alpha g, \beta g)$ and $\{\alpha,\beta\}g = \{\alpha g, \beta g\}$ for all α, $\beta \in \Omega$ and $g \in G$. A G-orbit in Ω^2 is said to be a *G-orbital*. Since the G-orbitals are orbits for a G-action, they form a partition of Ω^2 (Lemma 2.2). The set $\Delta_0 = \{(\alpha,\alpha) \mid \alpha \in \Omega\}$ is always a union of orbitals and Δ_0 is an orbital if and only if G is transitive. An orbital that is contained in Δ_0 is called a *diagonal orbital.*

If Δ is an orbital, then $\Delta^* = \{(\beta, \alpha) \mid (\alpha, \beta) \in \Delta\}$ is also an orbital and is called the *paired orbital* of Δ. An orbital Δ is said to be *self-paired* if $\Delta^* = \Delta$.

Example 2.26 Suppose that $\Omega = \underline{n}$, C_n is the cyclic group generated by the permutation $r = (1, 2, \ldots, n)$ and D_n is the dihedral group generated by r and t where

$$t = \begin{cases} (1, n)(2, n-1) \cdots (n/2, n/2 + 1) & \text{when } n \text{ is even;} \\ (1, n)(2, n-1) \cdots ((n-1)/2, (n-1)/2 + 2) & \text{when } n \text{ is odd.} \end{cases}$$

Then the group C_n has precisely n orbitals, namely

$$\begin{aligned} \Delta_0 &= \{(1,1), (2,2) \ldots, (n,n)\}; \\ \Delta_1 &= \{(1,2), (2,3) \ldots, (n,1)\}; \\ \Delta_2 &= \{(1,3), (2,4) \ldots, (n,2)\}; \\ &\vdots \\ \Delta_{n-2} &= \{(1,n-1), (2,n), \ldots, (n,n-2)\}; \\ \Delta_{n-1} &= \{(1,n), (2,1), \ldots, (n,n-1)\}. \end{aligned}$$

It is clear from this list, that Δ_1 is paired with Δ_{n-1}, while Δ_2 is paired with Δ_{n-2}. In general, Δ_i is paired with Δ_{n-i} for all $i \leqslant n-1$. Thus if n is even, then $\Delta_{n/2}$ is a self-paired non-diagonal C_n-orbital. If n is odd then there is no such self-paired non-diagonal C_n-orbital.

The permutation t permutes the C_n-orbitals Δ_i in such a way that t takes Δ_i to Δ_{n-i} for all $i \leqslant n-1$. Thus the D_n-orbitals in Ω are

$$\Delta_0,\ \Delta_1 \cup \Delta_{n-1},\ \Delta_2 \cup \Delta_{n-2}, \ldots, \Delta_{n/2-1} \cup \Delta_{n/2+1}, \Delta_{n/2}$$

when n is even, while the D_n-orbitals are

$$\Delta_0,\ \Delta_1 \cup \Delta_{n-1},\ \Delta_2 \cup \Delta_{n-2}, \ldots, \Delta_{(n-1)/2} \cup \Delta_{(n-1)/2+1}$$

when n is odd. Note that each of the D_n-orbitals is self-paired.

Lemma 2.27 *Suppose that Ω is a finite set and that $G \leqslant \mathsf{Sym}\,\Omega$. Then G admits a self-paired non-diagonal orbital if and only if $|G|$ is even.*

Proof If $|G|$ is even, then G contains an involution g. Since g is an element of order 2, there exist distinct elements α, $\beta \in \Omega$ such that $\alpha g = \beta$ and $\beta g = \alpha$. Then $(\alpha, \beta)g = (\beta, \alpha)$, and so (α, β) and (β, α)

are elements of the same non-diagonal orbital Δ. Thus $\Delta \cap \Delta^* \neq \emptyset$, and so $\Delta = \Delta^*$.

Conversely, suppose that Δ is a self-paired non-diagonal orbital of G and let $(\alpha, \beta) \in \Delta$. Then $\alpha \neq \beta$, and $(\beta, \alpha) \in \Delta^* = \Delta$, and hence there exists some $g \in G$ such that $(\alpha, \beta)g = (\beta, \alpha)$. Thus the disjoint cycle decomposition of g contains the transposition (α, β), and so g is an element of even order. Since the order of the element g divides $|G|$, we obtain that $|G|$ is even. □

If $G \leqslant \mathsf{Sym}\,\Omega$ is a transitive group and $\alpha \in \Omega$, then a G_α-orbit in Ω is said to be a *G-suborbit* relative to α. Note that $\{\alpha\}$ is always a suborbit relative to α, and so if $|\Omega| \geqslant 2$, then G has at least two suborbits.

Lemma 2.28 *Suppose that $G \leqslant \mathsf{Sym}\,\Omega$ is a transitive group, let $\Delta \subseteq \Omega^2$ be a G-orbital, and let $\alpha \in \Omega$. Then the set*

$$\Delta(\alpha) = \{\beta \in \Omega \mid (\alpha, \beta) \in \Delta\}$$

is a G-suborbit relative to α. Further, the correspondence $\Delta \mapsto \Delta(\alpha)$ is a bijection between the set of G-orbitals and the set of G-suborbits relative to α.

Proof If $\beta \in \Delta(\alpha)$, then $(\alpha, \beta) \in \Delta$, and hence, for $g \in G_\alpha$, $(\alpha, \beta)g = (\alpha, \beta g) \in \Delta$, and so $\beta g \in \Delta(\alpha)$. Hence $\Delta(\alpha)$ is invariant under G_α. If β_1, $\beta_2 \in \Delta(\alpha)$, then (α, β_1), $(\alpha, \beta_2) \in \Delta$, and hence there exists some $g \in G$ such that $(\alpha, \beta_1)g = (\alpha, \beta_2)$. In other words, $g \in G_\alpha$ and $\beta_1 g = \beta_2$. Hence $\Delta(\alpha)$ is a G-suborbit relative to α.

Consider the map $\psi\colon \Delta \mapsto \Delta(\alpha)$. We claim that ψ is injective. Suppose that $\Delta_1(\alpha) = \Delta_2(\alpha)$ for two orbitals Δ_1 and Δ_2 and let $\beta \in \Delta_1(\alpha)$. Then $(\alpha, \beta) \in \Delta_1 \cap \Delta_2$, and hence $\Delta_1 = \Delta_2$. Let us now verify that ψ is surjective. Let Γ be a G-suborbit relative to α and $\beta \in \Gamma$. Set $\Delta = (\alpha, \beta)G$. Then Δ is an orbital by definition, and the suborbit $\Delta\psi = \Delta(\alpha)$ contains β, and hence must be equal to Γ. Therefore ψ is also surjective, and so ψ is a bijection, as claimed. □

Example 2.29 Let $\Sigma = \underline{n}$ with $n \geqslant 3$, let $G = \mathsf{Sym}\,\Sigma$ and consider the G-action on $\Omega = \Sigma^{(2)}$. Since the set Ω^2 consists of ordered pairs of ordered pairs of elements of Σ, it is not easy to visualise the G-orbitals in Ω^2. In this case, we can more easily calculate the set of G_α-orbits for $\alpha = (1,2)$. In fact, G_α has 7 orbits in Ω^2:

$$\begin{aligned}
\Sigma_0 &= \{(1,2)\};\\
\Sigma_1 &= \{(2,1)\};\\
\Sigma_2 &= \{(1,i) \mid i \geqslant 3\};\\
\Sigma_3 &= \{(2,i) \mid i \geqslant 3\};\\
\Sigma_4 &= \{(i,1) \mid i \geqslant 3\};\\
\Sigma_5 &= \{(i,2) \mid i \geqslant 3\};\\
\Sigma_6 &= \{(i,j) \mid i,\ j \geqslant 3\}.
\end{aligned}$$

Therefore, by Lemma 2.28, there are 7 G-orbitals $\Delta_0, \ldots, \Delta_6$ in Ω^2, where Δ_i is the G-orbit of $((1,2),\beta)$ with β an arbitrary point in Σ_i.

The 2-homogeneous and 2-transitive groups form important families of permutation groups. A permutation group G acting on Ω is 2-*homogeneous* if G is transitive on $\Omega^{\{2\}}$, while G is 2-*transitive* if G is transitive on $\Omega^{(2)}$. It is immediate from the definition that a 2-transitive group is 2-homogeneous.

Suppose that G is imprimitive and $\{\Delta_1, \ldots, \Delta_\ell\}$ is a non-trivial system of imprimitivity for G. Let α_1, α_2, β_1, $\beta_2 \in \Omega$ such that $\alpha_1 \neq \alpha_2$, $\beta_1 \neq \beta_2$, α_1, α_2, $\beta_1 \in \Delta_1$, while $\beta_2 \in \Delta_2$. Then there is no element $g \in G$ such that $\{\alpha_1, \alpha_2\}g = \{\beta_1, \beta_2\}$ and hence G is not 2-homogeneous, and hence not 2-transitive. This proves the following lemma.

Lemma 2.30 *Let G be a permutation group on Ω such that $G \neq 1$. If G is 2-homogeneous or 2-transitive, then G is primitive.*

The condition that $G \neq 1$ in the previous lemma is necessary; indeed, the trivial group G acting on a two-element set is 2-homogeneous, as in this case $|\Omega^{\{2\}}| = 1$. However, this group G is imprimitive; in fact, it is intransitive.

For a transitive group G, we define the *rank* of G as the cardinality of the set of G-orbitals, or, equivalently (by Lemma 2.28), the cardinality of the set of G-suborbits. The 2-transitive permutation groups can be characterised using their rank and their suborbits.

Lemma 2.31 *If $|\Omega| \geqslant 2$, then the following are equivalent for a transitive group $G \leqslant \mathsf{Sym}\,\Omega$:*

(i) *G is 2-transitive;*

(ii) *G has rank 2;*
(iii) *for $\alpha \in \Omega$, the point stabiliser G_α is transitive on $\Omega \setminus \{\alpha\}$.*

Proof If G is 2-transitive, then $\Omega^{(2)}$ is a G-orbital, and hence G has two orbitals in Ω^2; namely, $\Omega^{(2)}$ and the diagonal orbital $\Delta_0 = \{(\alpha, \alpha) \mid \alpha \in \Omega\}$. Thus, G has rank 2. Now if G has rank 2 and $\alpha \in \Omega$, then, by Lemma 2.28, there are precisely two G-suborbits relative to α; that is, there are two G_α-orbits in Ω. Since $\{\alpha\}$ and $\Omega\setminus\{\alpha\}$ are G_α-invariant, they must be the two G_α-orbits. Thus (iii) must hold. Finally, if (iii) holds, then the number of G-suborbits is two, and so G has two orbitals in Ω^2, by Lemma 2.28. Since G is transitive, one of these orbitals is Δ_0, and so the other one must be $\Omega^2 \setminus \Delta_0 = \Omega^{(2)}$. Therefore G is transitive on $\Omega^{(2)}$, which implies that G is 2-transitive, and so (i) holds. □

Remark 2.32 We will prove in Section 3.5 that a finite 2-transitive permutation group has a unique minimal normal subgroup that is either elementary abelian or simple. The classification of the finite 2-transitive permutation groups is one of the most celebrated consequences of the finite simple group classification (see (Cameron 1999, Table 7.4) or (Dixon & Mortimer 1996, Section 7.7)). It is also known that a finite 2-homogeneous group is almost always 2-transitive. The exceptions were determined by (Kantor 1969, Proposition 3.1) and are explicitly listed in (Dixon & Mortimer 1996, Theorem 9.4B).

If $\Delta \subseteq \Omega^{(2)}$ then the pair $\mathfrak{G} = (\Omega, \Delta)$ is referred to as a *directed graph* or as a *digraph*, while if $\Delta \subseteq \Omega^{\{2\}}$, then $\mathfrak{G} = (\Omega, \Delta)$ is called an *undirected graph* or simply a *graph*. An element $\alpha \in \Omega$ is said to be a *vertex* of $\mathfrak{G}$ and Ω is called the *vertex set* of $\mathfrak{G}$, and is usually written as $\mathsf{V}(\mathfrak{G})$.

For a directed graph $\mathfrak{G} = (\Omega, \Delta)$, distinct vertices α, $\beta \in \mathsf{V}(\mathfrak{G})$ are said to be *adjacent* if either $(\alpha, \beta) \in \Delta$ or $(\beta, \alpha) \in \Delta$; a pair $(\alpha, \beta) \in \Delta$ is called an *arc* of $\mathfrak{G}$. For an undirected graph $\mathfrak{G} = (\Omega, \Delta)$, distinct points α, $\beta \in \Omega$ are *adjacent* if $\{\alpha, \beta\} \in \Delta$; a pair $\{\alpha, \beta\} \in \Delta$ is referred to as an *edge* of $\mathfrak{G}$. The set of edges of an undirected graph $\mathfrak{G}$ is denoted $\mathsf{E}(\mathfrak{G})$. An *arc* in an undirected graph $\mathfrak{G}$ is a pair (α, β) of vertices such that $\{\alpha, \beta\}$ is an edge in $\mathfrak{G}$. Hence in an undirected graph, (α, β) is an arc if and only if (β, α) is an arc. Note that our definition of directed and undirected graphs does not allow arcs of the form (α, α). Other authors allow such arcs and they often call them loops.

Example 2.33 Graphs and digraphs are often represented by diagrams

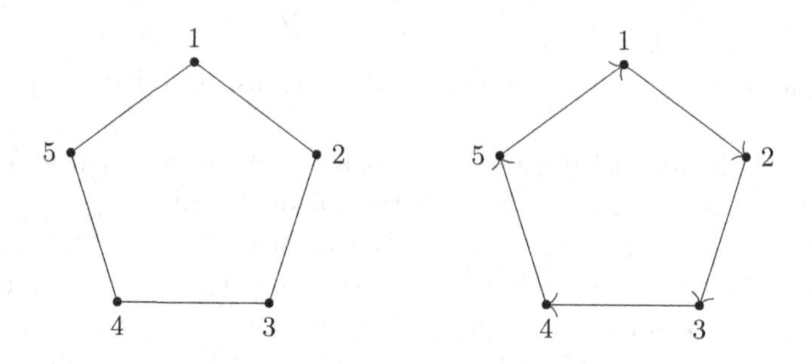

Fig. 2.1. Graphical representation of a graph and of a digraph

where vertices are represented by nodes, edges are represented by connections (lines) between the nodes, while arcs are represented by arrows. For instance the first diagram of Figure 2.1 represents the graph

$$\mathfrak{C}_5 = (\{1,\ldots,5\},\{\{1,2\},\{2,3\},\{3,4\},\{4,5\},\{5,1\}\}),$$

while the second diagram represents the digraph

$$\vec{\mathfrak{C}}_5 = (\{1,\ldots,5\},\{(1,2),(2,3),(3,4),(4,5),(5,1)\}).$$

Suppose now that $\mathfrak{G} = (\Omega,\Delta)$ is a directed graph or an undirected graph. A sequence of vertices $(\alpha_0,\alpha_1,\ldots,\alpha_k)$ is said to be a *path* if either $k=0$ or α_i is adjacent to α_{i+1} for all $i \in \{0,\ldots,k-1\}$. A *cycle* is a path of the form $(\alpha_0,\alpha_1,\ldots,\alpha_{k-1},\alpha_0)$. The sequence $(\alpha_0,\ldots,\alpha_k)$ is said to be a k*-arc*, if (α_i,α_{i+1}) is an arc in $\mathfrak{G}$ for all $i \in \{0,\ldots,k-1\}$ and $\alpha_{i+1} \neq \alpha_{i-1}$ for all $i = 1,\ldots,k-1$. Thus a 2*-arc* is a sequence (α,β,γ) such that (α,β) and (β,γ) are arcs and $\alpha \neq \gamma$. A 1-arc is simply an arc.

If $\mathfrak{G} = (\Omega,\Delta)$ is a directed graph, then a path $(\alpha_0,\ldots,\alpha_k)$ or a cycle $(\alpha_0,\ldots,\alpha_{k-1},\alpha_k)$ with $\alpha_k = \alpha_0$ is said to be *directed* if $(\alpha_i,\alpha_{i+1}) \in \Delta$ for all $i \in \{0,\ldots,k-1\}$. Thus in a digraph, a k-arc is a directed path.

If $g \in \mathsf{Sym}\,\Omega$ and $\mathfrak{G} = (\Omega,\Delta)$ is a graph or a digraph such that $\Delta g = \Delta$, then we say that g is an *automorphism* of $\mathfrak{G}$. The set of automorphisms of a graph or a digraph $\mathfrak{G}$ is a subgroup of $\mathsf{Sym}\,\Omega$ and is denoted $\mathsf{Aut}(\mathfrak{G})$. The (di)graph $\mathfrak{G}$ is said to be *vertex-transitive, edge-transitive, arc-transitive,* k*-arc-transitive* (for a fixed k), if $\mathsf{Aut}(\mathfrak{G})$ is transitive on the set of vertices, on the set of edges, on the set of arcs, or on the set of k-arcs, respectively. If G is a subgroup of $\mathsf{Aut}(\mathfrak{G})$, then we say that $\mathfrak{G}$ is G*-vertex-transitive*, G*-edge-transitive*, G*-arc-transitive*, (G,k)*-arc-transitive*, if G is transitive on the set of vertices, on the set of edges, on the set of arcs, or on the set of k-arcs, respectively.

Example 2.34 Consider the graphs $\mathfrak{C}_5$ and $\vec{\mathfrak{C}}_5$ defined in Figure 2.1. The cyclic group C_5 generated by the 5-cycle $(1,2,3,4,5)$ acts as automorphisms on both $\mathfrak{C}_5$ and $\vec{\mathfrak{C}}_5$. Identify the vertex set $\{1,2,3,4,5\}$ of these graphs with the additive group of the residue class ring $\mathbb{Z}/5\mathbb{Z}$ by writing the zero element of this ring as '5'. Then every k-arc in $\vec{\mathfrak{C}}_5$ has the form $(i,i+1,\ldots,i+k)$, whereas k-arcs in $\mathfrak{C}_5$ have the form of either $(i,i+1,\ldots,i+k)$ or $(i,i-1,\ldots,i-k)$ for some $k\in\{1,\ldots,5\}$. In particular in $\vec{\mathfrak{C}}_5$ there is a unique k-arc that originates from the vertex i for all i, while in $\mathfrak{C}_5$ there are two such k-arcs. Hence the group C_5 is transitive on the set of k-arcs in $\vec{\mathfrak{C}}_5$. In the case of $\mathfrak{C}_5$, the permutation $(2,5)(3,4)$ is an automorphism of $\mathfrak{C}_5$ and the group $D_5=\langle(1,2,3,4,5),(2,5)(3,4)\rangle$ is transitive on the set of k-arcs of $\mathfrak{C}_5$. In other words the graphs $\vec{\mathfrak{C}}_5$ and $\mathfrak{C}_5$ are (C_5,k)-arc-transitive and (D_5,k)-arc-transitive, respectively, for all $k\geqslant 1$.

The following lemma follows immediately from the definition of digraphs and orbitals.

Lemma 2.35 *Suppose that Ω is a set, $\Delta\subseteq\Omega^{(2)}$ and let $G\leqslant\mathsf{Sym}\,\Omega$. Let $\mathfrak{G}=(\Omega,\Delta)$. Then $G\leqslant\mathsf{Aut}(\mathfrak{G})$ if and only if Δ is a union of non-diagonal G-orbitals. Further, $G\leqslant\mathsf{Aut}(\mathfrak{G})$ and $\mathfrak{G}$ is G-arc-transitive if and only if Δ is a G-orbital.*

Suppose that $G\leqslant\mathsf{Sym}\,\Omega$ and $\Delta\subseteq\Omega^{(2)}$ is a non-diagonal orbital for G. Then the pair $\mathfrak{G}=(\Omega,\Delta)$ is said to be an *orbital digraph* for G. Considering non-diagonal orbitals in this definition is necessary, since our concept of digraphs does not allow loops. If $\mathfrak{G}=(\Omega,\Delta)$ is an orbital digraph, then there is a corresponding *orbital graph* $\mathfrak{G}=(\Omega,\overline{\Delta})$ where

$$\overline{\Delta}=\{\{\alpha,\beta\}\in\Omega^{\{2\}}\mid(\alpha,\beta)\in\Delta\}.$$

Using these new concepts of orbital graphs and orbital digraphs, we can draw the following corollary to Lemma 2.35.

Corollary 2.36 *Suppose that $G\leqslant\mathsf{Sym}\,\Omega$ and let $\Delta\subseteq\Omega^{(2)}$ be a non-diagonal G-orbital. Let $\mathfrak{G}=(\Omega,\Delta)$ and $\overline{\mathfrak{G}}=(\Omega,\overline{\Delta})$ be the orbital digraph and the orbital graph defined above, respectively. Then $\mathfrak{G}$*

is G-arc-transitive and $\overline{\mathfrak{G}}$ is G-edge-transitive. Further, $\overline{\mathfrak{G}}$ is G-arc-transitive if and only if Δ is a self-paired orbital.

Proof The graph $\mathfrak{G}$ is G-arc-transitive, by Lemma 2.35. Suppose that $\{\alpha, \beta\}$ and $\{\gamma, \delta\}$ are edges of $\overline{\mathfrak{G}}$. Then we may assume without loss of generality that (α, β), $(\gamma, \delta) \in \Delta$. Since Δ is a G-orbital, there is $g \in G$ such that $(\alpha, \beta)g = (\gamma, \delta)$, and hence $\{\alpha, \beta\}g = \{\gamma, \delta\}$. Therefore $\overline{\mathfrak{G}}$ is G-edge-transitive. Finally, the arc set of $\overline{\mathfrak{G}}$ is $\Delta \cup \Delta^*$, and hence $\overline{\mathfrak{G}}$ is G-arc-transitive if and only if $\Delta \cup \Delta^*$ is a G-orbit, which occurs if and only if $\Delta = \Delta^*$; that is, Δ is self-paired. □

For a graph or a digraph $\mathfrak{G}$ and a vertex $\alpha \in \mathsf{V}(\mathfrak{G})$, the *neighbourhood* $\mathfrak{G}(\alpha)$ is defined as

$$\mathfrak{G}(\alpha) = \{\beta \in \mathsf{V}(\mathfrak{G}) \mid (\alpha, \beta) \text{ is an arc in } \mathfrak{G}\}.$$

If $G \leqslant \mathsf{Aut}(\mathfrak{G})$, then G_α preserves $\mathfrak{G}(\alpha)$ setwise and hence the G_α-action on $\mathfrak{G}(\alpha)$ induces a subgroup of $\mathsf{Sym}\,\mathfrak{G}(\alpha)$. The following lemma relates the study of $(G, 2)$-arc-transitive graphs to the study of 2-transitive permutation groups.

Lemma 2.37 *Suppose that $\mathfrak{G}$ is a G-vertex transitive graph or digraph and let $\alpha \in \mathsf{V}(\mathfrak{G})$. Then*

(i) *$\mathfrak{G}$ is G-arc-transitive if and only if G_α is transitive on $\mathfrak{G}(\alpha)$.*
(ii) *$\mathfrak{G}$ is $(G, 2)$-arc-transitive if and only if G_α is 2-transitive on $\mathfrak{G}(\alpha)$.*

Proof (i) Suppose that $\mathfrak{G}$ is G-arc-transitive and let β_1, $\beta_2 \in \mathfrak{G}(\alpha)$. Then (α, β_1) and (α, β_2) are arcs in $\mathfrak{G}$, and hence there exists some $g \in G$ such that $(\alpha, \beta_1)g = (\alpha, \beta_2)$. Thus $\alpha g = \alpha$ and $\beta_1 g = \beta_2$. Therefore G_α is transitive on $\mathfrak{G}(\alpha)$.

Conversely, suppose that G is transitive on $\mathsf{V}(\mathfrak{G})$ and that G_α is transitive on $\mathfrak{G}(\alpha)$. Suppose that (α_1, β_1) and (α_2, β_2) are arcs in $\mathfrak{G}$. By vertex-transitivity, there are g_1, $g_2 \in G$, such that $\alpha_1 g_1 = \alpha$ and $\alpha_2 g_2 = \alpha$. Then $\beta_1 g_1$ and $\beta_2 g_2$ are elements of $\mathfrak{G}(\alpha)$, and hence there exists some $g \in G_\alpha$ such that $\beta_1 g_1 g = \beta_2 g_2$. Then $(\alpha_1, \beta_1) g_1 g g_2^{-1} = (\alpha_2, \beta_2)$, which shows that $\mathfrak{G}$ is G-arc-transitive.

(ii) Suppose first that $\mathfrak{G}$ is $(G, 2)$-arc-transitive and let α_1, α_2, β_1, $\beta_2 \in \mathfrak{G}(\alpha)$ be such that $\alpha_1 \neq \alpha_2$ and $\beta_1 \neq \beta_2$. Then $(\alpha_1, \alpha, \alpha_2)$ and $(\beta_1, \alpha, \beta_2)$ are 2-arcs, and hence there exists $g \in G$ such that

$(\alpha_1, \alpha, \alpha_2)g = (\beta_1, \alpha, \beta_2)$. This implies that $g \in G_\alpha$ and $(\alpha_1, \alpha_2)g = (\beta_1, \beta_2)$. Thus G_α is 2-transitive on $\mathfrak{G}(\alpha)$.

Suppose now that $\mathfrak{G}$ is G-vertex transitive and G_α is 2-transitive on $\mathfrak{G}(\alpha)$. Let $(\alpha_0, \alpha_1, \alpha_2)$ and $(\beta_0, \beta_1, \beta_2)$ be 2-arcs in $\mathfrak{G}$. By vertex-transitivity, there exist g_1, $g_2 \in G$ such that $\alpha_1 g_1 = \beta_1 g_2 = \alpha$. Therefore $\alpha_0 g_1$, $\alpha_2 g_1$, $\beta_0 g_2$, $\beta_2 g_2 \in \mathfrak{G}(\alpha)$ such that $\alpha_0 g_1 \neq \alpha_2 g_1$ and $\beta_0 g_2 \neq \beta_2 g_2$. By the 2-transitivity of G_α on $\mathfrak{G}(\alpha)$, there exists $g \in G_\alpha$ such that $(\alpha_0 g_1, \alpha_2 g_1)g = (\beta_0 g_2, \beta_2 g_2)$. Then $(\alpha_0, \alpha_1, \alpha_2)g_1 g g_2^{-1} = (\beta_0, \beta_1, \beta_2)$, and so $\mathfrak{G}$ is $(G, 2)$-arc-transitive. □

As finite 2-transitive groups are classified (Remark 2.32), Lemma 2.37 is particularly useful in the study of finite 2-arc-transitive graphs and it will be invoked several times in Section 12.7.

A graph or digraph $\mathfrak{G}$ is *connected* if either $|\mathsf{V}(\mathfrak{G})| = 1$ or, for all distinct α, $\beta \in \mathsf{V}(\mathfrak{G})$, there is a path $(\alpha, \alpha_1, \ldots, \alpha_k, \beta)$ from α to β. A graph or a digraph is *disconnected* when it is not connected. If $\mathfrak{G}$ is a digraph, then $\mathfrak{G}$ is said to be *strongly connected* if either $|\mathsf{V}(\mathfrak{G})| = 1$ or, for all distinct α, $\beta \in \mathsf{V}(\mathfrak{G})$, there is a directed path $(\alpha, \alpha_1, \ldots, \alpha_k, \beta)$ from α to β. For a graph or digraph $\mathfrak{G}$, we can define a relation $\sim$ on $\mathsf{V}(\mathfrak{G})$ by setting $\alpha \sim \beta$ if and only if there is a path from α to β. The relation $\sim$ is clearly an equivalence relation and an equivalence class of $\sim$ in $\mathsf{V}(\mathfrak{G})$ is said to be a *connected component* of $\mathfrak{G}$.

It is interesting that the most widely used classes of permutation groups can be characterised using orbital graphs or orbital digraphs. For instance, $G \leqslant \mathsf{Sym}\,\Omega$ is 2-transitive if and only if $(\Omega, \Omega^{(2)})$ is G-arc-transitive, while G is 2-homogeneous if and only if $(\Omega, \Omega^{\{2\}})$ is G-edge transitive. The following fundamental result due to D. G. Higman (Higman 1967, (1.12)) presents such a characterisation for primitive groups. It is worth noting the class of separating and the more recently introduced class of synchronising groups can also be characterised using orbital graphs; see (Araújo, Cameron & Steinberg 2017, Corollary 4.5 and Theorem 5.3).

Proposition 2.38 *A transitive permutation group G acting on Ω is primitive if and only if each of the non-diagonal orbital digraphs of G is connected.*

Proof Suppose that G is primitive and let $\mathfrak{G} = (\Omega, \Delta)$ be a non-diagonal orbital digraph for G. The relation $\sim$ on the vertex set of $\mathfrak{G}$ introduced before the proposition is an equivalence relation on Ω and it is preserved by G, in the sense that if $g \in G$ and α, $\beta \in \Omega$ such that $\alpha \sim \beta$, then $\alpha g \sim \beta g$. Thus the equivalence classes of $\sim$ form a G-invariant partition of Ω, and since G is primitive, this partition is either $\{\Omega\}$ or the partition into singletons. Since $\Delta \neq \emptyset$, $\mathfrak{G}$ has at least one arc, and so Ω must be the unique equivalence class of $\sim$. Hence $\mathfrak{G}$ is connected.

Suppose now that G is imprimitive and that $\{\Delta_1, \ldots, \Delta_s\}$ is a non-trivial G-invariant partition of Ω. Set

$$\Sigma = \{(\alpha, \beta) \mid \alpha \neq \beta \text{ and } \alpha,\ \beta \in \Delta_i \text{ for some } i\}$$

and set $\mathfrak{G} = (\Omega, \Sigma)$. Then the blocks $\Delta_1, \ldots, \Delta_s$ are the connected components of $\mathfrak{G}$, and so $\mathfrak{G}$ is not a connected graph. Further, $G \leqslant \mathsf{Aut}(\mathfrak{G})$ and hence, by Lemma 2.35, Σ is a union of non-diagonal G-orbitals. Suppose that Δ is a G-orbital contained in Σ. Then $\overline{\mathfrak{G}} = (\Omega, \Delta)$ is a G-orbital digraph that is not connected. □

Example 2.39 Consider the groups C_n and D_n acting on the set $\Omega = \underline{n}$ defined in Example 2.26. If n is a composite number and $n = ab$ with a, $b \geqslant 2$, then the orbital digraph $\mathfrak{G} = (\Omega, \Delta_a)$ is disconnected. Indeed, the arc set of $\mathfrak{G}$ is

$$\{(1, a+1), (2, a+2), \ldots, (n-1, a-1), (n, a)\}$$

and $\Sigma_i = \{i, i+a, \ldots, i+(b-1)a\}$, for $i = 1, \ldots, a$, are the connected components of $\mathfrak{G}$. Let us consider now the graph $\overline{\mathfrak{G}} = (\Omega, \Delta_a \cup \Delta_{n-a})$. It is an orbital graph for D_n. Since the arc set of $\overline{\mathfrak{G}}$ is the union of two paired orbitals for C_n, we can view $\overline{\mathfrak{G}}$ as an undirected graph in which two vertices are adjacent if and only if they are adjacent in $\mathfrak{G}$. Since connectedness does not depend on the orientation of the arcs, the graph $\overline{\mathfrak{G}}$ is also disconnected with the same connected components as $\mathfrak{G}$.

Consider now the example presented in Example 2.29. Let $\mathfrak{G} = (\Omega, \Delta_1)$ where Δ_1 is the G-orbital containing $((1,2),(2,1))$; that is,

$$\Delta_1 = \{((i,j),(j,i)) \mid i,\ j \in \Sigma \text{ and } i \neq j\}.$$

Then Δ_1 is self-paired and every vertex (i,j) of $\mathfrak{G}$ is adjacent only to the vertex (j,i), and so $\mathfrak{G}$ is a disconnected orbital digraph. Thus G is not primitive on Ω.

For finite primitive groups the orbital digraphs are strongly connected as shown by the following result. This need not hold for infinite groups; see Example 2.41.

Proposition 2.40 *Suppose that $G \leqslant \mathsf{Sym}\,\Omega$ is a primitive group and assume that for each $(\alpha,\beta) \in \Omega^{(2)}$, there exists $g \in G$ such that the disjoint cycle representation of g contains a finite cycle that contains α and β. Then for each non-diagonal orbital Δ for G, the digraph (Ω,Δ) is strongly connected. In particular, if G is finite and $\mathfrak{G}$ is an orbital digraph for a non-diagonal orbital for G, then $\mathfrak{G}$ is strongly connected.*

Proof Suppose that $\mathfrak{G} = (\Omega,\Delta)$ is an orbital digraph for some non-diagonal orbital Δ. Define the relation $\equiv$ on Ω by setting $\alpha \equiv \beta$ if and only if there is a directed cycle in $\mathfrak{G}$ that contains α and β. Then $\equiv$ is an equivalence relation such that the equivalence classes form a G-invariant partition $\mathcal{P}$ of Ω. Since G is primitive, either $\mathcal{P} = \{\{\omega\} \mid \omega \in \Omega\}$ or $\mathcal{P} = \{\Omega\}$. We show that the former option is impossible. Let $(\alpha,\beta) \in \Delta$ and $g \in G$ such that the cycle representation of g contains a finite cycle that contains the points α and β. Let $m \geqslant 1$ and $h = g^m$ such that $\alpha h = \beta$. Since $(\alpha,\beta) \in \Delta$, it follows that $(\alpha h^i, \alpha h^{i+1}) = (\alpha,\beta)h^i \in \Delta$ for all i. Moreover, since the g-cycle containing α and β is finite, this is also true for the h-cycle containing α and $\beta = \alpha h$, and so $\alpha h^n = \alpha$ for some $n \geqslant 1$. Then $(\alpha, \alpha h, \alpha h^2, \ldots, \alpha h^n = \alpha)$ is a directed cycle in $\mathfrak{G}$ containing $\alpha = \alpha h^n$ and $\beta = \alpha h$. Thus the G-invariant partition $\mathcal{P}$ determined by the relation $\equiv$ does not consist of singletons. This implies that $\mathcal{P} = \{\Omega\}$ and hence every pair of distinct points α, $\beta \in \Omega$ is contained in a directed cycle in $\mathfrak{G}$. Hence $\mathfrak{G}$ is strongly connected.

Suppose that G is finite. Then for $(\alpha,\beta) \in \Omega^{(2)}$, there is $g \in G$ such that $\alpha g = \beta$, and hence the disjoint cycle representation of g contains a finite cycle that starts as $(\alpha,\beta,\ldots)$. This proves the final assertion of the theorem. □

Example 2.41 Proposition 2.40 is not true for infinite groups. Consider for example the set G of order-preserving permutations of $\mathbb{Q}$. Then $G \leqslant \mathsf{Sym}\,\mathbb{Q}$. Further, by (Dixon & Mortimer 1996, Exercise 2.2.8), G is 2-homogeneous, but not 2-transitive on $\mathbb{Q}$. By Lemma 2.30, G is primitive on $\mathbb{Q}$ and G has two non-diagonal orbitals Δ and its pair Δ^* where

$$\Delta = \{(\alpha,\beta) \in \mathbb{Q}^{(2)} \mid \alpha < \beta\}.$$

Now if α, $\beta \in \mathbb{Q}$ such that $\alpha < \beta$, then $(\alpha,\beta) \in \Delta$ and $(\beta,\alpha) \in \Delta^*$, and hence all pairs of distinct vertices are adjacent in the digraphs $\mathfrak{G} = (\mathbb{Q},\Delta)$ and $\mathfrak{G}^* = (\mathbb{Q},\Delta^*)$; thus $\mathfrak{G}$ and $\overline{\mathfrak{G}}$ are connected. However, there is no directed path from β to α in $\mathfrak{G}$ and there is no directed path from α to β in $\mathfrak{G}^*$. Hence the digraphs $\mathfrak{G}$ and $\mathfrak{G}^*$ are not strongly connected.

3
Minimal normal subgroups of transitive permutation groups

In Chapter 3 we focus on the structure of transitive minimal normal subgroups of permutation groups. *In this chapter also, the results usually hold for general (not necessarily finite) groups, and if the group needs to be finite, this will be stated explicitly.* We emphasise that infinite groups may not have minimal normal subgroups (take, for instance, the infinite cyclic group). However we show in Theorem 3.18 that, for certain significant infinite families of infinite transitive groups with finite stabilisers, minimal normal subgroups do exist. First, in Section 3.1, we study centralisers of transitive permutation groups. Section 3.3 is devoted to the study of the holomorph. We prove in Theorem 3.10 that every permutation group with a regular normal subgroup can be embedded into the holomorph of that normal subgroup (Holomorph Embedding Theorem). The final sections, Sections 3.5 and 3.6, are the only places in the chapter where the groups are assumed to be finite. In Section 3.5, we give an elementary proof to Burnside's Theorem on the socles of finite 2-transitive permutation groups, and in Section 3.6 we study stabilisers in finite primitive groups with regular normal subgroups.

3.1 The centraliser of a transitive permutation group

In the investigation of primitive, quasiprimitive, and innately transitive groups, it is often important to determine the centraliser and the normaliser of a given transitive permutation group. We denote the centraliser of a subgroup H of G by $\mathsf{C}_G(H)$, namely

$$\mathsf{C}_G(H) = \{g \in G \mid h^g = h \text{ for all } h \in H\}.$$

In particular, the *centre* of G is $\mathsf{C}_G(G)$ and is denoted by $\mathsf{Z}(G)$. The results of this chapter provide the necessary tools for the determination of the centraliser $\mathsf{C}_{\mathsf{Sym}\,\Omega}(G)$ of a transitive subgroup G of $\mathsf{Sym}\,\Omega$.

Let us determine the centraliser of a coset action defined in (2.1). Suppose that G is a group and let H be a subgroup of G. Consider the right coset action ϱ_H of G on $[G\colon H]$. As noted before, this is a transitive action with point stabiliser H. Let us define an action λ_H of $\mathsf{N}_G(H)$ on the coset space $[G\colon H]$. For $h \in \mathsf{N}_G(H)$ and $Hg \in [G\colon H]$ let

$$Hg(h\lambda_H) = h^{-1}Hg = Hh^{-1}g.$$

The proof that λ_H is an action is standard and is left to the reader.

Lemma 3.1 *Using the notation above, the following hold:*

(i) $\mathsf{N}_G(H)\,\lambda_H = \mathsf{C}_{\mathsf{Sym}[G\colon H]}(G\varrho_H)$;
(ii) *the orbits* $H(\mathsf{N}_G(H)\,\varrho_H)$ *and* $H(\mathsf{C}_{\mathsf{Sym}[G\colon H]}(G\varrho_H))$ *coincide;*
(iii) $\ker\lambda_H = H$.

Proof (i) Let $h \in \mathsf{N}_G(H)$ and $g \in G$. We are required to show that $h\lambda_H$ and $g\varrho_H$ commute; that is, for all $g_1 \in G$,

$$Hg_1((h\lambda_H)(g\varrho_H)) = Hg_1((g\varrho_H)(h\lambda_H)).$$

This holds, since

$$Hg_1(h\lambda_H)(g\varrho_H) = Hh^{-1}g_1g = Hg_1(g\varrho_H)(h\lambda_H).$$

Thus $\mathsf{N}_G(H)\,\lambda_H \leqslant \mathsf{C}_{\mathsf{Sym}[G\colon H]}(G\varrho_H)$. Now let $c \in \mathsf{C}_{\mathsf{Sym}[G\colon H]}(G\varrho_H)$. Then, we have, for all $g,\ g_1 \in G$ that $Hg_1((g\varrho_H)c) = Hg_1(c(g\varrho_H))$. Let g_0 be an element of G such that the image of H under c is Hg_0^{-1}. First we show that g_0 normalises H. Let h be an arbitrary element of H. Then $c(h\varrho_H) = h\varrho_H c$, and so

$$Hg_0^{-1}h = Hc(h\varrho_H) = H(h\varrho_H)c = Hhc = Hc = Hg_0^{-1}.$$

That is, $Hg_0^{-1}hg_0 = H$, which shows that $h^{g_0} \in H$. Since this holds for all $h \in H$, we conclude that $g_0 \in \mathsf{N}_G(H)$, as claimed. Finally we show that $c = g_0\lambda_H$. Indeed, for $g \in G$,

$$Hgc = H(g\varrho_H)c = Hc(g\varrho_H) = Hg_0^{-1}g = Hg(g_0\lambda_H).$$

(ii) It follows from the definition of λ_H that the orbit $H(\mathsf{N}_G(H)\,\varrho_H)$ is equal to $H(\mathsf{N}_G(H)\,\lambda_H)$, and we just proved that the latter is equal to $H(\mathsf{C}_{\mathsf{Sym}[G\colon H]}(G\varrho_H))$.

(iii) If $h \in H$ then we have, for all $g \in G$, that

$$Hg(h\lambda_H) = Hh^{-1}g = Hg.$$

Thus $H \leqslant \ker \lambda_H$. Conversely if $h \in \ker \lambda_H$ then h fixes the trivial coset H, and so $H = H(h\lambda_H) = Hh^{-1}$. Thus $h \in H$. This shows that $\ker \lambda_H = H$. □

The following theorem gives a summary of the most important properties of the centralisers of transitive groups.

Theorem 3.2 *If G is a transitive permutation group acting on Ω, then the following hold.*

(i) *If $\omega \in \Omega$, then $\mathsf{C}_{\mathsf{Sym}\,\Omega}(G) \cong \mathsf{N}_G(G_\omega)/G_\omega$.*
(ii) $\omega\mathsf{C}_{\mathsf{Sym}\,\Omega}(G) = \omega\mathsf{N}_G(G_\omega) = \mathsf{Fix}_\Omega(G_\omega)$.
(iii) $\mathsf{C}_{\mathsf{Sym}\,\Omega}(G)$ *is semiregular on Ω.*
(iv) $\mathsf{C}_{\mathsf{Sym}\,\Omega}(G)$ *is transitive on Ω if and only if G is regular on Ω.*

Proof By Lemma 2.8 we may assume without loss of generality that Ω is the set of right cosets of a core-free subgroup H of G. Now (i) follows from Lemma 3.1, which also shows that $\omega\mathsf{C}_{\mathsf{Sym}\,\Omega}(G) = \omega\mathsf{N}_G(G_\omega)$. The rest of (ii) follows from part (i) and Lemma 2.19.

(iii) In order to prove that $C = \mathsf{C}_{\mathsf{Sym}\,\Omega}(G)$ is semiregular, we only need to show that if $c \in C$ and $\omega \in \Omega$ such that $\omega c = \omega$, then $c = 1$. Let $\omega' \in \Omega$. Since G is transitive, there is some $g \in G$ such that $\omega g = \omega'$. As $gc = cg$ and $\omega c = \omega$, we obtain that $\omega' c = \omega gc = \omega cg = \omega g = \omega'$. Therefore c fixes all points in Ω. Since c is an element of $\mathsf{Sym}\,\Omega$, it follows that $c = 1$ must hold, as claimed.

(iv) By (ii), $\omega\mathsf{C}_{\mathsf{Sym}\,\Omega}(G) = \mathsf{Fix}_\Omega(G_\omega)$, and hence $\mathsf{C}_{\mathsf{Sym}\,\Omega}(G)$ is transitive if and only if $\mathsf{Fix}_\Omega(G_\omega) = \Omega$. In turn, this holds if and only if $G_\omega = 1$, that is, the transitive group G is regular. □

A special case of Lemma 3.1 is when $H = 1$. In this case the left action $\lambda_1 \colon G \to \mathsf{Sym}\,G$ (defined at the beginning of the section) is a homomorphism and is called the *left regular action of G*. The action λ_1 is usually denoted by λ. For $g_1, g_2 \in G$ we write $g_1(g_2\lambda) = g_2^{-1}g_1$. Recall that ϱ denotes the right regular action defined in (2.2).

Corollary 3.3 *If G is a group, then $\mathsf{C}_{\mathsf{Sym}\,G}(G\varrho) = G\lambda$. Further, if H is a transitive subgroup of $\mathsf{Sym}\,G$ centralising $G\varrho$ then $H = G\lambda$.*

Proof The assertion that $\mathsf{C}_{\mathsf{Sym}\,G}(G\varrho) = G\lambda$ follows from Lemma 3.1(i). If H is a transitive subgroup of $\mathsf{Sym}\,G$ centralising $G\varrho$, then $H \leqslant \mathsf{C}_{\mathsf{Sym}\,G}(G\varrho) = G\lambda$. Consequently, $G\lambda = \mathsf{C}_{\mathsf{Sym}\,G}(G\varrho)$ is transitive. Thus by Theorem 3.2(iv) applied to $G\lambda$, we obtain that $G\lambda$ is regular on Ω. Thus by Lemma 2.11, $G\lambda = H$. □

3.2 Transitive minimal normal subgroups

If G is a group and N is a subgroup of G, then N is said to be a *minimal normal subgroup of* G if N is a normal subgroup of G and the only normal subgroup of G properly contained in N is the identity subgroup. In a finite group every non-trivial normal subgroup contains a minimal normal subgroup, but this is not necessarily true for infinite groups. For instance every subgroup of the additive group $\mathbb{Z}$ is cyclic of the form $\langle n\rangle$ for some $n \in \mathbb{Z}$ and the subgroup $\langle 2n\rangle$ is properly contained in $\langle n\rangle$ for all $n \in \mathbb{Z}$. Thus $\mathbb{Z}$ has no minimal normal subgroups. For subgroups H and L of a group G, the commutator subgroup $[H, L]$ of H and L is defined as the subgroup generated by all commutators $[h, l] = h^{-1}l^{-1}hl$ where $h \in H$ and $l \in L$.

Lemma 3.4 *Let N be a minimal normal subgroup of a group G and let H be a normal subgroup of G. Then the following hold.*

(i) *Either $N \leqslant H$ or $N \cap H = 1$. In the latter case $[N, H] = 1$.*
(ii) *Either N is abelian, or $\mathsf{Z}(N) = \mathsf{C}_G(N) \cap N = 1$.*

Proof (i) As $H \cap N$ is normal in G contained in N, either $H \cap N = 1$ or $H \cap N = N$. Since $[H, N] \leqslant H \cap N$, in the first case we also obtain that $[H, N] = 1$.

(ii) The centre $\mathsf{Z}(N)$ is invariant under all automorphisms of N, including the automorphisms that are induced by conjugation by the elements of G. Therefore $\mathsf{Z}(N)$ is normal in G and, as N is a minimal normal subgroup, either $\mathsf{Z}(N) = N$, in which case N is abelian, or $\mathsf{Z}(N) = 1$. □

Definition 3.5 Suppose that M is an abelian group. If r is a natural number, then we define $M^r = \{m^r \mid m \in M\}$ and note that $M^r \leqslant M$. If M is a non-trivial abelian group and $M^p = 1$ for some prime p, then M is said to be *elementary abelian.* If $M^r = M$ for all $r \in \mathbb{N}$, then M is said to be *divisible.*

Minimal normal subgroups, when they exist, have an important role in the theory of primitive, quasiprimitive, and innately transitive permutation groups. The next result shows that a permutation group can have at most two transitive minimal normal subgroups.

Theorem 3.6 *Let G be a permutation group acting on Ω, and suppose that G has a transitive minimal normal subgroup M.*

(i) *If M is abelian, then $\mathsf{C}_{\mathsf{Sym}\,\Omega}(M) = M$, M is regular on Ω, and M is the unique minimal normal subgroup of G. Further, M is either elementary abelian or divisible.*
(ii) *If M is the unique minimal normal subgroup of G, then $\mathsf{C}_G(M)$ is contained in M.*
(iii) *The number of transitive minimal normal subgroups of G is at most two, and if N is a second such subgroup, then $N \cong M$ is non-abelian, M and N are both regular on Ω, $\mathsf{C}_{\mathsf{Sym}\,\Omega}(N) = M$, $\mathsf{C}_{\mathsf{Sym}\,\Omega}(M) = N$, and M, N are the only minimal normal subgroups of G.*

Proof Suppose that M is a transitive minimal normal subgroup of a permutation group G, acting on a set Ω. If N is a minimal normal subgroup of G distinct from M, then Lemma 3.4 shows that $N \leqslant \mathsf{C}_G(M)$, and so $N \leqslant \mathsf{C}_{\mathsf{Sym}\,\Omega}(M)$.

(i) If M is abelian, then $M \leqslant \mathsf{C}_{\mathsf{Sym}\,\Omega}(M)$. Further, M is regular, by Lemma 2.4, and so, for $\omega \in \Omega$, $M_\omega = 1$ and $\mathsf{Fix}_\Omega(M_\omega) = \Omega$. By Lemma 2.8, the action of M on Ω is permutationally isomorphic to the action ϱ of M on M as defined in (2.2), and Corollary 3.3 gives that $\mathsf{C}_{\mathsf{Sym}\,M}(M\varrho) = M\lambda$. Since M is abelian, $M\lambda = M\varrho$, that is, to say, $M\varrho$ is self-centralising in $\mathsf{Sym}\,M$ and therefore $\mathsf{C}_{\mathsf{Sym}\,\Omega}(M) = M$. If N were a transitive minimal normal subgroup of G distinct from M, then, as $N \leqslant \mathsf{C}_{\mathsf{Sym}\,\Omega}(M)$, we would have that $N \leqslant M$, which is a contradiction, as M is already a minimal normal subgroup.

Let us now prove that M is elementary abelian or divisible. For $r \in \mathbb{N}$, consider the subgroup M^r defined in Definition 3.5. Note that M^r is invariant under conjugation by elements of G, and so M^r is normal in G. Since M is minimal normal either $M^r = M$ or $M^r = 1$. Assume that $M^r \neq M$ for some r and let r be minimal such that $M^r \neq M$. We claim that r is a prime. Suppose that r is not prime and $r = r_1 r_2$ with r_1, $r_2 \geqslant 2$. Then by the minimality of r we have that $M^{r_1} = M^{r_2} = M$, and hence $M^r = (M^{r_1})^{r_2} = M^{r_2} = M$, which is a contradiction. Hence r is a prime. By the minimality of M, we have $M^r = 1$ and so M is elementary abelian. The other alternative is that $M^r = M$ for all $r \geqslant 1$, and hence M is divisible, as claimed.

(ii) Suppose that M is the unique minimal normal subgroup of G and let $C = \mathsf{C}_G(M)$. As C is normal in G and M is the unique minimal normal subgroup of G, we find that either $M \leqslant C$ or $C = 1$. In the first case, M is abelian and $M = C$ follows from (i). Thus, in both cases, $C \leqslant M$, as claimed.

(iii) Suppose now that N is a second transitive minimal normal subgroup of G. Part (i) implies that M and N are non-abelian. As $N \leqslant \mathsf{C}_{\mathsf{Sym}\,\Omega}(M)$, Theorem 3.2(iii) shows that N is semiregular. As N is transitive, we obtain that N is regular. As $M \leqslant \mathsf{C}_{\mathsf{Sym}\,\Omega}(N)$ and M is transitive, the same argument implies that M is regular. Thus if M and N are distinct transitive minimal normal subgroups of G, then they both are regular, and so are $\mathsf{C}_{\mathsf{Sym}\,\Omega}(N)$ and $\mathsf{C}_{\mathsf{Sym}\,\Omega}(M)$. Now, by Lemma 2.11, $\mathsf{C}_{\mathsf{Sym}\,\Omega}(M) = N$ and $\mathsf{C}_{\mathsf{Sym}\,\Omega}(N) = M$. Any minimal normal subgroup of G distinct from M must be contained in $\mathsf{C}_{\mathsf{Sym}\,\Omega}(M) = N$. As N itself is a minimal normal subgroup, any minimal normal subgroup distinct from M must coincide with N. Therefore M and N must be the only minimal normal subgroups of G. Theorem 3.2(i) implies that $N = \mathsf{C}_{\mathsf{Sym}\,\Omega}(M) \cong \mathsf{N}_M(M_\omega)/M_\omega \cong M$. □

3.3 The holomorph

Suppose that G is a (not necessarily finite) group. Recall from (2.2) that ϱ denotes the right regular action of G; that is $\varrho\colon G \to \mathsf{Sym}\,G$ and, for $g_1,\ g_2 \in G$, we define $g_1(g_2\varrho) = g_1 g_2$. We consider $\mathsf{Aut}(G)$ as a permutation group on G. Clearly ϱ is faithful, and so is the $\mathsf{Aut}(G)$-action on G. We define the *holomorph* $\mathsf{Hol}\,G$ of G as the subgroup of $\mathsf{Sym}\,G$ generated by $G\varrho$ and $\mathsf{Aut}(G)$; that is,

$$\mathsf{Hol}\,G = \langle G\varrho, \mathsf{Aut}(G)\rangle\,.$$

Recall that the left regular action of a group G was defined before Lemma 3.1.

Lemma 3.7 *Let G be a group, let $g \in G$ and $a \in \mathsf{Aut}(G)$, let ϱ denote the right regular action of G, and let λ denote the left regular action of G. Then the conjugate $(g\varrho)^a$ coincides with $(ga)\varrho$ and, similarly, $(g\lambda)^a = (ga)\lambda$. Further, for any subgroup H of G, the following are equivalent:*

(i) $H\lambda$ *is* a*-invariant;*
(ii) H *is* a*-invariant;*
(iii) $H\varrho$ *is* a*-invariant.*

Proof If $h \in G$, then

$$ha^{-1}(g\varrho)a = (ha^{-1}g)a = (ha^{-1}a)(ga) = h(ga) = h((ga)\varrho).$$

Thus $a^{-1}(g\varrho)a = (ga)\varrho$ and the equation concerning λ can be proved similarly. The other assertions hold on considering these equations as g runs through the elements of H. □

If N and H are subgroups of a group G such that H normalises N, then let ξ_N^H denote the conjugation action of H on N. That is, $\xi_N^H \colon H \to \mathsf{Sym}\, N$ and, for $h \in H$ and $n \in N$, the image of n under the permutation $h\xi_N^H$ is n^h. Recall that $\mathsf{Inn}(G)$ denotes the group of all inner automorphisms of G.

Lemma 3.8 *If G is a group, then the following hold.*

(i) $G\varrho \cap \mathsf{Aut}(G) = 1$.
(ii) $\mathsf{Aut}(G)$ *normalises* $G\varrho$*, and hence* $\mathsf{Hol}\, G = (G\varrho)\mathsf{Aut}(G)$.
(iii) $(G\varrho)(\mathsf{Inn}(G)) = (G\varrho)(G\lambda) = (G\lambda)(\mathsf{Inn}(G))$ *and* $(G\varrho) \cap (G\lambda) = \mathsf{Z}(G)\varrho = \mathsf{Z}(G)\lambda$.
(iv) *Let* ι *be the element of* $\mathsf{Sym}\, G$ *that maps* $g \mapsto g^{-1}$ *for all* $g \in G$. *Then* $\iota \in (\mathsf{Sym}\, G)_1$, $\iota \in \mathsf{C}_{\mathsf{Sym}\, G}(\mathsf{Aut}(G))$, $|\iota| = 2$, *and* $(G\varrho)^\iota = G\lambda$.

Proof As $(G\varrho)_1 = 1$ and $\mathsf{Aut}(G)_1 = \mathsf{Aut}(G)$, we find that $G\varrho \cap \mathsf{Aut}(G) = 1$, and so (i) holds. The fact that $\mathsf{Aut}(G)$ normalises $G\varrho$ follows from Lemma 3.7, and this implies that $(G\varrho)\mathsf{Aut}(G) = \mathsf{Hol}\, G$. Thus (ii) holds.

For the proof of (iii) we note first that $G\lambda = \mathsf{C}_{\mathsf{Sym}\, G}(G\varrho)$ (Corollary 3.3), and so $\langle G\varrho, G\lambda\rangle = (G\varrho)(G\lambda)$. Set $\xi = \xi_G^G$. Then for $g,\ g_1 \in G$ we have

$$g_1(g\xi)(g^{-1}\varrho) = g^{-1}g_1gg^{-1} = g^{-1}g_1 = g_1(g\lambda).$$

Thus $(g\lambda)(g\varrho) = g\xi$ which shows that $G\lambda \leqslant (G\varrho)\mathsf{Inn}(G) \leqslant \mathsf{Hol}\, G$ and that $\mathsf{Inn}(G) \leqslant (G\varrho)(G\lambda)$. Hence $(G\varrho)(\mathsf{Inn}(G)) = (G\varrho)(G\lambda)$ holds, and similarly $(G\lambda)(\mathsf{Inn}(G)) = (G\varrho)(G\lambda)$. For $g \in \mathsf{Z}(G)$ we have that $g\varrho = g^{-1}\lambda$, and hence $\mathsf{Z}(G)\varrho = \mathsf{Z}(G)\lambda$, and also $\mathsf{Z}(G)\varrho \leqslant (G\varrho) \cap (G\lambda)$. Let $g_1,\ g_2 \in G$ such that $g_1\varrho = g_2\lambda$. Then $g_1 = 1(g_1\varrho) = 1(g_2\lambda) = g_2^{-1}$. Thus, for $g \in G$ we have that $gg_1 = g(g_1\varrho) = g(g_1^{-1}\lambda) = g_1g$. Hence

$g_1 \in \mathsf{Z}(G)$. This shows that $(G\varrho)\cap(G\lambda) \leqslant \mathsf{Z}(G)\varrho$, and so $(G\varrho)\cap(G\lambda) = \mathsf{Z}(G)\varrho = \mathsf{Z}(G)\lambda$, as claimed.

(iv) Clearly $\iota \in \mathsf{Sym}\, G$, $1\iota = 1$, and $|\iota| = 2$. Let $x,\ g \in G$ and $\alpha \in \mathsf{Aut}(G)$. Then $x\iota\alpha\iota = (x^{-1}\alpha)^{-1} = x\alpha$, and hence $\iota \in \mathsf{C}_{\mathsf{Sym}\, G}(\mathsf{Aut}(G))$. Further, $x\iota(g\varrho)\iota = (x^{-1}g)^{-1} = g^{-1}x = x(g\lambda)$. Hence $(G\varrho)^{\iota} \leqslant G\lambda$. A similar computation yields that $x\iota(g\lambda)\iota = x(g\varrho)$ for all $x \in G$, and so $(G\lambda)^{\iota} \leqslant G\varrho$, whence $G\lambda = (G\lambda)^{\iota^2} \leqslant (G\varrho)^{\iota} \leqslant G\lambda$. Therefore $(G\varrho)^{\iota} = G\lambda$. □

Lemma 3.8 implies that, for a group G, the holomorph $\mathsf{Hol}\, G$ is the semidirect product of $G\varrho$ and $\mathsf{Aut}(G)$; see Section 5.1 for the definition of semidirect products. It turns out that $\mathsf{Hol}\, G$ is equal to the full normaliser of the regular subgroup $G\varrho$.

Lemma 3.9 *If G is a group then $\mathsf{N}_{\mathsf{Sym}\, G}(G\varrho) = \mathsf{Hol}\, G$.*

Proof Lemma 3.8 shows that $\mathsf{Aut}(G) \leqslant \mathsf{N}_{\mathsf{Sym}\, G}(G\varrho)$, and so $\mathsf{Hol}\, G \leqslant \mathsf{N}_{\mathsf{Sym}\, G}(G\varrho)$. Therefore the conjugation action of $\mathsf{N}_{\mathsf{Sym}\, G}(G\varrho)$ on $G\varrho$ induces all automorphisms of $G\varrho$, and the kernel of this action is $\mathsf{C}_{\mathsf{Sym}\, G}(G\varrho) = G\lambda$ (Corollary 3.3). Let $x \in \mathsf{N}_{\mathsf{Sym}\, G}(G\varrho)$. Then x induces an automorphism a of G by conjugation such that $a \in \mathsf{N}_{\mathsf{Sym}\, G}(G\varrho)$. Thus $xa^{-1} \in \mathsf{N}_{\mathsf{Sym}\, G}(G\varrho)$ induces the identity automorphism of G by conjugation, that is, $xa^{-1} \in \mathsf{C}_{\mathsf{Sym}\, G}(G\varrho) = G\lambda$ (the equality follows from Corollary 3.3). Thus $xa^{-1} = y\lambda$ for some $y \in G$ and so $x = (y\lambda)a \in \mathsf{Hol}\, G$. □

Before the next results we recall, for subgroups N and H of a group G such that N is normalised by H, that the conjugation action of H on N is denoted by ξ_N^H.

The following embedding theorem will be an essential tool when working with groups that have a regular normal subgroup.

Theorem 3.10 (Holomorph Embedding Theorem) *Suppose that G is a permutation group on Ω, that M is a regular normal subgroup of G, and that ω_0 is a fixed element of Ω. Then there is a permutational embedding (ϑ, χ) where $\vartheta\colon \Omega \to M$ is a bijection and $\chi\colon G \to \mathsf{Hol}\, M$ such that, for all $m \in M$, $m\chi = m\varrho$ and, for all $g \in G_{\omega_0}$, $g\chi = g\xi_M^{G_{\omega_0}}$.*

Proof As M is a regular subgroup of G, for $\omega \in \Omega$, there is a unique $m \in M$ such that $\omega_0 m = \omega$. Let us define the bijection $\vartheta\colon \Omega \to M$ as $\omega\vartheta = m$. Let $g \in G$. Since M is transitive, Lemma 2.11 shows that there is $m \in M$ and $h \in G_{\omega_0}$ such that $g = mh$. Further, if $m' \in M$ and $h' \in G_{\omega_0}$ is such that $g = mh = m'h'$, then, as M is regular, $m^{-1}m' = hh'^{-1} \in M \cap G_{\omega_0} = 1$. Thus $m = m'$ and $h = h'$, which shows that this decomposition of the element g is unique. Set $\xi = \xi_M^{G_{\omega_0}}$. Since h normalises M, conjugation by h induces the automorphism $h\xi$. Define χ as $g\chi = (m\varrho)(h\xi)$. We claim that χ is a monomorphism. Let $m \in M$, and let g_1, $g_2 \in G$ such that $g_1 = m_1h_1$ and $g_2 = m_2h_2$ where m_1, $m_2 \in M$ and h_1, $h_2 \in G_{\omega_0}$. Then

$$\begin{aligned} m(g_1\chi)(g_2\chi) &= m(m_1\varrho)(h_1\xi)(m_2\varrho)(h_2\xi) \\ &= ((mm_1)^{h_1}m_2)^{h_2} = m^{h_1h_2}m_1^{h_1h_2}m_2^{h_2} \end{aligned}$$

and

$$m(g_1g_2)\chi = m(m_1h_1m_2h_2)\chi = m(m_1m_2^{h_1^{-1}}h_1h_2)\chi = m^{h_1h_2}m_1^{h_1h_2}m_2^{h_2}.$$

Thus $(g_1g_2)\chi = (g_1\chi)(g_2\chi)$, and hence χ is a homomorphism. If $g_1 = m_1h_1 \in \ker\chi$ then

$$m_1^{-1} = m_1^{-1}(g_1\chi) = m_1^{-1}(m_1\varrho)(h_1\xi) = 1(h_1\xi) = 1.$$

Thus $m_1 = 1$ must follow, and conjugation by h_1 must induce a trivial automorphism of M. Thus $h_1 \in \mathsf{C}_{\mathsf{Sym}\,\Omega}(M)$. As $\mathsf{C}_{\mathsf{Sym}\,\Omega}(M)$ is semiregular by Theorem 3.2(iii), and as $h_1 \in G_{\omega_0}$, we obtain that $h_1 = 1$. Thus $\ker\chi = 1$, as claimed.

It remains to show that (ϑ, χ) is a permutational isomorphism between $G \leqslant \mathsf{Sym}\,\Omega$ and $G\chi \leqslant \mathsf{Sym}\,M$. Let $\omega \in \Omega$, $g \in G$. We must show that $\omega g\vartheta = \omega\vartheta(g\chi)$. Let n, n_1, $n_2 \in M$ and $h \in G_{\omega_0}$ such that $g = nh$, $\omega = \omega_0 n_1$ and $\omega g = \omega_0 n_2$. Then $\omega g\vartheta = n_2$ and $\omega\vartheta(g\chi) = n_1^h n^h$. Now

$$\omega_0 n_1^h n^h = \omega_0 n_1 nh = \omega g.$$

Since n_2 is the unique element of M such that $\omega g = \omega_0 n_2$ we find that $n_2 = n_1^h n^h$ and so $\omega g\vartheta = \omega\vartheta(g\chi)$. Therefore (ϑ, χ) is a permutational isomorphism. The definition of χ implies, for all $m \in M$, that $m\chi = m\varrho$ and, for all $g \in G_{\omega_0}$, that $g\chi = g\xi$. □

The previous lemma combined with Lemma 3.8(iv) gives the following important corollary.

Corollary 3.11 *Suppose that $G \leqslant \mathsf{Sym}\,\Omega$ such that G has two transitive minimal normal subgroups M_1 and M_2 and let $\omega_0 \in \Omega$. Then* *there exists an involution $\pi \in (\mathsf{Sym}\,\Omega)_{\omega_0}$ such that $\pi \in \mathsf{C}_{\mathsf{Sym}\,\Omega}(G_{\omega_0})$ and $M_1^\pi = M_2$.*

Proof Let us use the subgroup M_1, which is regular by Theorem 3.6(iii), to construct the permutational embedding (ϑ, χ) of Theorem 3.10. Then $\vartheta\colon \Omega \to M_1$, $\chi\colon G \to \mathsf{Sym}(M_1)$ are such that $\omega_0\vartheta = 1$, $M_1\chi = M_1\varrho$, and $G_{\omega_0}\chi \leqslant \mathsf{Aut}(M_1)$. By Lemma 3.4(i), $M_2 \leqslant \mathsf{C}_G(M_1)$, and hence $M_2\chi \leqslant \mathsf{C}_{\mathsf{Sym}\,M_1}(M_1\varrho)$ which implies by Corollary 3.3 that $M_2\chi \leqslant M_1\lambda$. As $M_2\chi$ is transitive and $M_1\lambda$ is regular, we must have $M_2\chi = M_1\lambda$. Then Lemma 3.8(iv) implies that the element $\iota \in \mathsf{Sym}\,M_1$ that maps $m \mapsto m^{-1}$ is an involution, stabilising 1, centralising $\mathsf{Aut}(M_1)$, and swapping $M_1\varrho$ and $M_1\lambda$. Thus the corresponding element π of $\mathsf{Sym}\,\Omega$ is as required. □

Next we characterise primitive, quasiprimitive and innately transitive groups with regular normal subgroups. We emphasise that G is arbitrary here; in particular; G is not necessarily finite.

Theorem 3.12 *Suppose that M is a regular normal subgroup of a permutation group G, and let $\omega \in \Omega$.*

(i) *The group G is innately transitive if and only if M is a minimal normal subgroup of G.*
(ii) *The group G is quasiprimitive if and only if M is a minimal normal subgroup of G and either $G_\omega\xi_M^{G_\omega} \cap \mathsf{Inn}(M) = 1$ or $\mathsf{Inn}(M) \leqslant G_\omega\xi_M^{G_\omega}$.*
(iii) *The group G is primitive if and only if no proper, non-trivial subgroup of M is normalised by G_ω.*

Proof Choose $\omega \in \Omega$, and use Theorem 3.10 to construct a permutational embedding (ϑ, χ) where $\vartheta\colon \Omega \to M$ is a bijection such that $\omega\vartheta = 1$, and $\chi\colon G \to \mathsf{Hol}\,M$ is a monomorphism such that $G_\omega\chi \leqslant \mathsf{Aut}(M)$, for $m \in M$, $m\chi = m\varrho$, and, for $g \in G_\omega$, $g\chi = g\xi_M^{G_\omega}$.

(i) Clearly, if M is a minimal normal subgroup, then, as M is transitive, G is innately transitive. Assume now that G is innately transitive. Then G has a transitive minimal normal subgroup N. By Lemma 3.4, either $N \leqslant M$ or $M \cap N = [M, N] = 1$. If $N \leqslant M$ then, as M is regular and N is transitive, Corollary 2.10(ii) implies that $M = N$, and hence M is a minimal normal subgroup of G. Assume now that $M \cap N = [M, N] = 1$. Then $N\chi$ is transitive and centralises $M\chi = M\varrho$.

Also, by Corollary 3.3, $N\chi = M\lambda$, where λ denotes the left regular action. Thus $G\chi$ contains $(M\varrho)(M\lambda) = (M\varrho)(\mathsf{Inn}(M)) = (M\lambda)(\mathsf{Inn}(M))$ (by Lemma 3.8), and it follows that $G\chi = (M\lambda)(G_\omega\xi_M^{G_\omega})$. We claim that M is a minimal normal subgroup of G. Suppose that K is a non-trivial normal subgroup of G contained in M. Then $K\chi$ is a non-trivial normal subgroup of $G\chi$ contained in $M\chi$. Further, by the definition of χ, $K\chi = K\varrho$ and, as we noted above, $M\chi = M\varrho$. Therefore $G_\omega\chi$ normalises $K\varrho$. On the other hand, $K\lambda$ is a normal subgroup of $M\lambda$, and Lemma 3.7 implies that $K\lambda$ is normalised by $G_\omega\chi$. Since $G\chi = (M\lambda)(G_\omega\chi)$, we obtain that $K\lambda$ is normal in $G\chi$, and, since $N = M\lambda$ is minimal normal in $G\chi$, we find that $K\lambda = M\lambda$. Since λ is one-to-one, this shows that $K = M$. Therefore M is a minimal normal subgroup of G, as claimed.

(ii) If M is not a minimal normal subgroup, then there exists a non-trivial normal subgroup M_1 of G properly contained in M. As M is regular, M_1 cannot be transitive by Corollary 2.10(ii). Therefore if G is quasiprimitive, then M must be a minimal normal subgroup of G, and hence G must be innately transitive. Let $Z = \mathsf{Z}(M)$, the centre of M. Then Z is invariant under all automorphisms of M, and so it is invariant under conjugation by G. Hence Z is a normal subgroup of G. As M is a minimal normal subgroup of G, $Z = M$ and M is abelian, or $Z = 1$. If M is abelian, then $\mathsf{Inn}(M) = 1$, and so $G_\omega\xi_M^{G_\omega} \cap \mathsf{Inn}(M) = 1$ holds. Thus we may assume that $Z = 1$, and so M is non-abelian. Suppose that $\mathsf{C}_G(M) \neq 1$, or equivalently, that $\mathsf{C}_{G\chi}(M\varrho) \neq 1$. Note that $\mathsf{C}_{G\chi}(M\varrho) \trianglelefteq G\chi$ and so, as $G\chi$ is quasiprimitive, $\mathsf{C}_{G\chi}(M\varrho)$ is transitive. Then Corollary 3.3 implies that $\mathsf{C}_{G\chi}(M\varrho) = M\lambda$. Thus $(M\lambda)(M\varrho) \leqslant G\chi$, and it follows from Lemma 3.8(iii) that $\mathsf{Inn}(M) \leqslant G\chi$. Since $\mathsf{Inn}(M)$ lies in the stabiliser of the identity, we obtain that $\mathsf{Inn}(M) \leqslant G_\omega\chi = G_\omega\xi_M^{G_\omega}$.

This leaves us to consider the case $\mathsf{C}_G(M) = 1$, or equivalently, $\mathsf{C}_{G\chi}(M\varrho) = 1$. Suppose that $a \in G_\omega\xi_M^{G_\omega}$ and let $m \in M$ such that a is conjugation by m. As $m\varrho \in G$, $m\lambda = a(m\varrho)^{-1} \in G\chi$. On the other hand, $m\lambda \in \mathsf{C}_{G\chi}(M\varrho) = 1$, and hence $m\lambda = 1$, which implies that $m = 1$ and, in turn, that $a = 1$. Thus $G_\omega\xi_M^{G_\omega} = 1$ in this case.

Conversely, assume that M is a minimal normal subgroup of G and either $G_\omega\xi_M^{G_\omega} \cap \mathsf{Inn}(M) = 1$ or $\mathsf{Inn}(M) \leqslant G_\omega\xi_M^{G_\omega}$. Suppose first that $G_\omega\xi_M^{G_\omega} \cap \mathsf{Inn}(M) = 1$. Then $G_\omega\chi \cap \mathsf{Inn}(M) = 1$, and so $G\chi \cap \mathsf{Inn}(M) = 1$. Then, as for $m \in M$, $m\xi_M^M = (m\varrho)(m\lambda)$, we obtain that $M\lambda \cap G\chi = 1$. Since $M\lambda = \mathsf{C}_{\mathsf{Sym}\,M}(M\varrho)$ we find that $\mathsf{C}_{G\chi}(M\varrho) = 1$, and hence

$\mathsf{C}_G(M) = 1$. If N is a non-trivial normal subgroup of G, then $\mathsf{C}_G(M) = 1$ and Lemma 3.4 imply that $M \leqslant N$ and hence N is transitive. Assume now that $\mathsf{Inn}(M) \leqslant G_\omega \xi_M^{G_\omega}$. As $G_\omega \xi_M^{G_\omega} = G_\omega \chi$, Lemma 3.8(iii) implies that $M\lambda \leqslant G\chi$. Now, as $M\lambda = \mathsf{C}_{\mathsf{Sym}\,M}(M\varrho)$, $M\lambda$ is a normal subgroup of $G\chi$. If N is a non-trivial normal subgroup of G, then either $M \leqslant N$ or $N \leqslant \mathsf{C}_G(M)$. In the former case N is transitive. In the latter, $N\chi \leqslant M\lambda$, and since $M\lambda$ is minimal normal, by Corollary 3.11, we find that $N\chi = M\lambda$. Hence $N\chi$ is transitive, and so is N. Thus G is quasiprimitive.

(iii) If Δ is a block for M containing ω, then $\Delta\vartheta$ is a block for $M\chi = M\varrho$ containing 1. Thus Lemma 2.18 shows that $\Delta\vartheta$ is a subgroup of M. As the action of M on Ω is permutationally isomorphic to that of $M\varrho$ on M, for every subgroup $\Delta\varrho$ of $M\varrho$, we have, by Lemma 2.18, that Δ is a block for M. As M is transitive, Lemma 2.11 shows that $G = MG_\omega$, and so such an M-block Δ is also a G-block if and only if $\Delta\vartheta$ is invariant under $G_\omega\chi$. However, for $g \in G_\omega$, the permutation $g\chi$ coincides with $g\xi_M^{G_\omega}$. Thus Δ is a G-block if and only if $\Delta\varrho$ is a subgroup of M normalised by G_ω. Now assertion (iii) follows. □

Corollary 3.13 *Suppose that G is a permutation group on Ω such that either G has an abelian transitive minimal normal subgroup or G has two transitive minimal normal subgroups. Then G is primitive.*

Proof Let M be a transitive minimal normal subgroup of G. By Theorem 3.6, in both cases, M is regular. Fix an $\omega \in \Omega$. Then Lemma 2.11 implies that $G = MG_\omega$. Let ϱ denote the right regular representation of M and let (ϑ, χ) be the permutational embedding constructed in Theorem 3.10. Then $\vartheta\colon \Omega \to M$ is a bijection, and $\chi\colon G \to \mathsf{Hol}\,M$ is a permutational embedding, such that $\omega\vartheta = 1$, $M\chi = M\varrho$, and $G_\omega\chi \leqslant \mathsf{Aut}(M)$.

Suppose first that M is an abelian minimal normal subgroup of G. If, in addition, K is a subgroup of M, then it is a normal subgroup of M. If K is normalised by G_ω, then, as $G = MG_\omega$, we obtain that K is a normal subgroup of G. As M is minimal normal in G, either $K = 1$ or $K = M$. Thus no non-trivial, proper subgroup of M can be normalised by G_ω. Therefore G is primitive, in this case, by Theorem 3.12(iii).

Suppose next that G has two transitive minimal normal subgroups M and N. Let λ denote the left regular representation of M. By Theorem 3.6, $N\chi = \mathsf{C}_{\mathsf{Sym}\,M}(M\chi) = \mathsf{C}_{\mathsf{Sym}\,M}(M\varrho) = M\lambda$ (Corollary 3.3). Therefore it follows from Lemma 3.8 that $\mathsf{Inn}(M) \leqslant G\chi$. Now if K is a subgroup of M normalised by G_ω, then $K\vartheta$ is invariant under $G_\omega\chi = G_\omega\xi_M^{G_\omega}$. As $\mathsf{Inn}(M) \leqslant G_\omega\xi_M^{G_\omega}$, we obtain that K is normal in M, and, as $G = MG_\omega$, it is normal in G. As M is minimal normal, we obtain that either $K = 1$ or $K = M$. It now follows from Theorem 3.12(iii) that G is primitive. □

3.4 Characteristically simple groups

A subgroup N of a group G is said to be *characteristic* if N is invariant under all automorphisms of G. A characteristic subgroup of G is normal in G. Examples for characteristic subgroups of G include the centre $\mathsf{Z}(G)$ and the *derived subgroup* G', that is, the subgroup generated by all the *commutators* $[x, y] = x^{-1}y^{-1}xy$. The group G is said to be *characteristically simple* if its only characteristic subgroups are 1 and G. The *socle* of G is defined as the subgroup generated by all the minimal normal subgroups of G, and is denoted by $\mathsf{Soc}(G)$. If G has no minimal normal subgroups, for instance G is an infinite cyclic group, then $\mathsf{Soc}\, G = 1$. The socle of a group is always a characteristic subgroup.

Given an index set I and groups G_i, $i \in I$, the *unrestricted direct product* $\mathsf{Prod}_i\, G_i$ of the G_i is the group whose underlying set is the set of vectors $(g_i \mid i \in I)$ where the i-th coordinate is an element of G_i. Multiplication is performed componentwise; that is,

$$(g_i \mid i \in I)(h_i \mid i \in I) = (g_i h_i \mid i \in I).$$

The *restricted direct product* $\prod_i G_i$ is the subgroup of $\mathsf{Prod}_i\, G_i$ containing elements $(g_i \mid i \in I)$ such that $g_i = 1$ for all but finitely many i. In this book we will usually study the case when I is finite, and in this case the restricted and unrestricted direct products coincide.

The next result describes the structure of minimal normal subgroups and the socle; we refer to (Robinson 1996) for its proof.

Lemma 3.14 *Let G be a group.*

(i) *If G is a characteristically simple group that contains a minimal normal subgroup, then G can be written as a restricted direct product $\prod_{i \in I} T_i$ where the T_i are pairwise isomorphic simple groups. Conversely, a restricted direct product of pairwise isomorphic simple groups is characteristically simple.*

(ii) *The minimal normal subgroups of G are characteristically simple groups.*

(iii) *The socle* $\mathsf{Soc}\, G$ *of* G *can be written as a restricted direct product* $\prod_i N_i$ *where the* N_i *are minimal normal subgroups of* G*. Further, if* G *does not have an abelian minimal normal subgroup, then the* N_i *are the only minimal normal subgroups of* G*.*

(iv) *The socle of a group is a (possibly trivial) restricted direct product of simple groups.*

Proof See (Robinson 1996, 3.3.15) for part (i), and the discussion before this result for part (ii) and for the first statement of part (iii). The second statement of part (iii) follows from (Robinson 1996, 3.3.16), and part (iv) follows from parts (i) and (iii). □

Theorem 3.12 states that if G is a permutation group with a regular normal subgroup M then G is innately transitive if and only if M is a minimal normal subgroup. Since such a minimal normal subgroup is characteristically simple (see Lemma 3.14(ii)), the class of characteristically simple groups plays an important role in the theory of innately transitive, quasiprimitive, and primitive permutation groups. It follows from Lemma 3.14(i), that a finite abelian characteristically simple group M is isomorphic to a direct product of finitely many copies of a cyclic group of order p, for some prime p. Thus such a group can be considered as a vector space over the field $\mathbb{F}_p$ of p elements. The automorphism group of M coincides with the group of linear transformations $\mathsf{GL}(M) = \mathsf{GL}_k(p)$. A subgroup K of $\mathsf{GL}_k(p)$ is said to be *irreducible* if there are no proper and non-trivial K-invariant subspaces in M.

Theorem 3.15 *Suppose that* G *is a finite permutation group with an abelian minimal normal subgroup* M*. Then* G *is primitive if and only if* $G_\omega \xi_M^{G_\omega}$ *is an irreducible subgroup of* $\mathsf{GL}(M)$*.*

Proof The claim that $G_\omega \xi_M^{G_\omega}$ is irreducible is equivalent to the assertion that no proper non-trivial subgroup of M is normalised by G_ω. Thus the theorem follows from Theorem 3.12. □

Extending the notation introduced Section 2.5, for an element g of a permutation group acting on Ω, we denote by $\mathsf{Fix}_\Omega(g)$ the set of fixed points of g. That is,

$$\mathsf{Fix}_\Omega(g) = \{\omega \in \Omega \mid \omega g = \omega\}.$$

A permutation group $G \leqslant \mathsf{Sym}\,\Omega$ is said to be a *Frobenius group*, if G is not regular on Ω, but $|\mathsf{Fix}_\Omega(g)| \leqslant 1$ for all $g \in G\backslash\{1\}$. Frobenius groups have been widely studied. The following is a combination of results by Frobenius (1902) and Thompson (1959); see (Dixon & Mortimer 1996, Section 3.4) for more details concerning Frobenius groups.

Theorem 3.16 *Let $G \leqslant \mathsf{Sym}\,\Omega$ be a finite Frobenius group. Then the set $\{g \in G \mid |\mathsf{Fix}_\Omega(g)| = 0\} \cup \{1\}$ is a non-trivial nilpotent regular normal subgroup of G.*

The nilpotent normal subgroup in Theorem 3.16 is usually referred to as a *Frobenius kernel*. Using the existence and nilpotence of Frobenius kernels, we can state and prove the following structure theorem for finite primitive groups with abelian stabilisers.

Proposition 3.17 *Let $G \leqslant \mathsf{Sym}\,\Omega$ be a finite primitive permutation group such that the point stabilisers are abelian. Then G is either a cyclic group of prime order, or G is a Frobenius group with an elementary abelian Frobenius kernel which is a minimal normal subgroup of G.*

Proof Let $\alpha \in \Omega$. If G is regular, then $G_\alpha = 1$. On the other hand, G_α is a maximal subgroup of G, and so G must be a cyclic group of prime order. Thus the proposition holds when G is regular, and so we may assume in the rest of the proof that G is not regular.

We claim that G is a Frobenius group. Suppose that $g \in G$ such that g stabilises distinct points α, $\beta \in \Omega$. We will prove that $g = 1$. Assume to the contrary that $g \neq 1$. Then, since G_α and G_β are abelian, $G_\alpha \leqslant \mathsf{C}_G(g)$ and $G_\beta \leqslant \mathsf{C}_G(g)$. On the other hand, as G is primitive, G_α and G_β are maximal subgroups of G, and so either $G_\alpha = \mathsf{C}_G(g)$ or $\mathsf{C}_G(g) = G$. If $\mathsf{C}_G(g) = G$, then g lies in the centre of G. However, this implies that the cyclic group $\langle g \rangle$ is normal in G, and so $\langle g \rangle$ has to be transitive by Corollary 2.21. Since g fixes α, this is impossible. Hence $G_\alpha = \mathsf{C}_G(g)$ and we obtain similarly that $G_\beta = \mathsf{C}_G(g)$. In particular, $G_\alpha = G_\beta$. Therefore $|\mathsf{Fix}_\Omega(G_\alpha)| \geqslant 2$ and Corollary 2.21 implies that $\mathsf{Fix}_\Omega(G_\alpha) = \Omega$. Since G is a permutation group, $G_\alpha = 1$, which is a contradiction, since G is not regular. Hence $g = 1$ must hold, and so G is a Frobenius group. Therefore the claim is proved.

By Theorem 3.16, $N = \{g \in G \mid |\mathsf{Fix}_\Omega(g)| = 0\} \cup \{1\}$ is a non-trivial nilpotent regular normal subgroup of G. Let M be a minimal normal subgroup of G contained in N. Then M is a direct product of pairwise isomorphic finite simple groups by Lemma 3.14. Since M is nilpotent, a simple direct factor of M must be a cyclic group of order p for some prime p. Thus M is an elementary abelian p-group. Further, by Corollary 2.21, M is a transitive subgroup of the regular group N, and so $M = N$. □

We finish this section with a theorem that shows that groups with FCR-plinth occur naturally; namely, quasiprimitive groups with finite point stabilisers satisfy this condition. Our argument follows the proof of (Neumann, Praeger & Smith 2017, Observation 1.1 and Theorem 1.4). See also (Smith 2015) for related results concerning primitive groups with finite stabilisers.

Theorem 3.18 *Let G be a permutation group acting on an infinite set Ω, let $\omega \in \Omega$, and suppose that G_ω is finite.*

(i) *If every infinite normal subgroup of G is transitive, then G has a minimal normal subgroup.*
(ii) *If G is quasiprimitive, then G has a unique minimal normal subgroup that is an FCR-group.*

Proof (i) If G contains non-trivial finite normal subgroups, then such a subgroup of least order is a minimal normal subgroup. Therefore we may assume without loss of generality that all non-trivial normal subgroups of G are infinite. Let M be a non-trivial (hence infinite) normal subgroup of G such that the order $|M_\omega|$ is minimal possible. Such a normal subgroup M exists, since G_ω is finite and $H_\omega \leqslant G_\omega$ holds for all subgroups H of G. We claim that M is a minimal normal subgroup of G. Suppose that N is a non-trivial normal subgroup of G contained in M. Then $N_\omega \leqslant M_\omega$, and so the minimality of $|M_\omega|$ implies that $N_\omega = M_\omega$, and hence $N = M$, by Lemma 2.11. Therefore M is a minimal normal subgroup of G.

(ii) Suppose now that G is quasiprimitive. Then G satisfies the conditions of part (i), and so G has a minimal normal subgroup M. Since M is transitive, Lemma 2.11 implies that $G = MG_\omega$. We claim that M is FCR. Let T be a normal subgroup of M such that the elements of the G_ω-orbit $T^{G_\omega} = \{T^h \mid h \in G_\omega\}$ are pairwise disjoint (that is, for all h_1, $h_2 \in G_\omega$, either $T^{h_1} = T^{h_2}$ or $T^{h_1} \cap T^{h_2} = 1$) and the cardinality $|T^{G_\omega}|$ is maximal possible. Such T exists, as, for example, $T = M$ satisfies the first condition and $|H^{G_\omega}| \leqslant |G_\omega|$ for any subgroup H of G.

Suppose that $T^{G_\omega} = \{T_1, \ldots, T_q\}$ with $T = T_1$ and note that T_i

is a normal subgroup of M for all $i = 1, \ldots, q$. Hence the product $T_1 \cdots T_q$ is a non-trivial subgroup of M normalised by $G = MG_\omega$. By the minimality of M, we have that $M = T_1 \cdots T_q$. We claim that $M = T_1 \times \cdots \times T_q$. If $i \neq j$, then $T_i \cap T_j = 1$, and so the normal subgroups T_i and T_j commute. In particular, for all i, T_i centralises the product $T_1 \cdots T_{i-i} T_{i+1} \cdots T_q$. We are required to show that $T_i \cap (T_1 \cdots T_{i-1} T_{i+1} \cdots T_q) = 1$. Without loss of generality, we show this for $i = 1$. Set $Z_1 = T_1 \cap (T_2 \cdots T_q)$. We seek a contradiction by assuming that $Z_1 \neq 1$. Then $Z_1 \leqslant \mathsf{Z}(T_1)$, and hence Z_1 is abelian. Now $Z_1 \trianglelefteq M$ and the number of G-conjugates of Z_1 is not greater than $|G_\omega|$, which is finite. Assume that $Z_1, \ldots, Z_s$ are the G-conjugates of Z_1 and set $Z = Z_1 \cdots Z_s$. Then Z is a non-trivial normal subgroup of G contained in M, and so $Z = M$. Then M is a product of abelian normal subgroups and hence, by Fitting's Theorem (Robinson 1996, 5.2.8), M is nilpotent. Thus the centre $\mathsf{Z}(M)$ is a non-trivial normal subgroup of G contained in M. By the minimality of M, $\mathsf{Z}(M) = M$, and so M is abelian. Let $a \in M \setminus \{1\}$ and let M_0 be the subgroup of M generated by the finite set $\{a^h \mid h \in G_\omega\}$. Then M_0 is normal in G and is contained in M. Hence $M_0 = M$, which implies that M is finitely generated. By the Fundamental Theorem of Finitely Generated Abelian Groups (Robinson 1996, 4.2.10), $M = F \times X$ where F is a free abelian group isomorphic to $\mathbb{Z}^r$ for some r, X is a finite group, and X is characteristic in M. Hence X is normal in G, and minimality implies that either $X = M$ or $X = 1$. The case $X = M$ is impossible, as X is finite and M is infinite. Thus $M \cong \mathbb{Z}^r$. However, in this case $M^2 = \{a^2 \mid a \in M\}$ is a non-trivial normal subgroup of G properly contained in M, which contradicts the minimality of M. Thus $Z_1 = 1$, as claimed, and hence $M = T_1 \times \cdots \times T_q$.

Next we claim that the T_i are simple. It suffices to show that T is simple. Let U be a non-trivial normal subgroup of T. Set $N = \mathsf{N}_{G_\omega}(T)$ and let $\{U_1, \ldots, U_r\}$ be the (finite) set of N-conjugates of U such that $U = U_1$. Let I be a maximal subset (by inclusion) of the index set $\{1, \ldots, r\}$ such that $V = \bigcap_{i \in I} U_i \neq 1$. Since V is an intersection of normal subgroups of T, the subgroup V is normal in $T = T_1$. Further, $M = T_1 \times \cdots \times T_q$, and so V is normal in M. We claim that the distinct G_ω-conjugates of V are pairwise disjoint. It suffices to show, for $h \in G_\omega$, that either $V = V^h$ or $V \cap V^h = 1$. Let $h \in G_\omega$. If $T^h \neq T$, then $V^h \leqslant T_i$ for some $i \neq 1$. Since $T \cap T_i = 1$, we obtain $V \cap V^h = 1$. Hence, we may assume that $T^h = T$; that is, $h \in N = \mathsf{N}_{G_\omega}(T)$. If h stabilises the set $\{U_i \mid i \in I\}$ then the definition of V implies that

$V^h = V$. Suppose that h does not stabilise $\{U_i \mid i \in I\}$ and suppose without loss of generality that $U_i^h = U_j$ with $i \in I$ and $j \notin I$. Then $V \cap V^h \leqslant V \cap U_j = (\bigcap_{i \in I} U_i) \cap U_j$ and the maximality of I implies that $V \cap U_j = 1$. Hence $V \cap V^h = 1$. Therefore the G_ω-conjugates of V are pairwise disjoint.

If $h_i \in G_\omega$ is such that $T_i = T^{h_i}$, then $V^{h_i} \leqslant T_i$, and so each T_i contains at least one G_ω-conjugate of V. In particular, the number of G_ω-conjugates of V is at least q. Since distinct G_ω-conjugates of V are disjoint, the maximality of q implies that V has precisely q G_ω-conjugates $V_1, \ldots, V_q$, where $V = V_1$, and we may assume that $V_i \leqslant T_i$ for all i. Then the argument above implies that $M = V_1 \times \cdots \times V_q$. Since also $M = T_1 \times \cdots \times T_q$, and $V_i \leqslant T_i$, it follows that $V = T$, and since $V \leqslant U \leqslant T$, we obtain that $U = T$. Therefore T is simple as claimed, and therefore M is FCR.

Finally, we show that M is the unique minimal normal subgroup of G. If G has another minimal normal subgroup, then Theorem 3.6 implies that G has two minimal normal subgroups M and N and that both are regular, and hence transitive. Thus, $G = NG_\omega$ and so $G/N = NG_\omega/N \cong G_\omega/(N \cap G_w)$ is finite. On the other hand $M \cap N = 1$, and so $M \cong MN/N \leqslant G/N$ and M is infinite (since transitive). This is a contradiction, and so M is the unique minimal normal subgroup of G, as claimed. □

In Theorem 7.9 we will determine the possible O'Nan–Scott classes of the quasiprimitive permutation groups with finite stabilisers.

3.5 The socles of finite 2-transitive permutation groups

In this section we prove Burnside's Theorem that the socle of a finite 2-transitive permutation group is either elementary abelian or simple. This theorem is often proved using Frobenius kernels (as in (Cameron 1999, Theorem 4.3)) and can also be derived from the O'Nan–Scott Theorem for primitive permutation groups (as in (Dixon & Mortimer 1996, Theorem 4.1B)). Here we present an elementary proof that was inspired by the argument sketched on Peter Cameron's website (Cameron 2001).

Lemma 3.19 *Suppose that G is a finite 2-transitive group on a set Ω and let M be a regular normal subgroup of G. Then M is an elementary abelian p-group.*

Proof By Theorem 3.10, we may assume without loss of generality that $\Omega = M$, that $G = (M\varrho) \rtimes H$ where $H \leqslant \mathsf{Aut}(M)$, and that H is the stabiliser of $1 \in M$. Since G is 2-transitive, by Lemma 2.31, H is transitive on the set of non-identity elements of M. Hence all non-identity elements of M have the same order, and so there is a prime p, such that the order of all non-identity elements of M is p. Hence M is a group of exponent p, and so M is a finite p-group. The centre of M is a non-trivial subgroup of M which is invariant under H, hence $\mathsf{Z}(M) = M$. Thus M is an elementary abelian p-group. □

Proposition 3.20 *Let G be a finite 2-transitive permutation group and suppose that M is a minimal normal subgroup of G such that M is imprimitive. Then M is an elementary abelian p-group for some prime p and M is regular.*

Proof Suppose that M is a minimal normal subgroup of G. Since M is a non-trivial normal subgroup of a primitive group, by Corollary 2.21, M is transitive. Let Δ be a minimal non-trivial block of imprimitivity for M.

Claim 1. For $g \in G$, the image Δg is also a minimal block of imprimitivity for M.

Proof of Claim 1. Let $g \in G$. If $m \in M$, then $(\Delta g)m = \Delta m^{g^{-1}} g$ and hence

$$\Delta g \cap \Delta g m = \Delta g \cap \Delta m^{g^{-1}} g = (\Delta \cap \Delta m^{g^{-1}})g.$$

Since Δ is a block for M, and $m^{g^{-1}} \in M$, the intersection $\Delta \cap \Delta m^{g^{-1}}$ is either equal to Δ or is empty. Thus $\Delta g \cap \Delta g m$ is either equal to Δg or is empty, and so Δg is a block for M. If Σ is an M-block properly contained in Δg, then the same argument shows that Σg^{-1} is an M-block properly contained in Δ. By the minimality of Δ we obtain that $|\Sigma g^{-1}| = |\Sigma| = 1$. Thus Δg is a minimal block as claimed. □

Set $\mathcal{B} = \{\Delta g \mid g \in G\}$. Then by Claim 1, $\mathcal{B}$ is a set of non-trivial minimal blocks for M.

Claim 2. If γ, δ are distinct elements of Ω, then there is a unique block $\Sigma \in \mathcal{B}$ such that γ, $\delta \in \Sigma$.

Proof of Claim 2. If γ, δ are contained in Σ_1, $\Sigma_2 \in \mathcal{B}$, then γ, $\delta \in \Sigma_1 \cap \Sigma_2$. However, $\Sigma_1 \cap \Sigma_2$ is a block for M which contains the distinct elements γ and δ, and hence must be non-trivial. Therefore the minimality of Σ_1 and Σ_2 implies that $\Sigma_1 = \Sigma_2$. Thus there exists at most one block $\Sigma \in \mathcal{B}$ such that $\gamma,\ \delta \in \Sigma$. Suppose that $\alpha,\ \beta \in \Delta$ are distinct elements. Since G is 2-transitive, there exists $g \in G$ such that $\alpha g = \gamma$ and $\beta g = \delta$. Thus $\gamma,\ \delta \in \Delta g$, and Claim 1 implies that $\Delta g \in \mathcal{B}$. □

Claim 3. If $\gamma,\ \delta$ are distinct points of Ω, then $M_{\gamma,\delta} = 1$.

Proof of Claim 3. Suppose that α and β are distinct elements of Δ. Since G is 2-transitive on Ω, there exists $g \in G$ such that $\alpha g = \gamma$ and $\beta g = \delta$. Then $M_{\gamma,\delta} = (M_{\alpha,\beta})^g$ and it suffices to show that $M_{\alpha,\beta} = 1$. Suppose that $\omega \in \Omega$ such that $\omega \notin \Delta$. By Claim 2, there exist a unique $\Sigma_1 \in \mathcal{B}$ and a unique $\Sigma_2 \in \mathcal{B}$ such that $\alpha,\ \omega \in \Sigma_1$ and $\beta,\ \omega \in \Sigma_2$. Since Σ_1 and Σ_2 are blocks that contain α or β, the stabiliser $M_{\alpha,\beta}$ stabilises Σ_1 and Σ_2 setwise, and so $M_{\alpha,\beta}$ stabilises $\Sigma_1 \cap \Sigma_2$. By Claim 2, $|\Sigma_1 \cap \Sigma_2| \leqslant 1$, and so $\Sigma_1 \cap \Sigma_2 = \{\omega\}$. Hence $M_{\alpha,\beta}$ stabilises ω for all $\omega \in \Omega \setminus \Delta$. Further, $M_{\alpha,\beta}$ is conjugate to $M_{\alpha,\omega}$ and, by the argument above, $M_{\alpha,\beta} \leqslant M_{\alpha,\omega}$. Thus $M_{\alpha,\omega} = M_{\alpha,\beta}$. Considering the unique block Σ_1 that contains α and ω, we have $\Sigma_1 \neq \Delta$, and the argument above shows that $M_{\alpha,\beta} = M_{\alpha,\omega}$ stabilises all points in $\Omega \setminus \Sigma_1$. Since $\Omega = (\Omega \setminus \Delta) \cup (\Omega \setminus \Sigma_1) \cup \{\alpha\}$, $M_{\alpha,\beta}$ stabilises all points in Ω. Hence $M_{\alpha,\beta} = 1$. □

Set $n = |\Omega|$. Let F denote the set of fixed-point-free elements of M.

Claim 4. $|F| = n - 1$ and F is a G-conjugacy class. Further, if α and β are distinct elements of Ω, there is a unique element of $h \in F$ such that $\alpha h = \beta$.

Proof of Claim 4. We prove this claim using a well-known counting argument. Set

$$X = \{(\gamma, g) \in \Omega \times M \mid \gamma g = \gamma\} = \{(\gamma, g) \in \Omega \times M \mid g \in M_\gamma\}.$$

Recall, for an element $m \in M$, that we denote by $\mathsf{Fix}_\Omega(m)$ the set of fixed points of m. Counting $|X|$ in two different ways and using the Orbit-Stabiliser Theorem (Corollary 2.9),

$$|X| = \sum_{m \in M} |\mathsf{Fix}_\Omega(m)| = \sum_{\gamma \in \Omega} |M_\gamma| = |\Omega||M_\alpha| = |M|. \tag{3.1}$$

Now for all elements $m \in M$ we have, by Claim 3, that $|\mathsf{Fix}_\Omega(m)| \in \{0, 1, n\}$. In fact, $|\mathsf{Fix}_\Omega(m)| = n$ if and only if $m = 1$, $|\mathsf{Fix}_\Omega(m)| = 0$ if and only if m is fixed-point-free, and $|\mathsf{Fix}_\Omega(m)| = 1$ if and only if m is a non-trivial element of a point stabiliser. Suppose that $L = \{m \in M \mid$

$|\mathsf{Fix}_\Omega(m)| = 1\}$. Then, by equation (3.1), we obtain that

$$|L| + |F| + 1 = |M| = \sum_{m \in M} |\mathsf{Fix}_\Omega(m)| = |L| + n,$$

and so $|F| = n - 1$.

Let us now show that F is a G-conjugacy class. Since $M \trianglelefteq G$ and since each of the conjugates of a fixed-point-free element is fixed-point-free, F is a union of G-conjugacy classes. Let $h \in F$ and let $\gamma = \alpha h$. Then $\gamma \neq \alpha$ by the definition of F. Since $\alpha \neq \beta$, then there exists $g \in G$ such that $\alpha g = \alpha$ and $\gamma g = \beta$. Setting $h' = h^g$, we have that $\alpha h' = \beta$. This shows that for all $\beta \neq \alpha$, there is an element $h' \in F$ such that $\alpha h' = \beta$. Since $n - 1 = |F| = |\Omega \setminus \{\alpha\}|$, this element h' must be unique. Also, h' is a G-conjugate to h, and so the elements of F are pairwise G-conjugates. □

Claim 5. n is a power of a prime p and $F \cup \{1\}$ is a Sylow p-subgroup of M. Further, $F \cup \{1\}$ acts regularly on Ω.

Proof of Claim 5. Let p be a prime divisor of n. Since M is transitive, $|M|$ is divisible by n and hence by p. Let P be a Sylow p-subgroup of M and suppose that $h \in P$ is an element of order p. Then the disjoint cycle representation of h contains cycles of length one or p. By Claim 3, h may contain at most one cycle of length one. However, if h contained a cycle of length one, we would obtain that $n = |\Omega| \equiv 1 \pmod{p}$ which is a contradiction as we assume that $p \mid n$. Thus $h \in F$. Hence for each prime p dividing n, there is at least one element of F of order p. As, by Claim 4, the elements of F are G-conjugate, all elements of F have the same order p. In particular p is the only prime that divides n, which gives that n is a power of p.

We claim that P is a transitive subgroup of M. By the argument in the previous paragraph, for a point α, the stabiliser M_α does not contain an element of order p, and hence $|M_\alpha|$ is not divisible by p. Hence $P \cap M_\alpha = 1$, and P is semiregular. However, since M is transitive, the p-power n divides $|M|$ and hence divides the order $|P|$ of its Sylow p-subgroup. Thus the orbits of the semiregular subgroup P have size at least (and hence equal to) n. Hence P is a regular subgroup of M. In particular $P \setminus \{1\}$ consists of fixed-point-free elements and so is contained in F. Since $|P| = n$ and by Claim 4, $|F| = n - 1$, it follows that $P = F \cup \{1\}$. □

Now we can prove the main assertion of Proposition 3.20. By Claim 4, the set F is a G-conjugacy class and hence G normalises $P = F \cup \{1\}$

which is a subgroup, by Claim 5. Since M is minimal normal, $M = P$. Thus M is regular, by Claim 5. A minimal normal p-group is elementary abelian, and hence M is an elementary abelian group. □

Now we can prove Burnside's Theorem announced in the introduction of this section.

Theorem 3.21 *Suppose that G is a finite 2-transitive permutation group and M is a minimal normal subgroup of G. Then M is the unique minimal normal subgroup of G and either M is regular and is an elementary abelian p-group for some prime p, or M is a non-abelian simple group.*

Proof If M is not the unique minimal normal subgroup of G, then, by Theorem 3.6, the only other minimal normal subgroup is $\mathsf{C}_G(M)$, and both M and $\mathsf{C}_G(M)$ are regular non-abelian normal subgroups of the finite 2-transitive permutation group G. This is a contradiction, by Lemma 3.19.

If M is regular, then Lemma 3.19 implies that M is elementary abelian. If M is imprimitive, then, by Proposition 3.20, M is elementary abelian and regular. Hence we may assume that M is primitive and non-regular. By Theorem 3.6, M is non-abelian, and so by Lemma 3.14, $M = T_1 \times \cdots \times T_k$ where the T_i are pairwise isomorphic non-abelian simple groups and they are the minimal normal subgroups of M. Since M is primitive, Corollary 2.21 implies that the minimal normal subgroups $T_1, \ldots, T_k$ are transitive. Thus by Theorem 3.6, $k \leqslant 2$; further, if $k = 2$, then T_1 and T_2 are regular.

We need to rule out the case $k = 2$. Seeking a contradiction, assume that $k = 2$. Then Theorem 3.10 implies that M can be viewed as a subgroup of $\mathsf{Hol}\, T_1$. Let H denote the stabiliser in M of $1 \in T_1$. Then $\mathsf{Inn}(T_1) \leqslant H \leqslant \mathsf{Aut}(T_1)$, and, since $|M| = |T_1|^2 = |T_1||\mathsf{Inn}(T_1)|$, we find that $H = \mathsf{Inn}(T_1)$. Since G is 2-transitive on Ω, the stabiliser G_1 of the identity is transitive on $\Omega \setminus \{1\}$. The subgroup H is normal in G_1, which by Lemma 2.20 implies that the H-orbits that are different from $\{1\}$ have the same size. On the other hand, since $H = \mathsf{Inn}(T_1)$, the size of the H-orbit of $t \in T_1$ is $|T_1 \colon \mathsf{C}_{T_1}(t)|$ which means that $|\mathsf{C}_{T_1}(t)|$ is independent of t whenever $t \neq 1$. We show that this is impossible. Argue by contradiction that $d = |\mathsf{C}_{T_1}(t)|$ for all $t \in T_1 \setminus \{1\}$ independently of t. Suppose that p is a prime dividing $|T_1|$ and assume that p^s is the largest p-power such that $p^s \mid |T_1|$. Let P be a Sylow

p-subgroup of T_1, and let $x \in \mathsf{Z}(P)\setminus\{1\}$. Then $P \leqslant \mathsf{C}_{T_1}(x)$, and hence $d = |\mathsf{C}_{T_1}(x)|$ is divisible by $|P| = p^s$. Since this is true for each of the primes that divide $|T_1|$, we obtain that $d = |T_1|$, but this implies that T_1 is abelian, which is a contradiction.

Thus we have proved that $k = 2$ is impossible, and hence M is a non-abelian simple group. □

3.6 The point stabiliser of a primitive group with a regular normal subgroup

The aim of this section is to obtain structural information about the point stabiliser of a finite primitive group with a regular non-abelian minimal normal subgroup. It has several consequences regarding the structure of finite almost simple primitive groups (Corollary 3.23) and twisted wreath products (Theorem 6.11).

Theorem 3.22 *Suppose that G is a finite primitive permutation group with a regular normal subgroup M. Then M is a minimal normal subgroup, and either*

(i) *M is an elementary abelian q-group, for some prime q; or*
(ii) *M is non-abelian, and each non-trivial normal subgroup of a point stabiliser G_α is insoluble.*

Proof Let G be a finite primitive group with a non-abelian regular normal subgroup M, and let G_α be a point stabiliser. By Corollary 2.22, M is a minimal normal subgroup of G. Assume that G_α contains a non-trivial soluble normal subgroup. Then G_α contains a soluble minimal normal subgroup P which is an elementary abelian p-group for some prime p (Lemma 3.14). Since the action of G is faithful, $\mathsf{Core}_G G_\alpha = 1$, and so $\mathsf{N}_G(P) < G$. On the other hand, G_α is maximal in G (Corollary 2.15), which implies that $\mathsf{N}_G(P) = G_\alpha$. Hence $\mathsf{C}_M(P) \leqslant \mathsf{N}_M(P) = M \cap G_\alpha = 1$ (since M is regular), so $\mathsf{C}_M(P) = \mathsf{N}_M(P) = 1$. As M is normal in G, P acts by conjugation on M, and, since $\mathsf{C}_M(P) = 1$, the only P-orbit with length 1 in this action is $\{1\}$. Thus p divides $|M| - 1$, and hence $p \nmid |M|$.

Suppose that q is a prime divisor of $|M|$. Then the number of Sylow q-subgroups of M is a divisor of M, and hence this number is not divisible by p. Thus there is a Sylow q-subgroup Q of M normalised by P. We claim that Q is the unique Sylow q-subgroup of M with this property. Assume that $u \in M$ is such that Q^u is also normalised by P. If $h \in P$, then

$$Q^{uhu^{-1}} = ((Q^u)^h)^{u^{-1}} = (Q^u)^{u^{-1}} = Q,$$

and hence uPu^{-1} also normalises Q. Hence both P and uPu^{-1} are Sylow p-subgroups of $\mathsf{N}_{MP}(Q)$, and so there is $v \in \mathsf{N}_{MP}(Q)$ such that $vuPu^{-1}v^{-1} = P$. Since by definition $v \in MP$ we may assume that $v \in M$. Thus $vu \in \mathsf{N}_G(P) \cap M = \mathsf{N}_M(P) = 1$, as shown above. Hence $u = v^{-1} \in \mathsf{N}_{MP}(Q)$, which shows that $Q^u = Q$. Therefore Q is the unique Sylow q-subgroup of M normalised by P.

If $g \in G_\alpha$, then g normalises P. Let $y \in P$ and define $x := gyg^{-1} \in P$. Then x normalises Q, whence

$$Q^{gy} = Q^{xg} = Q^g.$$

Thus Q^g is normalised by P, and hence by uniqueness, $Q^g = Q$. We conclude that G_α normalises Q. Then Theorem 3.12(iii) implies that $M = Q$, and since M is a minimal normal subgroup, M is an elementary abelian q-group, as in part (i). Thus, if M is non-abelian, then G_α has no non-trivial soluble normal subgroup. □

The proof of the following result relies on the classification of finite simple groups, as it uses the fact that the outer automorphism group of a finite simple group is soluble. This result is usually referred to as *Schreier's Conjecture* and was proved using the finite simple group classification; see (Dixon & Mortimer 1996, p. 133).

Corollary 3.23 *Suppose that G is a finite primitive permutation group acting on Ω with a unique minimal normal subgroup T such that T is non-abelian and simple. Then T is not regular.*

Proof Suppose that G is as in the theorem. First we note that $G_\alpha \neq 1$, since G_α is a maximal subgroup of G. Since the centraliser of a normal subgroup is normal, $\mathsf{C}_G(T)$ is a normal subgroup of G. The facts that T is the unique minimal normal subgroup of G and that $\mathsf{Z}(T) = 1$ imply that $\mathsf{C}_G(T) = 1$. Hence G can be viewed as a subgroup of $\mathsf{Aut}(T)$. Suppose that T is regular; that is, $G_\alpha \cap T = 1$. Then $G_\alpha \cong G_\alpha/(G_\alpha \cap T) \cong TG_\alpha/T$, which shows that G_α can be embedded into $\mathsf{Out}(T)$. By Schreier's Conjecture, $\mathsf{Out}(T)$ is soluble, which is a contradiction, by Theorem 3.22. □

It was observed by Smith (Smith 2015) that Theorem 3.22 and Corollary 3.23 do not hold for infinite simple groups. We are grateful to Dr. Simon Smith for showing us the following infinite example where these results fail.

Example 3.24 Let p_1, p_2, q be odd primes such that $q \nmid (p_1 - 1)$ and $q \nmid (p_2 - 1)$. By a result of Obraztsov (Obraztsov 1996, Theorem C) (see also (Smith 2015, Theorem 4.1)) there exists an infinite simple group T with the following properties:

(i) every proper non-trivial subgroup of T is isomorphic to one of the groups $\mathbb{Z}$, C_{p_1}, or C_{p_2};
(ii) T admits an outer automorphism β of order q such that β has no non-trivial fixed points in T; that is, $\{t \in T \mid t\beta = t\} = \{1\}$.

Set $G = T \rtimes \langle \beta \rangle$ and $H = \langle \beta \rangle$. We claim that H is a maximal subgroup of G. Suppose that M is a subgroup of G such that $H < M \leqslant G$. Set $T_0 = M \cap T$. The subgroup T_0 is normalised by H, and, since $H < M$, we have that $T_0 \neq 1$. Suppose that $T_0 \neq T$. Then T_0 is a proper non-trivial subgroup of T, and so it is isomorphic to $\mathbb{Z}$, C_{p_1}, or to C_{p_2}. Further, β induces an automorphism on T_0. Noting that $\mathsf{Aut}(\mathbb{Z}) \cong C_2$, that $\mathsf{Aut}(C_{p_i}) \cong C_{p_i - 1}$ and that $|\beta| = q$ is an odd prime such that $q \nmid p_1 - 1$ and $q \nmid p_2 - 1$, we find that β must induce the trivial automorphism on T_0. Since β has no non-trivial fixed points in T, this is impossible, and hence $T_0 = T$. Thus $M = G$, and so H is maximal as claimed. Now set $\Omega = [G\colon H]$ and consider G as a permutation group on Ω acting by right multiplication. Then G is primitive (by Lemma 2.14), almost simple (by definition), and $\mathsf{Soc}\, G$ is regular (since $\mathsf{Soc}\, G = T$ and $T \cap H = 1$).

4
Finite direct products of groups

This book has a special focus on groups that act on cartesian products of sets. In this chapter we present some elementary results about direct products of groups, as such direct products play an important role in our investigation. The main theorems of this chapter are Theorems 4.8, 4.16, 4.22, 4.24, and 4.30. The first, sometimes attributed to Goursat, describes the subgroup structure of the direct product of two groups, while the second, often referred to as Scott's Lemma, describes subdirect subgroups in a direct product of non-abelian simple groups. In Section 4.7 we study the cartesian decompositions that arise most naturally for innately transitive permutation groups with an FCR-plinth. We call such decompositions normal. This chapter contains the first major results in this book that concern cartesian decompositions; namely, we prove in Theorems 4.22 and 4.24 that if G is innately transitive with non-abelian FCR-plinth M, then the normal cartesian decompositions for G can be completely characterised using certain direct decompositions of M that restrict to decompositions of a point stabiliser M_ω. Section 4.8 contains a study of the factorisations of non-abelian FCR-groups using diagonal subgroups. The main result of this section is Theorem 4.30 which shows that a direct power of a non-abelian simple group T admits certain factorisations using diagonal subgroups if and only if T has uniform automorphisms. This result may have several applications in the study of permutation groups with an FCR-plinth.

Most of the results in the chapter apply to arbitrary (not necessarily finite) groups. If a finiteness condition is required, this is specified.

4.1 Direct products

As mentioned above, direct products of groups, as defined in Section 3.4, play a vital role in this book. We briefly fix our conventions concerning them. *From now on, we treat only direct products of finitely many groups.*

Let $G_1, \ldots, G_k$ be groups that act on the pairwise disjoint sets $\Omega_1, \ldots, \Omega_k$, respectively, and set $G = G_1 \times \cdots \times G_k$. Then the direct product G has two natural actions. Firstly, G acts on the disjoint union $\Sigma = \Omega_1 \cup \cdots \cup \Omega_k$ as follows. If $g = (g_1, \ldots, g_k) \in G$ and $\omega \in \Sigma$ then there is precisely one $i \in \underline{k}$ such that $\omega \in \Omega_i$ and we let $\omega g = \omega g_i$. It is easy to see that the Ω_i are G-invariant subsets under this action, and so this action is intransitive whenever $k > 1$. Thus if $k > 1$ we call this action the *intransitive action* of G. The intransitive action of G is faithful if and only if, for each i, the action of G_i on Ω_i is faithful. Recall the definition of a transitive constituent after Lemma 2.2. The proof of the following lemma is easy and is left to the reader.

Lemma 4.1 *Let $k \geqslant 2$, and let G_i, Ω_i $(1 \leqslant i \leqslant k)$, G and Σ be as above. If G_i is faithful and transitive on Ω_i for each $i \leqslant k$, then the intransitive action of G on Σ is faithful, the G-orbits are $\Omega_1, \ldots, \Omega_k$, and the transitive constituents of G are $G^{\Omega_i} = G/G_{(\Omega_i)} \cong G_i$ for $i \leqslant k$.*

Conversely, every intransitive permutation group is permutationally isomorphic to a subgroup of the direct product of its transitive constituents.

Secondly, G acts on the cartesian product $\Pi = \Omega_1 \times \cdots \times \Omega_k$ as follows. If $g = (g_1, \ldots, g_k) \in G$ and $\omega = (\omega_1, \ldots, \omega_k) \in \Pi$ then

$$\omega g = (\omega_1 g_1, \ldots, \omega_k g_k).$$

This action of G on Π is called the *product action* of the direct product. The product action may be transitive for any k.

Lemma 4.2 *Using the notation of the previous paragraph, the product action of G on Π is transitive if and only if G_i is transitive on Ω_i for all $i \in \underline{k}$.*

Proof Assume, for some i, that G_i is intransitive on Ω_i, and that ω_1, ω_2 are from different G_i-orbits. Let γ_1 and γ_2 be elements of Π such that, for $j = 1,\ 2$, the i-th coordinate of γ_j is ω_j. Then the definition of the product action implies that γ_1 and γ_2 must be in different G-orbits.

Conversely, assume that all of the G_i are transitive, and choose points $(\omega_1, \ldots, \omega_k)$, $(\omega'_1, \ldots, \omega'_k) \in \Pi$. Then, for each $i \in \underline{k}$, there is an element $g_i \in G_i$ such that $\omega_i g_i = \omega'_i$. Therefore

$$(\omega_1, \ldots, \omega_k)(g_1, \ldots, g_k) = (\omega_1 g_1, \ldots, \omega_k g_k) = (\omega'_1, \ldots, \omega'_k),$$

and so G is transitive on Π. □

The direct product of groups $G_1, \ldots, G_k$, defined above, is sometimes called their *external* direct product. If G is a group and $G_1, \ldots, G_k$ are subgroups of G such that the map $G_1 \times \cdots \times G_k \to G$ given by

$$(g_1, \ldots, g_k) \longmapsto g_1 \cdots g_k \quad \text{for all} \quad g_i \in G_i$$

is an isomorphism, then we say that G is an *(internal) direct product* of $G_1, \ldots, G_k$ and that the set $\mathcal{G} = \{G_1, \ldots, G_k\}$ of subgroups is an *(internal) direct decomposition* of G. In such a situation it will be usual to identify G with $G_1 \times \cdots \times G_k$. In particular we usually do not distinguish between internal and external direct products, and we omit the word 'internal' for both direct products and direct decompositions. We will represent elements either as tuples $(g_1, \ldots, g_k)$, or as words $g_1 \cdots g_k$, whichever is most natural in the context. We allow direct decompositions with only one subgroup, and in this case we say that the direct decomposition is *trivial.*

Suppose that $\mathcal{G} = \{G_1, \ldots, G_k\}$ is a direct decomposition of a group G. Let I be a subset of $\underline{k}$ and set $J = \underline{k} \setminus I$. Define

$$G_I = \prod_{i \in I} G_i \quad \text{and} \quad G_J = \prod_{j \in J} G_j.$$

Then every element $g \in G$ can be written uniquely in the form $g = g_I g_J$ where $g_I \in G_I$ and $g_J \in G_J$. The reader can easily verify that the mapping $\sigma_I \colon g \mapsto g_I$ is a surjective homomorphism, which is often referred to as a *coordinate projection.* If I is a singleton $\{i\}$ then the projection σ_I will usually be denoted by σ_i. Coordinate projections will feature often in this work, and so we introduce further notation for them. If it is important to specify the direct factors of a direct decomposition, rather than their labels, we will sometimes denote the projection σ_I by $\sigma_{\mathcal{I}}$, where $\mathcal{I} = \{G_i \mid i \in I\}$. As the next lemma shows, under certain conditions, images under different projections may be isomorphic.

Lemma 4.3 *Let k be a positive integer, let $G_1, \ldots, G_k$ be groups and set $G = G_1 \times \cdots \times G_k$. Suppose that K is a subgroup of G, and that G_0 is a subgroup of $\mathsf{Aut}(G)$ such that $\{G_1, \ldots, G_k\}$ is invariant under* *the action of* G_0. *If* $I_1,\ I_2 \subseteq \underline{k}$ *and* $g \in G_0$ *is such that* $(G_{I_1})^g = G_{I_2}$, *then* $(K\sigma_{I_1})^g = K^g\sigma_{I_2}$.

Proof For $i = 1,\ 2$, set $J_i = \underline{k} \setminus I_i$, and let $x^g \in (K\sigma_{I_1})^g$. Then $x \in K\sigma_{I_1}$ and so there is $y \in G_{J_1}$ such that $xy \in K$. Then $(xy)^g \in K^g$. By assumption $(G_{I_1})^g = G_{I_2}$ and so $(xy)^g\sigma_{I_2} = x^g$. Thus $x^g \in K^g\sigma_{I_2}$, and hence $(K\sigma_{I_1})^g \leqslant K^g\sigma_{I_2}$. Conversely, if $x \in K^g\sigma_{I_2}$ then $xy \in K^g$ for some $y \in G_{J_2}$. As $(G_{I_2})^{g^{-1}} = G_{I_1}$, we obtain that $(xy)^{g^{-1}}\sigma_{I_1} = x^{g^{-1}}$ and so $x^{g^{-1}} \in K\sigma_{I_1}$ and $x \in (K\sigma_{I_1})^g$. Thus $(K\sigma_{I_1})^g = K^g\sigma_{I_2}$. □

4.2 Product action and intransitive cartesian decompositions

If G is a permutation group on Ω which preserves a cartesian decomposition $\mathcal{E} = \{\Gamma_1, \ldots, \Gamma_\ell\}$ of Ω (as defined in Definition 1.1), then G permutes the partitions Γ_δ. We denote the stabiliser $\{g \in G \mid \Gamma_\delta g = \Gamma_\delta\}$ in G of Γ_δ by G_{Γ_δ}. The δ*-component* of G is the permutation group induced by G_{Γ_δ} on Γ_δ. In Chapter 10 we will study cartesian decompositions that are acted upon intransitively by some group. This occurs, for instance, if $\mathcal{E}$ is inhomogeneous (that is, the $|\Gamma_\delta|$ do not all have the same cardinality). Here we show, for intransitive cartesian decompositions, that the action of G on Ω is permutationally isomorphic to that of a subgroup of a direct product in product action.

Let $\mathcal{E} = \{\Gamma_1, \ldots, \Gamma_\ell\}$ be a cartesian decomposition of a set Ω. Suppose that $\mathcal{E}_0$ is a proper subset of $\mathcal{E}$ and set $\mathcal{E}_1 = \mathcal{E} \setminus \mathcal{E}_0$. For $i \in \{0,1\}$, set $\Delta_i = \{\delta \mid \Gamma_\delta \in \mathcal{E}_i\}$ and let

$$\bar{\Omega}_i = \left\{ \bigcap_{\delta \in \Delta_i} \gamma_\delta \mid \gamma_\delta \in \Gamma_\delta \text{ for each } \delta \in \Delta_i \right\}.$$

By Lemma 2.16, each $\bar{\Omega}_i$ is a G-invariant partition of Ω.

If $\gamma \in \Gamma_\delta$ for some $\delta \in \Delta_i$, then γ is a union of blocks from $\bar{\Omega}_i$ and we set

$$\bar{\gamma} = \{\sigma \mid \sigma \in \bar{\Omega}_i,\ \sigma \subseteq \gamma\}, \quad \bar{\Gamma}_\delta = \{\bar{\gamma} \mid \gamma \in \Gamma_\delta\}. \tag{4.1}$$

For $i = 0, 1$, we let $\bar{\mathcal{E}}_i = \{\bar{\Gamma}_\delta \mid \delta \in \Delta_i\}$. Recall the definition of permutational isomorphism from Section 2.2.

Lemma 4.4 *Let* G *be a permutation group on a set* Ω *and suppose that* $\mathcal{E}$ *is a* G*-invariant cartesian decomposition of* Ω, *such that* G

leaves invariant a proper subset $\mathcal{E}_0$ of $\mathcal{E}$. Using the notation introduced above, the following hold.

(i) *$\{\bar{\Omega}_0, \bar{\Omega}_1\}$ is a G-invariant cartesian decomposition of Ω on which G acts trivially.*
(ii) *For $i = 0,\ 1$, $\bar{\mathcal{E}}_i$ is a G-invariant cartesian decomposition of $\bar{\Omega}_i$. Further, the pair $(\Gamma_i \mapsto \bar{\Gamma}_i, \mathrm{id})$ is a permutational isomorphism between the G-actions on $\mathcal{E}_i$ and on $\bar{\mathcal{E}}_i$.*

Proof (i) It follows, for $i = 0,\ 1$, from Lemma 2.16, that $\bar{\Omega}_i$ is a G-invariant partition of Ω. If $\sigma_i \in \bar{\Omega}_i$, then σ_i is an intersection of $|\Delta_i|$ blocks γ_δ taking precisely one for each $\delta \in \Delta_i$. Thus $\sigma_0 \cap \sigma_1$ is an intersection of ℓ blocks, one from each of the Γ_i in $\mathcal{E}$. As $\mathcal{E}$ is a cartesian decomposition, $|\sigma_0 \cap \sigma_1| = 1$ and so $\{\bar{\Omega}_0, \bar{\Omega}_1\}$ is a G-invariant cartesian decomposition whose members are G-invariant partitions. In other words, G acts trivially on $\{\bar{\Omega}_0, \bar{\Omega}_1\}$.

(ii) We prove this statement for $i = 0$, the proof for $i = 1$ being identical. Choose, for each $\delta \in \Delta_0$, an element $\gamma_\delta \in \Gamma_\delta$ and let $\bar{\gamma}_\delta \in \bar{\Gamma}_\delta$ be the corresponding element. By the definition of $\bar{\Omega}_0$, $\bigcap_{\delta \in \Delta_0} \gamma_\delta$ is an element of $\bar{\Omega}_0$. This shows that $\bar{\mathcal{E}}_0$ is a cartesian decomposition of $\bar{\Omega}_0$. The definition of $\bar{\Gamma}_\delta$ implies that $(\Gamma_\delta \mapsto \bar{\Gamma}_\delta, \mathrm{id})$ is a permutational isomorphism between the G-actions on $\mathcal{E}_0$ and $\bar{\mathcal{E}}_0$. □

For the analysis of embeddings into wreath products of permutation groups that preserve cartesian decompositions in Section 5.3, we need further details of the cartesian decompositions in Lemma 4.4. Let us define for $i = 0,\ 1$, the map $\bar{f}_i\colon \Omega \to \bar{\Omega}_i$ as $\omega \mapsto \sigma_i$ where σ_i is the unique element of $\bar{\Omega}_i$ that contains ω. We may recursively apply $\bar{f}_i$ to Γ_i and to $\mathcal{E}_i$, and obtain that $\Gamma_i \bar{f}_i = \bar{\Gamma}_i$ and $\mathcal{E}_i \bar{f}_i = \bar{\mathcal{E}}_i$. Define $\bar{f}\colon \Omega \to \bar{\Omega}_0 \times \bar{\Omega}_1$ as $\omega \mapsto (\omega \bar{f}_0, \omega \bar{f}_1)$. By the fact that $\{\bar{\Omega}_0, \bar{\Omega}_1\}$ is a cartesian decomposition and by (1.1), $\bar{f}$ is a bijection.

Recall that we established in (1.1) a bijection $f\colon \Omega \to \Gamma_1 \times \cdots \times \Gamma_\ell$ defined by $\omega f = (\gamma_1, \ldots, \gamma_\ell)$ where each γ_δ is the unique part in Γ_δ that contains ω. For $i \in \{0, 1\}$, set

$$\Omega_i = \prod_{\delta \in \Delta_i} \Gamma_\delta. \tag{4.2}$$

As $\mathcal{E}$ is a cartesian decomposition, the map $\psi_i\colon \bar{\Omega}_i \to \Omega_i$ defined as $\bigcap_{\delta \in \Delta_i} \gamma_\delta \mapsto (\gamma_\delta \mid \delta \in \Delta_i)$ is a bijection. Define $f_i\colon \Omega \to \Omega_i$ as the composition $\bar{f}_i \psi_i$. Hence $\omega f_i = (\gamma_\delta \mid \delta \in \Delta_i,\ \omega \in \gamma_\delta)$. Similarly, define

$\vartheta\colon \Omega \to \Omega_0 \times \Omega_1$ as $\omega\vartheta = (\omega\bar{f}_0\psi_0, \omega\bar{f}_1\psi_1) = (\omega f_0, \omega f_1)$. Recall the definition of a permutational embedding from Section 2.2.

Proposition 4.5 *Let G be a permutation group on a set Ω and suppose that $\mathcal{E}$ is a G-invariant cartesian decomposition of Ω, such that G leaves invariant a proper subset $\mathcal{E}_0$ of $\mathcal{E}$. Let $\mathcal{E}_1$, Ω_0, Ω_1, f_0, f_1 and ϑ be as above. Then the following hold.*

(i) *The map $\vartheta\colon \Omega \to \Omega_0 \times \Omega_1$ is a bijection.*
(ii) *For $i \in \{0,1\}$, define the map $\chi_i\colon G \to \mathsf{Sym}\,\Omega_i$ by the rule $(\omega f_i)(g\chi_i) = (\omega g)f_i$. Then χ_i is a well defined homomorphism and $\chi\colon G \to \mathsf{Sym}\,\Omega_0 \times \mathsf{Sym}\,\Omega_1$ defined by $g\chi = (g\chi_0, g\chi_1)$ is a monomorphism.*
(iii) *For $i = 0,\ 1$, the set $\mathcal{E}_i f_i$ is a cartesian decomposition of Ω_i and $G\chi_i$ is contained in the stabiliser W_i of $\mathcal{E}_i f_i$ in $\mathsf{Sym}\,\Omega_i$. Also, for each $\delta \in \Delta_i$, the δ-components of G and $G\chi_i$ are permutationally isomorphic.*
(iv) *The pair (ϑ, χ) is a permutational embedding of G on Ω into the group $\mathsf{Sym}\,\Omega_0 \times \mathsf{Sym}\,\Omega_1$ in its product action on $\Omega_0 \times \Omega_1$, and $G\chi \leqslant W_0 \times W_1$.*

Proof (i) This follows from the fact that the map $\omega \mapsto (\omega\bar{f}_0, \omega\bar{f}_1)$ is a bijection from Ω to $\bar{\Omega}_0 \times \bar{\Omega}_1$ and that $\psi_i\colon \bar{\Omega}_i \to \Omega_i$ are bijections.

(ii) Let $i \in \{0,1\}$. First note that f_i is a surjective map onto Ω_i. We prove that χ_i is a well defined homomorphism. Define an action of G on Ω_i by the rule $(\omega f_i)g = (\omega g)f_i$. We claim that this action is well defined. Let ω, $\omega' \in \Omega$ such that $\omega f_i = \omega' f_i$. For $\delta \in \Delta_i$, let γ_δ and γ_δ' be the parts of Γ_δ that contain ωg and $\omega' g$, respectively. Then $\gamma_\delta g^{-1}$ and $\gamma_\delta' g^{-1}$ are the parts of $\Gamma_\delta g^{-1}$ that contain ω and ω', respectively. As $\omega f_i = \omega' f_i$, we have that $\gamma_\delta g^{-1} = \gamma_\delta' g^{-1}$, which, in turn, implies that $\gamma_\delta = \gamma_\delta'$. Hence $(\omega g)f_i = (\omega' g)f_i$, and so $(\omega f_i)g$ is a well defined element of Ω_i. By the definition of $(\omega f_i)g$, this assignment defines an action of G on Ω_i. Further, the permutation of Ω_i induced by g is $g\chi_i$. Thus $\chi_i\colon g \mapsto g\chi_i$ is a homomorphism from G to $\mathsf{Sym}\,\Omega_i$.

If $g \in \ker\chi$ then, for each $\omega \in \Omega$ and each i, $\omega f_i = (\omega g)f_i$. Thus ω and ωg lie in the same part of each partition in $\mathcal{E}$, and hence $\omega = \omega g$. Since this holds for all $\omega \in \Omega$, we find that $g = 1$, and so χ is a monomorphism.

(iii) As noted before the proposition, $\Gamma_j \bar{f}_i = \bar{\Gamma}_j$ and $\mathcal{E}_i f_i = \bar{\mathcal{E}}_i$ for $i = 0,\ 1$. By Lemma 4.4, $\mathcal{E}_i \bar{f}_i$ is a cartesian decomposition of Ω_i, and, since ψ_i is a bijection, $\mathcal{E}_i\bar{f}_i\psi_i = \mathcal{E}_if_i$ is a cartesian decomposition of $\bar{\Omega}_i\psi_i = \Omega_i$. Moreover, it follows from part (ii) that $\mathcal{E}_if_i$ is invariant under $G\chi_i$. Hence $G\chi_i$ is contained in W_i.

The previous paragraph shows that, for g in the stabiliser G_{Γ_δ} of $\Gamma_\delta \in \mathcal{E}_i$, we have $(\Gamma_\delta f_i)(g\chi_i) = \Gamma_\delta f_i$, and that the permutation of $\Gamma_\delta f_i$ induced by $g\chi_i$ maps $\gamma f_i \mapsto \gamma g f_i$, for each $\gamma \in \Gamma_\delta$. It follows that the δ-component of G on Ω is permutationally isomorphic to the δ-component of $G\chi_i$ on Ω_i.

(iv) This follows since, for all $\omega \in \Omega$ and all $g \in G$, we have

$$\omega\vartheta\, g\chi = (\omega f_0, \omega f_1)g\chi = (\omega g f_0, \omega g f_1) = (\omega g)\vartheta.$$

□

Remark 4.6 Proposition 4.5 has an important and immediate application to inhomogeneous cartesian decompositions of a set Ω. Suppose that $G \leqslant \mathsf{Sym}\,\Omega$, that $\mathcal{E} = \{\Gamma_\delta \mid \delta \in \Delta\}$ is an inhomogeneous G-invariant cartesian decomposition of Ω, and consider the partition $\Delta = \dot{\bigcup}_{1 \leqslant i \leqslant s} \Delta_i$ of Δ such that, for all i, j and all $\delta \in \Delta_i$, $\delta' \in \Delta_j$, we have $|\Gamma_\delta| = |\Gamma_{\delta'}|$ if and only if $i = j$. For each i let $\mathcal{E}_i = \{\Gamma_\delta \mid \delta \in \Delta_i\}$ and define $\Omega_i = \prod_{\delta \in \Delta_i} \Gamma_\delta$. Let $f_i \colon \Omega \to \prod_{i \in \Delta_\delta} \Gamma_\delta$ be defined by the rule $\omega f_i = (\gamma_\delta \mid \delta \in \Delta_i)$ where γ_δ is the part in Γ_δ containing ω.

With the natural extensions of the definitions of ϑ, χ_0, χ_1, χ to the case with an arbitrary finite number of direct factors we obtain for each i, by Proposition 4.5, a $(G\chi_i)$-invariant homogeneous cartesian decomposition $\mathcal{E}_if_i$ of Ω_i, and a permutational embedding (ϑ, χ) of G on Ω into the direct product $\mathsf{Sym}\,\Omega_1 \times \cdots \times \mathsf{Sym}\,\Omega_s$ in its product action on $\Omega_1 \times \cdots \times \Omega_s$ such that $G\chi \leqslant W_1 \times \cdots \times W_s$ where W_i is the stabiliser of $\mathcal{E}_if_i$ in $\mathsf{Sym}\,\Omega_i$.

Thus for many applications and problems it is useful first to understand the situation for homogeneous cartesian decompositions, and then to approach the general case via the product action of a direct product. This is the philosophy behind Theorem 10.13.

4.3 Subgroups of direct products

The aim of this section is to characterise the subgroups of a direct product of two groups. Let G_1 and G_2 be groups, and let G denote their direct product $G_1 \times G_2$. Recall that σ_1 and σ_2 are the coordinate projections onto G_1 and G_2, respectively. Among the subgroups of G, the ones which are themselves direct products $H_1 \times H_2$ $(H_i \leqslant G_i)$ play an important role. The following lemma characterises such subgroups and will often be used without explicit reference. The proof is left to the reader.

Lemma 4.7 *Let $G = G_1 \times G_2$ be the direct product of the groups G_1 and G_2, and let H be a subgroup of G. Then the following are equivalent:*

(i) $H = (G_1 \cap H) \times (G_2 \cap H)$;
(ii) $H\sigma_1 = G_1 \cap H$;
(iii) $H\sigma_1 \leqslant H$;
(iv) $H\sigma_2 = G_2 \cap H$;
(v) $H\sigma_2 \leqslant H$.

For general subgroups H of $G_1 \times G_2$, we define the following connection between homomorphisms and subgroups.

(i) Let H be a subgroup of $G = G_1 \times G_2$, and, for $i = 1, 2$, set $K_i = G_i \cap H$. Easy argument shows that K_i is normal in $H\sigma_i$ for $i = 1, 2$. Define a map $\varphi_H \colon H\sigma_1 \to H\sigma_2/K_2$ by

$$x \mapsto yK_2 \quad \text{where } y \in G_2 \text{ is such that } xy \in H.$$

We will show in Theorem 4.8 below that φ_H is a well defined homomorphism from $H\sigma_1$ onto $H\sigma_2/K_2$.

(ii) Now assume that H_1 and H_2 are subgroups of G_1 and G_2, respectively, let N_2 be a normal subgroup of H_2, and let φ be a homomorphism from H_1 onto H_2/N_2. Define a subset H_φ of $G_1 \times G_2$ as follows:

$$H_\varphi = \{(x, y) \mid x \in H_1,\ y \in x\varphi\}.$$

We will show below that H_φ is a subgroup of G.

If H is a subgroup of a group G, and N is a normal subgroup of H, then the quotient H/N is said to be a *section* of G. If both $H \neq G$ and $N \neq 1$ then we call H/N a *proper section* of G. As the homomorphism φ above can be viewed as an isomorphism between $H_1/\ker\varphi$ and H_2/N_2, we usually refer to such a homomorphism as an *isomorphism between sections of G_1 and G_2*. The following result, usually attributed to Goursat, gives a characterisation of the subgroups of a direct product of two groups in terms of isomorphisms between their sections. Our proof follows that of (Schmidt 1994, Theorem 1.6.1).

Theorem 4.8 *Using the notation introduced above, the following hold.*

(i) *The map* φ_H *is a well defined homomorphism from* $H\sigma_1$ *onto* $H\sigma_2/K_2$ *with kernel* $K_1 = H \cap G_1$.

(ii) *The subset* H_φ *is a subgroup of* $G_1 \times G_2$ *with* $H_\varphi\sigma_1 = H_1$, $H_\varphi\sigma_2 = H_2$, $H_\varphi \cap G_1 = \ker\varphi$, *and* $H_\varphi \cap G_2 = N_2$.

(iii) *Let* φ *and* ψ *be isomorphisms between sections of* G_1 *and* G_2. *Then* $H_\varphi \leqslant H_\psi$ *if and only if the domain of* φ *is contained in the domain of* ψ *and* $x\varphi \subseteq x\psi$ *for all elements* x *in the domain of* φ.

(iv) *The correspondence* $\varphi \mapsto H_\varphi$ *is a bijection from the set of isomorphisms between sections of* G_1 *and* G_2 *to the set of subgroups of* $G_1 \times G_2$.

Proof (i) Suppose that H is a subgroup of $G_1 \times G_2$. First we show that φ_H is well defined. For $x \in H\sigma_1$ let y_1, $y_2 \in G_2$ such that xy_1, $xy_2 \in H$. Then y_1, $y_2 \in H\sigma_2$, and $y_1^{-1}y_2 \in G_2 \cap H$. Therefore y_1 and y_2 are in the same coset of $G_2 \cap H$. The proof that φ_H is a homomorphism is straightforward and is left to the reader. The kernel of φ_H consists of those $x \in H\sigma_1$ which are elements of H. Therefore $\ker\varphi_H = G_1 \cap H$.

(ii) Let x_1, $x_2 \in G_1$ and y_1, $y_2 \in G_2$ such that (x_1, y_1), $(x_2, y_2) \in H_\varphi$. As $y_1y_2 \in (x_1\varphi)(x_2\varphi) = (x_1x_2)\varphi$, we find that $(x_1x_2, y_1y_2) \in H_\varphi$, and so H_φ is closed under multiplication. Similarly H_φ is closed under taking inverses since $(x_1\varphi)^{-1} = (x_1^{-1})\varphi$. Thus H_φ is a subgroup. It follows from the definition of H_φ that $H_\varphi\sigma_1 = H_1$, $H_\varphi\sigma_2 = H_2$. An element $x \in G_1$ is an element $(x, 1)$ of H_φ if and only if $x\varphi$ is the trivial coset N_2. Thus $G_1 \cap H_\varphi = \ker\varphi$. Finally, an element $y \in G_2$ is an element $(1, y)$ of H_φ if and only if $y \in 1\varphi = N_2$; that is, $G_2 \cap H_\varphi = N_2$.

(iii) Let φ and ψ be homomorphisms as above and assume that $H_\varphi \leqslant H_\psi$. Then $H_\varphi\sigma_1 \leqslant H_\psi\sigma_1$ and so the domain of φ is contained in the domain of ψ. Further if x is an element of the domain of φ then

$$x\varphi = \{y \in G_2 \mid (x, y) \in H_\varphi\} \subseteq \{y \in G_2 \mid (x, y) \in H_\psi\} = x\psi.$$

Assume now that the domain of φ is contained in the domain of ψ and that $x\varphi \subseteq x\psi$ for all elements x in the domain of φ. Suppose that $x_1 \in G_1$ and $x_2 \in G_2$ such that $(x_1, x_2) \in H_\varphi$. Then x_1 lies in the domain of φ, and so x_1 lies in the domain of ψ also. Further, $x_2 \in x_1\varphi$, which implies that $x_2 \in x_1\psi$. Thus, the definition of H_ψ implies that $(x_1, x_2) \in H_\psi$, and so $H_\varphi \leqslant H_\psi$ as claimed.

(iv) It follows from (iii) that $H_\varphi = H_\psi$ if and only if $\varphi = \psi$. □

Let G be a group with a direct decomposition $\mathcal{G} = \{G_1, \ldots, G_k\}$ and let H be a subgroup of G. We say that H is a $\mathcal{G}$*-subdirect subgroup* of G if

$$H\sigma_i = G_i \quad \text{for all} \quad i \in \underline{k}.$$

If there is no danger of confusion, the group H will briefly be called *subdirect.* For example, if G is a permutation group on Ω and $\Omega_1, \ldots, \Omega_k$ are the G-orbits, then G is permutationally isomorphic to a subdirect subgroup of the direct product $G^{\Omega_1} \times \cdots \times G^{\Omega_k}$ of its transitive constituents; see the discussion after Lemma 2.2 and Lemma 4.1. Theorem 4.8 can be used to characterise subdirect subgroups of direct products.

Corollary 4.9 *Let G_1 and G_2 be groups and set $G = G_1 \times G_2$. There is a bijection between the set of surjective homomorphisms between quotients of G_1 and G_2 and the set of subdirect subgroups of G.*

Proof It follows from Theorem 4.8(ii) that H_φ is subdirect if and only if $H_1 = G_1$ and $H_2 = G_2$. Hence the required bijection is given by the restriction of the mapping $\varphi \mapsto H_\varphi$ to surjective homomorphisms between quotient groups of the form G_1/N_1 and G_2/N_2. □

If G is a group with a direct decomposition $\mathcal{G}$ and H is a subgroup of G, then, in some cases, the double projections $\sigma_\mathcal{I}(H)$, where $\mathcal{I}$ is a two-element subset of $\mathcal{G}$, give information concerning the subgroup H. Such a situation is exhibited by the following lemma and it will be useful in our analysis of permutation groups that preserve cartesian decompositions. A group G is said to be *perfect* if $G' = G$ where G' denotes the commutator subgroup. Recall that we use the notation $[a, b] = a^{-1}b^{-1}ab$.

Lemma 4.10 *Let k be a positive integer, let $G_1, \ldots, G_k$ be groups, and suppose, for $i \in \underline{k}$, that N_i is a perfect subgroup of G_i. Set $G = G_1 \times \cdots \times G_k$ and let K be a subgroup of G such that for all i_1, i_2 with $1 \leqslant i_1 < i_2 \leqslant k$, we have $N_{i_1} \times N_{i_2} \leqslant K\sigma_{\{i_1,i_2\}}$. Then $N_1 \times \cdots \times N_k \leqslant K$.*

Proof We prove by induction on m that, for all $i_1, \ldots, i_m$ such that

$2 \leqslant m \leqslant k$ and $1 \leqslant i_1 < \cdots < i_m \leqslant k$, we have

$$N_{i_1} \times \cdots \times N_{i_m} \leqslant K\sigma_{\{i_1,\ldots,i_m\}}.$$

Note that for $m = k$, this yields the required result. By assumption this condition holds for $m = 2$. Suppose that $3 \leqslant m \leqslant k$, and that the condition holds for $m - 1$. Let us prove it for m. Without loss of generality we show that $N_1 \times \cdots \times N_m \leqslant K\sigma_{\underline{m}}$. Let a and b be elements of N_1. Then, since $m \geqslant 3$, the inductive hypothesis applies to the projections $\sigma_{\{1,3,\ldots,m\}}$ and $\sigma_{\{1,2,4,\ldots,m\}}$ (which are taken to be $\sigma_{\{1,3\}}$ and $\sigma_{\{1,2\}}$, respectively, if $m = 3$), and we obtain

$$(a, 1, \ldots, 1) \in K\sigma_{\{1,3,\ldots,m\}} \quad \text{and} \quad (b, 1, \ldots, 1) \in K\sigma_{\{1,2,4,\ldots,m\}}.$$

Hence there are elements $c \in G_2$ and $d \in G_3$ such that

$$(a, c, 1, \ldots, 1),\ (b, 1, d, 1, \ldots, 1) \in K\sigma_{\underline{m}},$$

and so their commutator $([a, b], 1, \ldots, 1)$ is also an element of $K\sigma_{\underline{m}}$. Therefore $K\sigma_{\underline{m}}$ contains all commutators $[a, b]$ where a, $b \in N_1$. This amounts to saying that $N_1' = N_1 \leqslant K\sigma_{\underline{m}}$. Similar argument shows that $N_i \leqslant K\sigma_{\underline{m}}$ for $i = 2, \ldots, m$. The lemma now follows by induction. □

4.4 Strips

In this section we discuss some concepts that are related to diagonal subgroups of direct products. Let $G = G_1 \times \cdots \times G_k$ be a direct product and set $\mathcal{G} = \{G_1, \ldots, G_k\}$ so that $\mathcal{G}$ is a direct decomposition of G. Moreover assume that each G_i is non-trivial. The following definitions depend on $\mathcal{G}$. This is emphasised by including '$\mathcal{G}$' in the terminology, but in situations where no confusion arises it will be usual to omit the prefix '$\mathcal{G}$'. Recall, for $i \in \underline{k}$, the projection map $\sigma_i \colon G \to G_i$ given by $\sigma_i \colon (g_1, \ldots, g_k) \mapsto g_i$. Then for all $x \in G$ we have

$$x = (x\sigma_1)(x\sigma_2) \ldots (x\sigma_k).$$

We say that H is a $\mathcal{G}$*-strip* of G, if H is non-trivial and, for each $i \in \underline{k}$, either

$$H \cap \ker \sigma_i = 1 \quad \text{or} \quad H\sigma_i = 1.$$

Note that $H \cap \ker \sigma_i = 1$ if and only if the restriction of σ_i to H is an isomorphism between H and $H\sigma_i$. For $i \in \underline{k}$ and a strip H, we say that H *covers* G_i if $H\sigma_i$ is non-trivial; the $\mathcal{G}$*-support* $\mathrm{Supp}_{\mathcal{G}} H$ of H is defined as

$$\mathrm{Supp}_{\mathcal{G}} H = \{G_i \in \mathcal{G} \mid H \text{ covers } G_i\}.$$

If $|\mathrm{Supp}_{\mathcal{G}} H| > 1$, then we say that the strip H is *non-trivial.* Two strips, H and L, are *disjoint* if their supports are disjoint. Note that disjoint strips commute. A $\mathcal{G}$-strip H is said to be *full,* if $H\sigma_i = G_i$ for all $G_i \in \mathrm{Supp}_{\mathcal{G}} H$. A $\mathcal{G}$*-diagonal subgroup* is a $\mathcal{G}$-strip H such that $H\sigma_i = G_i$ for all i. In particular $\mathrm{Supp}_{\mathcal{G}} H = \mathcal{G}$ for $\mathcal{G}$-diagonal subgroups H.

Let us consider an example. Let G_1, G_2, G_3 be groups and, for $i = 1, 2, 3$, let H_i be a subgroup of G_i. Let $\varphi_2\colon H_1 \to H_2$ be an isomorphism. Set $G = G_1 \times G_2 \times G_3$. Then the subgroup

$$H = \{(x, x\varphi_2, 1) \mid x \in H_1\}$$

is a strip in G. Indeed, for $i = 1, 2, 3$, let τ_i denote the restriction of σ_i to H. Then the kernel of τ_1 consists of the elements $(1, x_2, x_3)$ such that $(1, x_2, x_3) \in H$. Since $(1, x_1, x_2) \in \ker\tau_1$, the definition of H implies that $1\varphi_2 = x_2$ and, as φ_2 is a bijection, we have that $x_2 = 1$. Since the third coordinate of an element of H is 1, we also have $x_3 = 1$. Therefore $\ker\tau_1 = 1$ and similar argument shows that $\ker\tau_2 = 1$. Hence $H \cong H\sigma_1 \cong H\sigma_2$. It follows from the definition of H that $H\sigma_3 = 1$, therefore H is a strip. Since $\mathrm{Supp}\, H = \{G_1, G_2\}$, the subgroup H is a non-trivial strip. Further H is a full strip if and only if $H_1 = G_1$ and $H_2 = G_2$. Since $G_3 \notin \mathrm{Supp}\, H$, we have that H is not a diagonal subgroup in G, but it is a diagonal subgroup in $G_1 \times G_2$.

The strip in this example was constructed using a bijection between two members of a direct decomposition. The following useful lemma shows that this is a typical occurrence. Since, by our discussion above, every strip is a diagonal subgroup in a possibly smaller group, we state this result for diagonal subgroups.

Lemma 4.11 *Suppose that G is a group with a finite direct decomposition $\mathcal{G} = \{G_1, \ldots, G_k\}$ and let H be a strip in G with $\mathrm{Supp}\, H = \mathcal{G}$. Then, for $i \in \underline{k}$, there is an isomorphism $\alpha_i\colon H\sigma_1 \to H\sigma_i$ such that*

$$H = \{(h\alpha_1, \ldots, h\alpha_k) \mid h \in H\sigma_1\}. \tag{4.3}$$

Proof Let τ_1 denote the restriction of σ_1 to H. As H is a strip, τ_1 is an isomorphism between H and $H\sigma_1$. Set $\alpha_i = \tau_1^{-1}\sigma_i$ for each i.

Let X denote the set on the right-hand side of (4.3) and suppose that $x \in H$. Then, for $i \in \underline{k}$, we have $x\sigma_i = x\tau_1\alpha_i$. Hence

$$x = (x\sigma_1, \ldots, x\sigma_k) = (x\tau_1\alpha_1, \ldots, x\tau_1\alpha_k).$$

This shows that $x \in X$. Conversely, let $x = (h\alpha_1, \ldots, h\alpha_k)$ be an element of X where $h \in H\sigma_1$. Then there is some $h_0 \in H$ such that $h = h_0\tau_1$, and so, for $i \in \underline{k}$, $h_0\sigma_i = h_0\tau_1\alpha_i = h\alpha_i$. Therefore

$$x = (h\alpha_1, \ldots, h\alpha_k) = (h_0\sigma_1, \ldots, h_0\sigma_k) = h_0,$$

which shows that $x \in H$. Therefore $H = X$, as required. □

4.5 Normalisers in direct products

Here we present several results about normalisers in direct products.

Lemma 4.12 *Let G be a group and let A, B be subgroups of G, such that $A \lhd B$ and $\mathsf{N}_G(A)/A$ is abelian. Then $\mathsf{N}_G(A) \leqslant \mathsf{N}_G(B)$.*

Proof As B normalises A, it is a subgroup of $\mathsf{N}_G(A)$. By an isomorphism theorem, there is a one-to-one correspondence between the set of subgroups of $\mathsf{N}_G(A)/A$ and the set of those subgroups of $\mathsf{N}_G(A)$ that contain A. Further, under this correspondence, normal subgroups correspond to normal subgroups. Since $\mathsf{N}_G(A)/A$ is abelian, each of its subgroups is normal. This shows that B must also be normal in $\mathsf{N}_G(A)$. □

Lemma 4.13 *Let k be a positive integer, let $G_1, \ldots, G_k$ be groups, set $G = G_1 \times \cdots \times G_k$, and let H be a subgroup of G. Then the following are valid.*

(i) $\mathsf{N}_G(H) \leqslant \prod_i \mathsf{N}_{G_i}(H\sigma_i)$. *Moreover if* $H = \prod_i H\sigma_i$ *then* $\mathsf{N}_G(H) = \prod_i \mathsf{N}_{G_i}(H\sigma_i)$.

(ii) *For $i \in \underline{k}$, let H_i be a subgroup of G_i such that the following hold:*

 (a) $H_1 \times \cdots \times H_k \lhd H$;

 (b) $\mathsf{N}_G(H_1 \times \cdots \times H_k)/(H_1 \times \cdots \times H_k)$ *is abelian;*

 (c) $\mathsf{N}_{G_i}(H\sigma_i) = \mathsf{N}_{G_i}(H_i)$.

 Then $\mathsf{N}_G(H) = \mathsf{N}_G(H_1 \times \cdots \times H_k)$.

Proof As part (i) is straightforward, we only prove part (ii). It follows from Lemma 4.12 that $\mathsf{N}_G(H_1 \times \cdots \times H_k) \leqslant \mathsf{N}_G(H)$. On the other hand part (i) and condition (c) imply that

$$\mathsf{N}_G(H) \leqslant \prod_{i=1}^{k} \mathsf{N}_{G_i}(H\sigma_i) = \prod_{i=1}^{k} \mathsf{N}_{G_i}(H_i) = \mathsf{N}_G(H_1 \times \cdots \times H_k).$$

Therefore equality holds. □

Next we determine the normaliser of a strip.

Lemma 4.14 *Let k be a positive integer, let $G_1, \ldots, G_k$ be groups, let H_1 be a subgroup of G_1, for $i = 2, \ldots, k$ let $\varphi_i \colon \mathsf{N}_{G_1}(H_1) \to G_i$ be an injective homomorphism such that $\mathsf{N}_{G_1}(H_1)\varphi_i = \mathsf{N}_{G_i}(H_1\varphi_i)$, and let $H = \{(h, h\varphi_2, \ldots, h\varphi_k) \mid h \in H_1\}$ be a non-trivial strip in $G_1 \times \cdots \times G_k$. Then*

$$\begin{aligned} &\mathsf{N}_{G_1 \times \cdots \times G_k}(H) \\ &= \{(t, c_2(t\varphi_2), \ldots, c_k(t\varphi_k)) \mid t \in \mathsf{N}_{G_1}(H_1),\ c_i \in (\mathsf{C}_{G_1}(H_1))\varphi_i\}. \end{aligned} \quad (4.4)$$

Proof Denote the right-hand side of equation (4.4) by N and set $G = G_1 \times \cdots \times G_k$. If $(t, c_2(t\varphi_2), \ldots, c_k(t\varphi_k)) \in N$ then for all $h \in H_1$

$$(h, h\varphi_2, \ldots, h\varphi_k)^{(t, c_2(t\varphi_2), \ldots, c_k(t\varphi_k))} = (h^t, h^t\varphi_2, \ldots, h^t\varphi_k) \in H.$$

Hence $N \leqslant \mathsf{N}_G(H)$. Let us prove that the other inclusion also holds. Suppose that $(t_1, \ldots, t_k) \in \mathsf{N}_G(H)$. Then for all $h \in H_1$ we have that

$$(h, h\varphi_2, \ldots, h\varphi_k)^{(t_1, \ldots, t_k)} = (h^{t_1}, (h\varphi_2)^{t_2}, \ldots, (h\varphi_k)^{t_k}) \in H,$$

and so $h^{t_1} \in H_1$, and $(h\varphi_i)^{t_i} \in H_1\varphi_i$, for $i \in \{2, \ldots, k\}$. This shows that $t_1 \in \mathsf{N}_{G_1}(H_1)$, and, for all i, $t_i \in \mathsf{N}_{G_i}(H_1\varphi_i)$. Therefore t_i is an element in $\mathsf{N}_{G_1}(H_1)\varphi_i$. As the element on the right-hand side of the last displayed equation is an element of H, we also obtain that $h^{t_1}\varphi_i = (h\varphi_i)^{t_i}$ for all $h \in H_1$. This amounts to saying that $h^{t_1} = h^{t_i\varphi_i^{-1}}$ for all $h \in H_1$, and hence $a_i := t_1(t_i\varphi_i^{-1})^{-1} \in \mathsf{C}_{G_1}(H_1)$. Therefore $t_i\varphi_i^{-1} = a_i^{-1}t_1$, and so $t_i = c_i(t_1\varphi_i)$ where $c_i = (a_i^{-1})\varphi_i \in (\mathsf{C}_{G_1}(H_1))\varphi_i$, for all $i = 2, \ldots, k$. Thus

$$(t_1, \ldots, t_k) = (t_1, c_2(t_1\varphi_2), \ldots, c_k(t_k\varphi_k)),$$

and so $\mathsf{N}_G(H) \subseteq N$, as required. □

It is worth specialising the previous lemma to the common case in which the simple direct factors are non-abelian simple groups and the strip is full.

Corollary 4.15 *Let $M = T_1 \times \cdots \times T_k$ be a direct product of non-abelian simple groups $T_1, \ldots, T_k$. Suppose that H is a full strip of M. Then*

$$\mathsf{N}_M(H) = H \times \prod_{T_i \notin \mathsf{Supp}\, H} T_i. \tag{4.5}$$

Proof Set $\mathcal{T} = \{T_1, \ldots, T_k\}$. It is clear that $M = M_1 \times M_2$ where

$$M_1 = M\sigma_{\mathsf{Supp}\, H} \quad \text{and} \quad M_2 = M\sigma_{\mathcal{T}\setminus\mathsf{Supp}\, H},$$

and $H = (H \cap M_1) \times (H \cap M_2)$ with $H \cap M_2 = 1$. By Lemma 4.14 applied to $M_1 = \prod_{T_i \in \mathsf{Supp}\, H} T_i$ with subgroup $H \cap M_1 = H$, the group H is self-normalising in M_1, that is $\mathsf{N}_{M_1}(H \cap M_1) = H$. Then, by Lemma 4.13(i) applied with $k = 2$ to $M = M_1 \times M_2$ with subgroup $H = (H \cap M_1) \times (H \cap M_2)$, we see that $\mathsf{N}_M(H) = \mathsf{N}_{M_1}(H \cap M_1) \times \mathsf{N}_{M_2}(H \cap M_2) = H \times M_2$. □

4.6 Scott's Lemma

The aim of this section is to prove that a subdirect subgroup in a direct product of finitely many non-abelian simple groups is a direct product of strips. This is known as Scott's Lemma and appeared, for instance, in (Scott 1980, Lemma, Appendix on maximal subgroups).

Suppose that G is a group with a direct decomposition $\mathcal{G}$, and K is a subgroup of G. If X is a strip in G then we say that X is *involved* in K if $K = X \times K\sigma_{\mathcal{G}\setminus\mathsf{Supp}\, X}$. For example, Corollary 4.15 exhibits subgroups involving strips in the case where $\mathcal{G}$ consists of non-abelian simple groups.

The result often referred to as Scott's Lemma is part (iii) of the following theorem.

Theorem 4.16 *Let k be a positive integer, let M be a direct product $T_1 \times \cdots \times T_k$ where the T_i are non-abelian simple groups, and let H be a subgroup of M.*

(i) *Suppose that H is a non-trivial subdirect subgroup of the product $\prod_{T_i \in \mathsf{Supp}\, H} T_i$ and that, for some $m \leqslant k$, $H\sigma_m = T_m$. Then there exists a full strip X involved in H covering T_m.*

(ii) *Suppose that each T_i is isomorphic to a fixed non-abelian simple group T such that no proper subgroup and no proper section of T is isomorphic to T. If $H\sigma_m = T_m$ for some $m \leqslant k$, then there exists a full strip X involved in H covering T_m.*

(iii) *If H is a subdirect subgroup of M, then H is a direct product of pairwise disjoint full strips of M.*

(iv) *If H is a non-trivial normal subgroup of M then H is the direct product of some of the T_i.*

Proof We start with (i) and (ii). Suppose without loss of generality that $m = 1$ in cases (i) and (ii), so that $T_1 \in \operatorname{Supp} H$ and $H\sigma_1 = T_1$. Let X be a normal subgroup of H such that $X\sigma_1 = T_1$ and $|\operatorname{Supp} X|$ is minimal among such subgroups. (Such a subgroup exists as H satisfies $H\sigma_1 = T_1$ and is a normal subgroup of H.)

Suppose that $i \neq 1$ is such that $X\sigma_i \neq 1$; that is, $T_i \in \operatorname{Supp} X$. Then $X \cap \ker \sigma_i$ is a proper subgroup of X. As $X \cap \ker \sigma_i$ is normal in X we deduce that $(X \cap \ker \sigma_i)\sigma_1$ is a normal subgroup of $X\sigma_1 = T_1$. Further, $\operatorname{Supp}(X \cap \ker \sigma_i) \subseteq \operatorname{Supp} X \setminus \{T_i\} \subset \operatorname{Supp} X$. Therefore, the minimality of $|\operatorname{Supp} X|$ implies that $(X \cap \ker \sigma_i)\sigma_1 \neq T_1$. Thus $(X \cap \ker \sigma_i)\sigma_1$ is trivial since T_1 is simple, and so $X \cap \ker \sigma_i \leqslant X \cap \ker \sigma_1$. In fact, as $\ker \sigma_i$ is normal in M, we find that $X \cap \ker \sigma_i$ is a normal subgroup of $X \cap \ker \sigma_1$ and this holds in statements (i) and (ii).

(i) To complete the proof of part (i), suppose that H is a subdirect subgroup of $\prod_{T_j \in \operatorname{Supp} H} T_j$. Now $X\sigma_i$ is a non-trivial normal subgroup of $H\sigma_i = T_i$, so $X\sigma_i = T_i$ and $X/(X \cap \ker \sigma_i) \cong T_i$. As $X \cap \ker \sigma_i$ is normal in $X \cap \ker \sigma_1$, we obtain that $(X \cap \ker \sigma_1)/(X \cap \ker \sigma_i)$ is a normal subgroup of $X/(X \cap \ker \sigma_i) \cong T_i$. As $X \cap \ker \sigma_1$ is a proper subgroup of X and the group T_i is simple, it follows that $X \cap \ker \sigma_1 = X \cap \ker \sigma_i$. As $X/(X \cap \ker \sigma_1) \cong T_1$, and $X/(X \cap \ker \sigma_i) \cong T_i$, we have $T_1 \cong T_i$. We have now proved that, for all $i \in \underline{k}$, either $X\sigma_i = 1$, or $X\sigma_i = T_i \cong T_1$ and $X \cap \ker \sigma_i = X \cap \ker \sigma_1$. In particular this implies that $(X \cap \ker \sigma_1)\sigma_i = 1$ for all i, whence $X \cap \ker \sigma_1$ is trivial and the restriction of σ_1 to X is an injection. We proved also that if $X\sigma_i$ is non-trivial, then $X \cap \ker \sigma_i = X \cap \ker \sigma_1 = 1$, and hence the restriction of σ_i to X is an injection also. Therefore X is a full strip as claimed and X covers T_1. By Corollary 4.15, X is self-normalising in $\prod_{T_i \in \operatorname{Supp} X} T_i$. Therefore, since X is normal in H, we have $H\sigma_{\operatorname{Supp} X} = X$, and then, by Lemma 4.7, $H = X \times (H\sigma_{\mathcal{G} \setminus \operatorname{Supp} X})$, proving part (i).

(ii) Suppose that $T_i \cong T_1$ for all i and that no proper subgroup and no proper section of T_1 is isomorphic to T_1. If $X\sigma_i = 1$ for all $i \neq 1$, then $X = T_1$ and part (ii) holds. Thus we may assume that $X\sigma_i \neq 1$ for some $i \neq 1$, We have shown that $X \cap \ker \sigma_i$ is a normal subgroup of $X \cap \ker \sigma_1$. Suppose that $X \cap \ker \sigma_i \neq X \cap \ker \sigma_1$. Then $(X \cap \ker \sigma_1)\sigma_i$ is a non-trivial normal subgroup of $X\sigma_i$ and

$$X\sigma_i/(X \cap \ker \sigma_1)\sigma_i \cong (X/(X \cap \ker \sigma_i))/((X \cap \ker \sigma_1)/(X \cap \ker \sigma_i))$$
$$\cong X/(X \cap \ker \sigma_1) \cong T_1.$$

Thus $(X \cap \ker \sigma_1)\sigma_i$ is a non-trivial proper normal subgroup of $X\sigma_i$ and consequently $X\sigma_i$ is not simple. In particular $X\sigma_i \neq T_i$. However this implies that $X\sigma_i/(X \cap \ker \sigma_1)\sigma_i$ is a proper section of $T_i \cong T_1$ which is isomorphic to T_1, contradicting our assumption in (ii). Thus $X \cap \ker \sigma_i = X \cap \ker \sigma_1$ and so $X\sigma_i \cong X/(X \cap \ker \sigma_i) = X/(X \cap \ker \sigma_1) \cong T_1$. Since $T_i \cong T_1$ has no proper subgroup isomorphic to T_1 it follows that $X\sigma_i = T_i$. We have now proved that, for all $i \in \underline{k}$, either $X\sigma_i = 1$, or $X\sigma_i = T_i \cong T_1$ and $X \cap \ker \sigma_i = X \cap \ker \sigma_1$. Now we obtain as in part (i) that if $X\sigma_i \neq 1$, then the restriction of σ_i to X is an injection, and so X is a full strip that covers T_1. We also obtain using the argument in part (i) that X is involved in H.

(iii) We deduce this assertion from part (i) by induction on k. If $k = 1$, then $H = T_1 = M$, and so (iii) holds. Suppose that $k \geqslant 2$. Then part (i) implies that $H = X \times (H \cap \ker \sigma_{\mathrm{Supp}\, X})$ where X is a full strip involved in H. If $\mathrm{Supp}\, X = \underline{k}$, then $H = X$ and we are done. Hence we may assume that $|\mathrm{Supp}\, X| \leqslant k - 1$. Then $\ker \sigma_{\mathrm{Supp}\, X} \neq 1$, and $\ker \sigma_{\mathrm{Supp}\, X}$ is the direct product of some of the T_i. Further, $H \cap \ker \sigma_{\mathrm{Supp}\, X}$ is a subdirect subgroup of $\ker \sigma_{\mathrm{Supp}\, X}$ and, by the inductive hypothesis, it is a direct product of pairwise disjoint full strips, say, $X_1, \ldots, X_t$. Since, for all $i \in \underline{t}$, X is disjoint from X_i, we obtain that H is the direct product of the pairwise disjoint full strips $X, X_1, \ldots, X_t$.

(iv) Let us suppose that H is a non-trivial normal subgroup of M. Then H is a normal subgroup of $\prod_{T_i \in \mathrm{Supp}\, H} T_i$. For each $T_i \in \mathrm{Supp}\, H$, $H\sigma_i$ is a non-trivial normal subgroup of T_i, and so $H\sigma_i = T_i$ since T_i is simple. Thus H is a subdirect subgroup of $\prod_{T_i \in \mathrm{Supp}\, H} T_i$, and so by part (iii) it is a direct product of pairwise disjoint full strips. Suppose that X is one of these strips. Then $X = H\sigma_{\mathrm{Supp}\, X}$, and so X is a normal subgroup of $\prod_{T_j \in \mathrm{Supp}\, X} T_j$. However, Corollary 4.15 shows that X is self-normalising in $\prod_{T_j \in \mathrm{Supp}\, X} T_j$, and hence $X = \prod_{T_j \in \mathrm{Supp}\, X} T_j$. Since X is a strip, $\mathrm{Supp}\, X$ must have only one element and $X = T_j$ for some j. Now the result follows. □

Corollary 4.17 *Suppose that N is a non-abelian characteristically simple normal subgroup of a group G, such that $N = T_1 \times \cdots \times T_k$ where the T_i are non-abelian simple groups. Then G acts on the set $\{T_1, \ldots, T_k\}$ by conjugation. Further N is a minimal normal subgroup of G if and only if G is transitive on $\{T_1, \ldots, T_k\}$.*

Proof By Theorem 4.16(iv), the set $\{T_1, \ldots, T_k\}$ is the set of minimal normal subgroups of N, and so this set is invariant under $\mathsf{Aut}(N)$. As conjugation by an element of G induces an automorphism of N, this set is also invariant under conjugation by elements of G. If $I \subseteq \underline{k}$ such that $\{T_i \mid i \in I\}$ is a G-orbit, then $\prod_{i \in I} T_i$ is a normal subgroup of G. Thus N is a minimal normal subgroup of G if and only if G is transitive on $\{T_1, \ldots, T_k\}$. □

4.7 Normal cartesian decompositions

In this section we develop a connection between cartesian decompositions of sets and a direct decomposition concept for groups. Before Lemma 4.2 we defined the product action of a direct product of permutation groups. Here we extend this concept by defining a class of cartesian decompositions that characterise the product action.

Definition 4.18 Suppose that M is a transitive permutation group on a set Ω, and $\mathcal{E} = \{\Gamma_1, \ldots, \Gamma_\ell\}$ is an M-invariant cartesian decomposition of Ω such that $M_{(\mathcal{E})} = M$. That is, for each $i \in \underline{\ell}$, the set Γ_i is an M-invariant partition of Ω. For $i \in \underline{\ell}$, let $\overline{M_i}$ denote the kernel of the M-action on Γ_i, and set

$$M_i = \bigcap_{j \neq i} \overline{M_j}. \tag{4.6}$$

The cartesian decomposition $\mathcal{E}$ is said to be *M-normal* if the set $\{M_1, \ldots, M_\ell\}$ is a direct decomposition of M.

If G is a permutation group acting on Ω and $\mathcal{E}$ is a G-invariant cartesian decomposition of Ω, then $\mathcal{E}$ is said to be *normal* if G has a transitive normal subgroup M such that $\mathcal{E}$ is M-normal.

A general construction for examples is given in Example 4.23. In the next lemma we summarise some basic properties of M-normal cartesian decompositions.

Lemma 4.19 *If $\mathcal{E} = \{\Gamma_1, \ldots, \Gamma_\ell\}$ is an M-normal cartesian decomposition of Ω, then, for all $i \in \underline{\ell}$, we have that $M = M_i \times \overline{M_i}$. Consequently M_i is faithful on Γ_i and $M^{\Gamma_i} = (M_i)^{\Gamma_i}$.*

Proof Since $\mathcal{E}$ is M-normal, we have that $M = M_1 \times \cdots \times M_\ell$. We claim, for $i \in \ell$, that $\overline{M_i} = M_1 \times \cdots \times M_{i-1} \times M_{i+1} \times \cdots \times M_\ell$. Assume without loss of generality that $i = 1$. By the definition of $\overline{M_1}$, we have, for $i \geqslant 2$, that $M_i \leqslant \overline{M_1}$. Thus $M_2 \times \cdots \times M_\ell \leqslant \overline{M_1}$. Hence it suffices to show that $M_1 \cap \overline{M_1} = 1$. If $x \in M_1 \cap \overline{M_1}$, then $x \in \bigcap_{j \geqslant 1} \overline{M}_j$, and so $x \in \overline{M}_j$ for all $j \in \underline{\ell}$. Thus x acts trivially on Γ_j for all $j \in \underline{\ell}$. Then Definition 1.1 of a cartesian decomposition implies that x acts trivially on Ω, and so, as M is a permutation group, $x = 1$. Hence $M_1 \cap \overline{M_1} = 1$, which implies that $M_1 \times \overline{M_1} = M$, as claimed. □

If $\mathcal{E} = \{\Gamma_1, \ldots, \Gamma_\ell\}$ is an M-normal cartesian decomposition for a transitive group M, and the M_i are as in (4.6), then the subgroups M_i and $\overline{M_i}$ will usually be denoted by M^{Γ_i} and $\overline{M^{\Gamma_i}}$, respectively. The group M^{Γ_i} may be viewed either as a permutation group on Ω, or as a permutation group on Γ_i, whichever is convenient.

Lemma 4.20 *Suppose that M is a transitive permutation group acting on Ω, $\mathcal{E}$ is an M-normal cartesian decomposition of Ω, and let $\Gamma \in \mathcal{E}$. Then the following all hold.*

(i) *The group M^Γ is transitive on Γ.*
(ii) *For $\gamma \in \Gamma$ and $\omega \in \gamma$, we have $(M^\Gamma)_\gamma = M_\omega \cap M^\Gamma$.*
(iii) *If $\omega \in \Omega$, then $M_\omega = \prod_{\Gamma' \in \mathcal{E}} \left(M_\omega \cap M^{\Gamma'} \right)$.*
(iv) *The partition Γ is the set of $\overline{M^\Gamma}$-orbits.*

Proof Let $\mathcal{E} = \{\Gamma_1, \ldots, \Gamma_\ell\}$. Without loss of generality we will prove parts (i), (ii), and (iv) in the case $\Gamma = \Gamma_1$.

(i) Since Γ_1 is a system of imprimitivity for M and M is transitive, Lemma 2.12 shows that M is transitive on Γ_1. Since M is the direct product $M^{\Gamma_1} \times \overline{M^{\Gamma_1}}$ and $\overline{M^{\Gamma_1}}$ acts trivially on Γ_1, we must have that M^{Γ_1} is transitive on Γ_1.

(ii) Let $\omega \in \Omega$ and $\gamma_1 \in \Gamma_1$ such that $\omega \in \gamma_1$. As the partition Γ_1 is a system of imprimitivity in Ω for the M-action, we obtain that $M_\omega \cap M^{\Gamma_1} \leqslant (M^{\Gamma_1})_{\gamma_1}$. Suppose that $m \in (M^{\Gamma_1})_{\gamma_1}$ and that $\{\omega\} = \gamma_1 \cap \gamma_2 \cdots \cap \gamma_\ell$ for some $\gamma_2 \in \Gamma_2, \ldots, \gamma_\ell \in \Gamma_\ell$. Since $m \in M^{\Gamma_1}$, the element m stabilises $\gamma_2, \ldots, \gamma_\ell$, and, by assumption, m also stabilises

γ_1. Hence m stabilises ω, and so $(M^{\Gamma_1})_{\gamma_1} \leqslant M_\omega \cap M^{\Gamma_1}$. Therefore $(M^{\Gamma_1})_{\gamma_1} = M_\omega \cap M^{\Gamma_1}$.

(iii) Assume that $\{\omega\} = \gamma_1 \cap \cdots \cap \gamma_\ell$ as in part (ii). Then

$$(M^{\Gamma_1})_{\gamma_1} \times \cdots \times (M^{\Gamma_\ell})_{\gamma_\ell} \leqslant M_\omega.$$

Let $m \in M_\omega$ and write m as a product $m = m_1 m_2 \ldots m_\ell$ where $m_i \in M^{\Gamma_i}$ for $i = 1, \ldots, \ell$. Then $\{\omega\} = \{\omega\}m = (\gamma_1 \cap \cdots \cap \gamma_\ell)m = \gamma_1 m_1 \cap \cdots \cap \gamma_\ell m_\ell$. Thus, for each i, ω lies in the part $\gamma_i m_i$ of Γ_i, and hence $\gamma_i m_i = \gamma_i$. Hence $m = m_1 \cdots m_\ell \in (M^{\Gamma_1})_{\gamma_1} \times \cdots \times (M^{\Gamma_\ell})_{\gamma_\ell} \leqslant M_\omega$ and so

$$M_\omega = (M^{\Gamma_1})_{\gamma_1} \times \cdots \times (M^{\Gamma_\ell})_{\gamma_\ell} = (M_\omega \cap M^{\Gamma_1}) \times \cdots \times (M_\omega \cap M^{\Gamma_\ell}).$$

(iv) Suppose, as above, that $\{\omega\} = \gamma_1 \cap \cdots \cap \gamma_\ell$ for some $\omega \in \Omega$ and $\gamma_1 \in \Gamma_1, \ldots, \gamma_\ell \in \Gamma_\ell$. As Γ_1 is a block-system for the action of M on Ω, the block γ_1 is stabilised by

$$K_1 = \left(M_\omega \cap M^{\Gamma_1}\right) \times \overline{M^{\Gamma_1}} = (M^{\Gamma_1})_{\gamma_1} \times \overline{M^{\Gamma_1}},$$

and it follows from part (ii) and Lemma 4.19 that K_1 is equal to the stabiliser of γ_1 in M. By Lemma 2.14 it follows that $\gamma_1 = \omega K_1$. As $\overline{M^{\Gamma_1}} \lhd M$, the set Σ of $\overline{M^{\Gamma_1}}$-orbits is also a system of imprimitivity for M, by Lemma 2.20. It suffices to prove that the block $\sigma \in \Sigma$ containing ω is equal to γ_1. If $\omega' \in \gamma_1 = \omega K_1$, then there is some $m \in K_1$, such that $\omega m = \omega'$. Write m as the product $m_1 \cdots m_\ell$ where $m_i \in M^{\Gamma_i}$ for $i = 1, \ldots, \ell$. Then, as $m_1 \in M_\omega$, we have $\omega' = \omega m = \omega m_2 \cdots m_\ell$. Since $m_2 \cdots m_\ell \in \overline{M^{\Gamma_1}}$, we obtain $\omega' \in \sigma$. Therefore $\gamma_1 \subseteq \sigma$. On the other hand, as $\overline{M^{\Gamma_1}} \leqslant K_1$, it follows that $\sigma \subseteq \omega K_1 = \gamma_1$. Thus $\sigma = \gamma_1$ and the two block systems Σ and Γ_1 coincide, as claimed. □

Lemma 4.21 *Suppose that G is a permutation group on Ω, $\mathcal{E}$ is a G-invariant cartesian decomposition of Ω, and M is a normal subgroup of G such that $\mathcal{E}$ is M-normal. Then G acts by conjugation on the set $\{M^\Gamma \mid \Gamma \in \mathcal{E}\}$ and the G-actions on $\mathcal{E}$ and on $\{M^\Gamma \mid \Gamma \in \mathcal{E}\}$ are permutationally isomorphic.*

Proof Suppose that $\Gamma_1, \Gamma_2 \in \mathcal{E}$ and $g \in G$. Then $\Gamma_1 g = \Gamma_2$ if and only if $(\overline{M^{\Gamma_1}})^g = \overline{M^{\Gamma_2}}$ and, by equation (4.6), this implies that $(M^{\Gamma_1})^g = M^{\Gamma_2}$. This shows that the pair (φ, id_G) where $\varphi \colon \Gamma \mapsto M^\Gamma$ is a permutational isomorphism. □

Recall Definition 1.2 that if G is a direct product of finitely many simple groups, then G is said to be an FCR-group (finitely completely reducible). By Lemma 3.14, a minimal normal subgroup of a finite group is an FCR-group. The next result shows that, if G has a transitive non-abelian minimal normal subgroup M which is a non-abelian FCR-group then each normal cartesian decomposition is M-normal. By Lemma 3.14, if G is finite, then each non-abelian minimal normal subgroup has this form.

Theorem 4.22 *Let G be a permutation group on Ω, and suppose that G has a transitive, non-abelian, minimal normal* FCR*-subgroup M of the form $M = T_1 \times \cdots \times T_k$, with each T_i non-abelian simple. Let $\mathcal{E}$ be a normal G-invariant cartesian decomposition of Ω. Then $\mathcal{E}$ is M-normal and G is transitive on $\mathcal{E}$.*

Proof Let N be a normal subgroup of G, such that $\mathcal{E}$ is N-normal, let $N_i = N^{\Gamma_i}$ and $\overline{N_i} = \overline{N^{\Gamma_i}}$, where $\mathcal{E} = \{\Gamma_1, \ldots, \Gamma_\ell\}$. By the definition of N-normal, $N = N_1 \times \cdots \times N_\ell$. By Lemma 3.4, either $M \leqslant N$ or $M \cap N = 1$ and, in the latter case, $[M, N] = 1$. Suppose first that $M \leqslant N$. We claim that

$$M = (N_1 \cap M) \times \cdots \times (N_\ell \cap M). \tag{4.7}$$

Note that $N = N_i \times \overline{N_i}$ and $M \trianglelefteq N$. Let T be one of the simple direct factors of M. For $i \in \underline{\ell}$ let σ_i be the projection map $N \to N_i$. As T is a non-trivial subgroup of N, there exists i such that $T\sigma_i$ is non-trivial, whence $T\sigma_i \cong T$ as T is a non-abelian simple group. We claim that $T \leqslant N_i$. Choose $x \in T\sigma_i$ with $x \neq 1$. As M is a normal subgroup of G, the subgroup T^x is also a minimal normal subgroup of M, and Theorem 3.6 shows that either $T = T^x$ or $[T, T^x]$ is trivial. If the latter, then

$$1 = [T, T^x]\sigma_i = [T\sigma_i, (T\sigma_i)^x] = [T\sigma_i, T\sigma_i];$$

but the last term is non-trivial as $T\sigma_i$ is a non-abelian simple group. Hence the former holds, that is $T = T^x$. As $T\sigma_i \cong T$, if $t, t' \in T$ then $t^x = t'$ if and only if $(t\sigma_i)^x = t'\sigma_i$. As $T\sigma_i$ is non-abelian simple, conjugation by x induces a non-trivial automorphism of $T\sigma_i$. Thus x must induce a non-trivial automorphism of T as well. However if $T\sigma_j$ were non-trivial for some $j \neq i$, then x would centralise $T\sigma_j \cong T$ and so, as shown by a similar argument, conjugation by x would induce a trivial automorphism of T. Thus $T\sigma_j$ is trivial for all $j \neq i$, and

$T \leqslant N_i$ as claimed. Thus each simple direct factor of M is contained in some N_i and it follows that (4.7) holds. In this case, therefore, $\mathcal{E}$ is M-normal, so $M = \prod_{\Gamma \in \mathcal{E}} M^{\Gamma}$. Since G is transitive on the simple direct factors of M, by Corollary 4.17, it follows that G is transitive by conjugation on the M^{Γ_i}, and hence Lemma 4.21 implies that G is transitive on $\mathcal{E}$.

Thus we may assume that $M \cap N = 1$. As M and N centralise each other, we obtain that they are semiregular (Theorem 3.2) and, as each is transitive, we find that each is regular. As the full centraliser of M in $\mathsf{Sym}\,\Omega$ is semiregular, and N is a regular subgroup of this centraliser, we conclude that $N = \mathsf{C}_{\mathsf{Sym}\,\Omega}(M)$, and the same argument implies that $M = \mathsf{C}_{\mathsf{Sym}\,\Omega}(N)$. Since M and N are isomorphic regular subgroups, Lemma 2.6 implies M and N are permutationally isomorphic, and so there is a permutational isomorphism $\chi\colon N \to M$. For each i set $M_i = N_i\chi$ and $\overline{M_i} = \overline{N_i}\chi$. Thus $M = M_1 \times \cdots \times M_\ell$. Also the $\overline{M_i}$-orbits are the same as the $\overline{N_i}$-orbits, and by Lemma 4.20(iv) the $\overline{M_i}$-orbits form the partition Γ_i. Thus $M^{\Gamma_i} = (M_i)^{\Gamma_i}$. If K_i is the kernel of the action of M_i on Γ_i, then $K_i\chi^{-1}$ also acts trivially on Γ_i, and hence lies in $N_{(\Gamma_i)} = \overline{N_i}$. However $K_i\chi^{-1} \leqslant N_i$, so $K_i\chi^{-1} \leqslant N_i \cap \overline{N_i} = 1$, whence also $K_i = 1$. Hence $M^{\Gamma_i} = M_i$, and $M = \prod_{i=1}^{\ell} M^{\Gamma_i}$, so $\mathcal{E}$ is M-normal. As M is a minimal normal subgroup of G, the conjugation action by G on the M^{Γ_i} is transitive by Corollary 4.17, and hence G is transitive on $\mathcal{E}$ (see Lemma 4.21). □

Finally we describe the set of normal cartesian decompositions for an innately transitive permutation group G. First we show how to construct such cartesian decompositions. Theorem 4.24 implies that our construction is as general as possible.

Example 4.23 Suppose that G is a permutation group with a transitive normal subgroup M such that M is the direct product $M = \prod_{i=1}^{\ell} M_i$ where the M_i form a G-conjugacy class. Suppose that $\omega \in \Omega$ such that

$$M_\omega = (M_\omega \cap M_1) \times \cdots \times (M_\omega \cap M_\ell). \tag{4.8}$$

For $i = 1, \ldots, \ell$ let $\overline{M_i} = \prod_{j \neq i} M_j$, and let Γ_i denote the set of $\overline{M_i}$-orbits in Ω.

Claim. $\mathcal{E} = \{\Gamma_1, \ldots, \Gamma_\ell\}$ *is an M-normal cartesian decomposition of Ω.*

Proof of the claim. We note that M_i and $\overline{M_i}$ are normal subgroups of M such that $M = M_i \times \overline{M_i}$. For $i \in \underline{\ell}$, let Γ_i denote the set of

$\overline{M_i}$-orbits in Ω. Since M is transitive and $\overline{M_i}$ is normal in M, each of the Γ_i is an M-invariant partition of Ω (Lemma 2.20). Further, since, for all i, M is transitive on Γ_i and $\overline{M_i}$ acts trivially on Γ_i, we find that M_i must be transitive on Γ_i. Moreover, G permutes transitively by conjugation the subgroups M_i and $\overline{M_i}$. Therefore G permutes transitively the partitions Γ_i in such a way that the G-actions on the M_i and on the Γ_i are permutationally isomorphic.

Choose, for all i, the block $\gamma_i \in \Gamma_i$ such that $\omega \in \gamma_i$. Then γ_i is an $\overline{M_i}$-orbit stabilised by $(M_i)_\omega \times \overline{M_i}$. Hence γ_i is also an $\left((M_i)_\omega \times \overline{M_i}\right)$-orbit. Since $M_\omega \leqslant (M_i)_\omega \times \overline{M_i}$, the correspondence between the over-groups of M_ω and the M-blocks containing ω given in Lemma 2.14 implies that $M_{\gamma_i} = (M_i)_\omega \times \overline{M_i}$. In particular

$$M_{\gamma_1} \cap \cdots \cap M_{\gamma_\ell} = (M_1)_\omega \times \cdots \times (M_\ell)_\omega = M_\omega.$$

Let $\gamma = \gamma_1 \cap \cdots \cap \gamma_\ell$ and note that $\omega \in \gamma$. Suppose that $\alpha \in \gamma$. Then there is $m \in M$ such that $\omega m = \alpha$, and so $m \in M_{\gamma_1} \cap \cdots \cap M_{\gamma_\ell}$. Consequently, $m \in M_\omega$ which gives that $\alpha = \omega m = \omega$. Thus $|\gamma| = 1$. Suppose now that $\gamma_i' \in \Gamma_i$ for all i. Since M_i is transitive on Γ_i while acting trivially on Γ_j with $j \neq i$, there exist for all i, $m_i \in M_i$ such that $\gamma_i' m_i = \gamma_i$ and $\gamma_j' m_i = \gamma_j'$ if $j \neq i$. Hence $(\gamma_1' \cap \cdots \cap \gamma_\ell')(m_1 \cdots m_\ell) = \gamma_1 \cap \cdots \cap \gamma_\ell = \gamma$. Thus $|\gamma_1' \cap \cdots \cap \gamma_\ell'| = |\gamma| = 1$, which gives that $\mathcal{E}$ is a cartesian decomposition of Ω. As was already shown above, $\mathcal{E}$ is G-invariant.

It remains to show that $\mathcal{E}$ is M-normal. Let K_i be the kernel of the M-action on Γ_i. Clearly $\overline{M_i} \leqslant K_i$. If $\overline{M_i} < K_i$ then, there exists a non-trivial $x \in M_i$ such that x acts trivially on Γ_i. On the other hand x acts trivially on each Γ_j with $j \neq i$, and hence x must act trivially on Ω. As G is a permutation group, this is impossible, which gives that $K_i = \overline{M_i}$. Now applying Definition 4.18, we find that $\mathcal{E}$ is M-normal. Further the argument in this paragraph also shows that $M^{\Gamma_i} = M_i$ for all i. □

Theorem 4.24 *Let G be an innately transitive group on Ω with a non-abelian* FCR*-plinth M of the form $M = T_1 \times \cdots \times T_k$ with each T_i simple, and let $\omega \in \Omega$. Then there is a bijection between the set of G-invariant normal cartesian decompositions of Ω and the set of G-invariant partitions $\{P_1, \ldots, P_\ell\}$ of $\mathcal{T} = \{T_1, \ldots, T_k\}$ satisfying*

$$M_\omega = M_\omega \sigma_{P_1} \times \cdots \times M_\omega \sigma_{P_\ell}, \tag{4.9}$$

where σ_{P_i} is the projection onto the direct product of the factors that *are contained in P_i. Further each G-invariant normal cartesian decomposition arises as in Example 4.23.*

Proof First we note that Theorem 4.22 implies that any G-invariant normal cartesian decomposition is M-normal. Hence the set of G-invariant normal cartesian decompositions coincides with the set of such M-normal decompositions.

If $\mathcal{P} = \{P_1, \ldots, P_\ell\}$ is a G-invariant partition of $\{T_1, \ldots, T_k\}$ such that (4.9) is valid, then let $\mathcal{E}(\mathcal{P})$ denote the G-invariant normal cartesian decomposition constructed in Example 4.23. Conversely, if $\mathcal{E} = \{\Gamma_1, \ldots, \Gamma_\ell\}$ is a G-invariant M-normal cartesian decomposition of Ω, then let $A_i = \{T_j \mid T_j \leqslant M^{\Gamma_i}\}$ where M^{Γ_i} is defined as before Lemma 4.20. By the definition of normal cartesian decomposition, the A_i form a G-invariant partition of $\mathcal{T}$ and let us denote this partition by $\mathcal{P}(\mathcal{E})$.

We show that the maps $\mathcal{E} \mapsto \mathcal{P}(\mathcal{E})$ and $\mathcal{P} \mapsto \mathcal{E}(\mathcal{P})$ are inverses of each other. If $\mathcal{E} = \{\Gamma_1, \ldots, \Gamma_\ell\}$ is an M-normal cartesian decomposition of Ω, then let $\mathcal{P}(\mathcal{E}) = \{P_1, \ldots, P_\ell\}$. Since $\mathcal{E}$ is G-invariant, we find that $\mathcal{P}(\mathcal{E})$ is G-invariant (Lemma 4.21). Further, by Lemma 4.20(iii), $\mathcal{P}(\mathcal{E})$ satisfies (4.9). Since Lemma 4.20(iv) implies that Γ_i is the set of M^{Γ_i}-orbits, we find that $\mathcal{E}(\mathcal{P}(\mathcal{E})) = \mathcal{E}$.

Suppose now that $\mathcal{P}$ is a G-invariant partition of $\mathcal{T}$ such that (4.9) is valid. Then the last comment in Example 4.23 shows that $\mathcal{P}(\mathcal{E}(\mathcal{P})) = \mathcal{P}$. This shows that the compositions of the maps $\mathcal{E} \mapsto \mathcal{P}(\mathcal{E})$ and $\mathcal{P} \mapsto \mathcal{E}(\mathcal{P})$ are identities which gives that they are bijections between the sets in the statement of the theorem. □

Corollary 4.25 *Suppose that G is an innately transitive group on Ω with a regular and non-abelian* FCR*-plinth M of the form $M = T_1 \times \cdots \times T_k$ with each T_i simple and $k \geqslant 2$. Then G preserves a non-trivial normal cartesian decomposition of Ω.*

Proof As M is regular, $1 = M_\omega = \prod_i (T_i \cap M_\omega)$, and so equation (4.9) is trivially satisfied with the direct decomposition $\{T_1, \ldots, T_k\}$ of M. As M is not simple, $k \geqslant 2$, and so G preserves a non-trivial normal cartesian decomposition of Ω. □

4.8 Uniform automorphisms and factorisations of direct products

In this section we prove several results on factorisations of direct products using strips. These results can also be found in (Praeger & Schneider 2017b).

An automorphism α of a group G is uniform, if the map $\tilde{\alpha}\colon g \mapsto g^{-1}(g\alpha)$ is surjective. If G is a finite group, then a map $G \to G$ is surjective if and only if it is injective. Thus, if G is a finite group, then $\alpha \in \mathsf{Aut}(G)$ is not uniform if and only if the map $g \mapsto g^{-1}(g\alpha)$ is not injective; that is, $g^{-1}(g\alpha) = h^{-1}(h\alpha)$ for some distinct $g,\ h \in G$. The last equation is equivalent to $hg^{-1} = (hg^{-1})\alpha$; that is, in this case, the element hg^{-1} is a non-identity fixed point of the automorphism α. It is a consequence of the finite simple group classification that every automorphism of a non-abelian finite simple group has non-identity fixed points. In fact the following stronger result holds; see (Gorenstein 1982, Theorem 1.48) or (Kurzweil & Stellmacher 2004, 9.5.3)

Lemma 4.26 *A finite non-solvable group has no uniform (that is, fixed-point-free) automorphisms.*

The situation is rather different for the class of infinite simple groups.

Example 4.27 Let $\mathbb{F}$ be the algebraic closure of the finite field $\mathbb{F}_p$ and consider the group $G = \mathsf{SL}_n(\mathbb{F})$ with $n \geqslant 2$. Then G is a connected algebraic group (see (Springer 2009, Exercise 2.2.2)). Further, the p-th powering map $(a_{i,j}) \mapsto (a_{i,j}^p)$ defines an automorphism $\varphi\colon G \to G$ known as the *Frobenius automorphism*. By Lang's Theorem (Springer 2009, Theorem 4.4.17), the map $G \to G$, $g \mapsto g^{-1}(g\varphi)$ is surjective. Since the centre $\mathsf{Z}(G)$ is invariant under automorphisms of G, φ induces an automorphism $\overline{\varphi}$ of the infinite simple group $\mathsf{PSL}_n(\mathbb{F}) \cong G/\mathsf{Z}(G)$ such that the map $\mathsf{PSL}_n(\mathbb{F}) \to \mathsf{PSL}_n(\mathbb{F})$ defined as $g \mapsto g^{-1}(g\overline{\varphi})$ is surjective. Thus $\overline{\varphi}$ is a uniform automorphism of the infinite simple group $\mathsf{PSL}_n(\mathbb{F})$. Similar examples can be constructed using other connected algebraic groups of Lie type.

The existence of factorisations of direct products with strips as factors is related to the existence of uniform automorphisms. It was proved in (Baddeley & Praeger 2003, Lemma 2.2) that a finite direct power of a finite simple group cannot be factorised into a product of two subgroups both of which are direct products of non-trivial full strips. We generalise this result for a larger class of groups. We start by proving the following lemma for two factors.

Lemma 4.28 *The following assertions are equivalent for a group G.*

(i) *There exist non-trivial full strips X and Y in $G \times G$ such that $G \times G = XY$.*
(ii) *G admits a uniform automorphism.*

Proof Suppose that $G \times G = XY$ with X and Y non-trivial full strips of $G \times G$. Then there exist α, $\beta \in \mathsf{Aut}(G)$ such that $X = \{(g, g\alpha) \mid g \in G\}$ and $Y = \{(g, g\beta) \mid g \in G\}$. If $g \in G$, then there exist g_1, $g_2 \in G$, such that $(g, 1) = (g_1^{-1}, g_1^{-1}\alpha)(g_2, g_2\beta)$. Thus $g = g_1^{-1}g_2$ and $1 = (g_1^{-1}\alpha)(g_2\beta)$, and so $g_2 = g_1\alpha\beta^{-1}$. Hence $g = g_1^{-1}(g_1\alpha\beta^{-1})$, which implies that the map $x \mapsto x^{-1}(x\alpha\beta^{-1})$ is surjective. That is, the automorphism $\alpha\beta^{-1}$ is uniform.

Conversely, assume that $\alpha \in \mathsf{Aut}(G)$ is uniform. Set $X = \{(g, g) \mid g \in G\}$ and $Y = \{(g, g\alpha) \mid g \in G\}$. Let $(x, y) \in G \times G$. Choose $h \in G$ such that $h^{-1}(h\alpha) = x^{-1}y$ (such an h exists, as α is uniform) and let $g = xh^{-1}$. Then $g(h\alpha) = xh^{-1}(h\alpha) = y$ and $gh = x$. Thus $(g, g)(h, h\alpha) = (x, y)$. Therefore $XY = G \times G$. □

Finite groups with uniform automorphisms do exist. For example, the automorphism $\alpha\colon x \mapsto x^{-1}$ of a finite abelian group G of odd order is uniform. In this case, we do in fact obtain a factorisation

$$G \times G = \{(g, g) \mid g \in G\}\{(g, g\alpha) \mid g \in G\}.$$

Lemma 4.29 *Suppose that T is a group and, for $i = 1, \ldots, d$, let α_i, $\beta_i \in \mathsf{Aut}(T)$. Consider the following subgroups X and Y of T^{2d}:*

$$\begin{aligned} X &= \{(t_1, t_1\alpha_1, t_2, t_2\alpha_2 \ldots, t_d, t_d\alpha_d) \mid t_i = T\} \text{ and} \\ Y &= \{(s_d\beta_d, s_1, s_1\beta_1, s_2, s_2\beta_2, \ldots, s_{d-1}, s_{d-1}\beta_{d-1}, s_d) \mid s_i \in T\}. \end{aligned}$$

Then the following are equivalent:

(i) *$XY = T^{2d}$;*
(ii) *the automorphism $\alpha_1\beta_1\alpha_2\beta_2 \cdots \alpha_d\beta_d$ of T is uniform.*

Proof Assume that $XY = T^{2d}$ and let $x \in T$. Then there exist

$$\begin{aligned} (t_1^{-1}, t_1^{-1}\alpha_1, t_2^{-1}, t_2^{-1}\alpha_2, \ldots, t_d^{-1}, t_d^{-1}\alpha_d) &\in X \text{ and} \\ (s_d\beta_d, s_1, s_1\beta_1, \ldots, s_{d-1}, s_{d-1}\beta_{d-1}, s_d) &\in Y \end{aligned}$$

such that $(x, 1, \ldots, 1)$ is equal to the product

$$(t_1^{-1}, t_1^{-1}\alpha_1, t_2^{-1}, t_2^{-1}\alpha_2, \ldots, t_d^{-1}, t_d^{-1}\alpha_d) \\ \cdot (s_d\beta_d, s_1, s_1\beta_1, \ldots, s_{d-1}, s_{d-1}\beta_{d-1}, s_d).$$

Comparing the entries for these two expressions, we obtain that

$$\begin{aligned} x &= t_1^{-1}(s_d\beta_d) \\ t_1 &= s_1\alpha_1^{-1} \\ t_2 &= s_1\beta_1 \\ t_2 &= s_2\alpha_2^{-1} \\ &\vdots \\ t_d &= s_{d-1}\beta_{d-1} \\ t_d &= s_d\alpha_d^{-1}. \end{aligned}$$

Hence $s_i = s_{i-1}\beta_{i-1}\alpha_i$ for $i = 2, \ldots, d$. Thus, by induction, $s_d = s_1\beta_1\alpha_2 \cdots \beta_{d-1}\alpha_d$. Then setting $s = s_1\alpha_1^{-1}$ and using the first two equations in the system above, we obtain

$$x = (s_1\alpha_1^{-1})^{-1}(s_1\beta_1\alpha_2 \cdots \beta_{d-1}\alpha_d\beta_d) = s^{-1}(s\alpha_1\beta_1 \ldots \alpha_d\beta_d).$$

Since we chose $x \in T$ arbitrarily, the automorphism $\alpha_1\beta_1 \ldots \alpha_d\beta_d$ of T is uniform.

Conversely, assume that $\alpha := \alpha_1\beta_1 \ldots \alpha_d\beta_d$ is a uniform automorphism of T. Let $x = (x_1, \ldots, x_{2d})$ be an arbitrary element of T^{2d}. Since α is uniform, there exists $s_0 \in T$ such that

$$s_0^{-1}(s_0\alpha) = \prod_{i=d}^{1} \left(\left(x_{2i}^{-1}\beta_i\alpha_{i+1} \cdots \alpha_d\beta_d \right) (x_{2i-1}\alpha_i\beta_i \cdots \alpha_d\beta_d) \right). \quad (4.10)$$

Let $s_d = s_0\beta_d^{-1}$ and $t_d = (s_d x_{2d}^{-1})\alpha_d^{-1}$. We next define sequence

$$s_{d-1},\ t_{d-1},\ s_{d-2}, \ldots, s_1,\ t_1$$

'backwards recursively' in the sense that we define, for $i = d-1, \ldots, 1$, the elements s_i and t_i assuming that we have defined the elements s_{i+1} and t_{i+1}:

$$\begin{aligned} s_i &= (t_{i+1}x_{2i+1})\beta_i^{-1}; \\ t_i &= (s_i x_{2i}^{-1})\alpha_i^{-1}. \end{aligned}$$

Now let

$$\begin{aligned} t &= (t_1^{-1}, t_1^{-1}\alpha_1, \ldots, t_d^{-1}, t_d^{-1}\alpha_d) \quad \text{and} \\ s &= (s_d\beta_d, s_1, s_1\beta_1, \ldots, s_{d-1}, s_{d-1}\beta_{d-1}, s_d). \end{aligned}$$

We claim that $ts = x$. The definition of t_i and s_i implies that

$$\begin{aligned} x_{2i} &= (t_i^{-1}\alpha_i)s_i \quad \text{and} \\ x_{2i+1} &= t_{i+1}^{-1}(s_i\beta_i) \end{aligned}$$

for all indices $2i$ and $2i+1$ between 2 and $2d$. This shows that the product ts agrees with x from the second coordinate onwards. It remains to show that ts agrees with x in the first coordinate also. Using the equations for t_i and s_i, we obtain by induction that

$$\begin{aligned} t_1 = (s_d\alpha_d^{-1}\beta_{d-1}^{-1}\cdots\beta_1^{-1}\alpha_1^{-1})(x_{2d}^{-1}\alpha_d^{-1}\beta_{d-1}^{-1}\cdots\beta_1^{-1}\alpha_1^{-1}) \\ \cdot(x_{2d-1}\beta_{d-1}^{-1}\cdots\beta_1^{-1}\alpha_1^{-1})\cdots(x_2^{-1}\alpha_1^{-1}). \end{aligned}$$

Applying α to the last equation and considering the definition of s_d, we obtain

$$\begin{aligned} t_1\alpha &= (s_d\beta_d)(x_{2d}^{-1}\beta_d)(x_{2d-1}\alpha_d\beta_d)\cdots(x_2^{-1}\beta_1\alpha_2\beta_2\cdots\alpha_d\beta_d) \\ &= s_0(x_{2d}^{-1}\beta_d)(x_{2d-1}\alpha_d\beta_d)\cdots(x_2^{-1}\beta_1\alpha_2\beta_2\cdots\alpha_d\beta_d) \end{aligned}$$

which gives (using (4.10) for the last equality in the next line)

$$\begin{aligned} & s_0^{-1}(t_1\alpha)(x_1\alpha) = s_0^{-1}(t_1\alpha)(x_1\alpha_1\beta_1\alpha_2\beta_2\cdots\alpha_d\beta_d) \\ & = (x_{2d}^{-1}\beta_d)(x_{2d-1}\alpha_d\beta_d)\cdots(x_2^{-1}\beta_1\alpha_2\beta_2\cdots\alpha_d\beta_d)(x_1\alpha_1\beta_1\alpha_2\beta_2\cdots\alpha_d\beta_d) \\ & = s_0^{-1}(s_0\alpha). \end{aligned}$$

This implies that $x_1 = t_1^{-1}s_0$ which, in turn, equals $t_1^{-1}(s_d\beta_d)$. Therefore ts agrees with x in its first coordinate also, and so $ts = x$. Since the choice of the element x was arbitrary, $XY = T^{2d}$. □

If M is the direct product $M = T_1 \times \cdots \times T_k$ and I is a subset of $\{T_1, \ldots, T_k\}$ or a subset of $\underline{k}$, then σ_I denotes the corresponding coordinate projection from M to $\prod_{T_i \in I} T_i$ or to $\prod_{i \in I} T_i$, respectively.

The proof of the following lemma uses some simple graph theoretic concepts. The graphs that occur in this proof are undirected graphs without multiple edges and loops. A graph which does not contain a cycle is said to be a *forest*, while a connected graph without a cycle is a *tree*. The *valency* of a vertex v in a graph is the number of vertices that are adjacent to v. A vertex v of valency one in a forest is said to be a *leaf*.

Theorem 4.30 *Suppose that T is a group that does not admit a uniform automorphism and that X, Y are direct products of pairwise disjoint non-trivial full strips in T^k with $k \geqslant 2$. Then $XY \neq T^k$.*

Proof By assumption, $X = X_1 \times \cdots \times X_r$ and $Y = Y_1 \times \cdots \times Y_s$, where the X_i and Y_i are non-trivial full strips. Suppose that $T^k = XY$. At the end of the proof, this will lead to a contradiction.

Let Γ be the graph with vertex set $\{X_1, \ldots, X_r, Y_1, \ldots, Y_s\}$ such that two vertices of Γ are adjacent if and only if the supports of the strips are not disjoint. First note that two strips that belong to X are disjoint, and the same is true for two strips that belong to Y. Hence Γ is a bipartite graph with bipartition $\{X_1, \ldots, X_r\} \cup \{Y_1, \ldots, Y_s\}$. We prove the result by proving a series of claims. Recall that $T^k = XY$. Suppose that $T_1, \ldots, T_k$ are the internal direct factors of T^k; that is $T_1, \ldots, T_k$ are normal subgroups of M such that $M = T^k = T_1 \times \cdots \times T_k$.

Claim 1. Each vertex of Γ lies on at least one edge.

Proof of Claim 1. If, say, X_1 lies on no edge of the graph Γ, then $\operatorname{Supp} X_1 \cap \operatorname{Supp} Y_i = \emptyset$ for each i, and so $\operatorname{Supp} X_1 \cap \operatorname{Supp} Y = \emptyset$. Thus if σ is the projection of M onto $\prod_{T_i \in \operatorname{Supp} X_1} T_i$, then $Y\sigma = 1$, and so

$$M\sigma = (XY)\sigma = X\sigma = X_1 \cong T,$$

contradicting the fact that X_1 is non-trivial. Thus X_1 lies on at least one edge and the same proof shows that each X_i and each Y_i lies on at least one edge. □

Claim 2. If X_{j_1}, Y_{j_2} are adjacent in Γ, then $|\operatorname{Supp} X_{j_1} \cap \operatorname{Supp} Y_{j_2}| = 1$.

Proof of Claim 2. This follows from Lemma 4.28. Indeed, if $T_{i_1}, T_{i_2} \in \operatorname{Supp} X_{j_1} \cap \operatorname{Supp} Y_{j_2}$ and $\sigma = \sigma_{\{i_1, i_2\}}$, then $X_{j_1}\sigma$ and $Y_{j_2}\sigma$ are non-trivial full strips in $T_{i_1} \times T_{i_2}$ such that $T_{i_1} \times T_{i_2} = (X\sigma)(Y\sigma) = (X_{j_1}\sigma)(Y_{j_2}\sigma)$. Since T does not admit a uniform automorphism, this is a contradiction, by Lemma 4.28. □

By Claim 2, an edge in Γ that connects X_{j_1} and Y_{j_2} can be labelled with T_i where $\{T_i\} = \operatorname{Supp} X_{j_1} \cap \operatorname{Supp} Y_{j_2}$.

Claim 3. Γ does not contain a cycle.

Proof of Claim 3. Suppose to the contrary that Γ contains a cycle, and choose a cycle $X_1, Y_1, \ldots, X_d, Y_d$ of shortest length $2d$. Then $2d \geqslant 4$ and the X_i and the Y_i are pairwise distinct. By reordering the T_i, we may also assume without loss of generality that

$$\begin{aligned}
\{T_1\} &= \operatorname{Supp} Y_d \cap \operatorname{Supp} X_1;\\
\{T_2\} &= \operatorname{Supp} X_1 \cap \operatorname{Supp} Y_1;\\
\{T_3\} &= \operatorname{Supp} Y_1 \cap \operatorname{Supp} X_2;\\
&\vdots\\
\{T_{2i-1}\} &= \operatorname{Supp} Y_{i-1} \cap \operatorname{Supp} X_i;\\
\{T_{2i}\} &= \operatorname{Supp} X_i \cap \operatorname{Supp} Y_i;\\
&\vdots\\
\{T_{2d-1}\} &= \operatorname{Supp} Y_{d-1} \cap \operatorname{Supp} X_d;\\
\{T_{2d}\} &= \operatorname{Supp} X_d \cap \operatorname{Supp} Y_d.
\end{aligned}$$

In particular, we have, for $i = 1, \ldots, d-1$, that

$$\begin{aligned}
\operatorname{Supp} X_i \cap \{T_1, \ldots, T_{2d}\} &= \{T_{2i-1}, T_{2i}\} \text{ and}\\
\operatorname{Supp} Y_i \cap \{T_1, \ldots, T_{2d}\} &= \{T_{2i}, T_{2i+1}\}.
\end{aligned}$$

Further, $\operatorname{Supp} Y_d \cap \{T_1, \ldots, T_{2d}\} = \{T_{2d}, T_1\}$. The factors T_{2i}, with $i = 1, \ldots, d$, are pairwise distinct, since the $\operatorname{Supp} X_i$ are pairwise disjoint. Similarly the factors T_{2i-1}, with $i = 1, \ldots, d$, are pairwise distinct since the $\operatorname{Supp} Y_i$ are pairwise disjoint. Suppose that $T_{2i} = T_{2j-1}$ where $1 \leqslant i,\ j \leqslant d$. Since $T_{2j-1} \in \operatorname{Supp} X_j$ and $T_{2i} \in \operatorname{Supp} X_i$, we have $i = j$. But then $T_{2i-1} \in \operatorname{Supp} Y_{i-1}$ while $T_{2i} \in \operatorname{Supp} Y_i$ contradicting the fact that $\operatorname{Supp} Y_{i-1} \cap \operatorname{Supp} Y_i = \emptyset$. Thus all the T_i are pairwise distinct. Let σ denote the projection $\sigma_{\underline{2k}}$. Then $X\sigma = X_1\sigma \times \cdots \times X_d\sigma$ and $Y\sigma = Y_1\sigma \times \cdots \times Y_d\sigma$. Further, for $i = 1, \ldots, d$, $X_i\sigma$ is a non-trivial full strip in $T_{2i-1} \times T_{2i}$, for $i = 1, \ldots, d-1$, $Y_i\sigma$ is a non-trivial full strip in $T_{2i} \times T_{2i+1}$, while $Y_d\sigma$ is a non-trivial full strip in $T_{2d} \times T_1$. Thus there exist isomorphisms $\alpha_i \colon T_{2i-1} \to T_{2i}$ (for $i = 1, \ldots, d$), $\beta_i \colon T_{2i} \to T_{2i+1}$ (for $i = 1, \ldots, d-1$) and $\beta_d \colon T_{2d} \to T_1$ such that

$$\begin{aligned}
X\sigma &= \{(t_1, t_1\alpha_1, t_2, t_2\alpha_2, \ldots, t_d, t_d\alpha_d) \mid t_i \in T_{2i-1}\} \text{ and}\\
Y\sigma &= \{(s_d\beta_d, s_1, s_1\beta_1, \ldots, s_{d-1}, s_{d-1}\beta_{d-1}, s_d) \mid s_i \in T_{2i}\}.
\end{aligned}$$

Since $XY = T^k$, we find that $(X\sigma)(Y\sigma) = T_1 \times \cdots \times T_{2d}$. By Lemma 4.29, the automorphism $\alpha_1\beta_1 \cdots \alpha_d\beta_d$ of T_1 is uniform. This is a contradiction. Hence Γ does not contain a cycle. □

Since Γ does not contain a cycle, the graph Γ is a forest with no isolated vertices. This verifies at once our next claim.

Claim 4. There are at least two leaves in Γ. □

Set $\mathcal{T} = \{T_1, \ldots, T_k\}$,

$$\mathcal{S}_1 = \bigcup_{i=1}^{r} \operatorname{Supp} X_i \quad \text{and} \quad \mathcal{S}_2 = \bigcup_{i=1}^{s} \operatorname{Supp} Y_i.$$

For $i = 1,\ 2$, let $a_i = |\mathcal{T} \setminus \mathcal{S}_i|$. Suppose without loss of generality that X_1 has valency one in Γ. Since X_1 is a non-trivial strip, $|\operatorname{Supp} X_1| \geqslant 2$, and hence there must be some $T_i \in \operatorname{Supp} X_1$ that is not covered by a strip in Y. Assume without loss of generality that $i = 1$. This implies that $T_1 \in \mathcal{T} \setminus \mathcal{S}_2$, and so $a_2 = |\mathcal{T} \setminus \mathcal{S}_2| \geqslant 1$.

Claim 5. Some Y_i has valency one, and hence $a_1 \geqslant 1$ as well as $a_2 \geqslant 1$.

Proof of Claim 5. Suppose to the contrary that every Y_i has valency at least two. Then a second strip of X, X_2 say, also has valency 1, so some $T_i \in \operatorname{Supp} X_2$ with $i \geqslant 2$ also lies in $\mathcal{T} \setminus \mathcal{S}_2$. Let Y_{s+1} be a full strip such that $\operatorname{Supp} Y_{s+1} = \mathcal{T} \setminus \mathcal{S}_2$. Since $|\mathcal{T} \setminus \mathcal{S}_2| \geqslant 2$, the strip Y_{s+1} is non-trivial. Set $\bar{Y} = Y \times Y_{s+1}$. Since $T^k = XY$, we have $T^k = X\bar{Y}$. However, the graph that corresponds to the factorisation $T^k = X\bar{Y}$ has no vertices of valency 1, which contradicts Claim 4 applied to the graph corresponding to the factorisation $T^k = X\bar{Y}$. Thus Y_i has valency one for some i, and hence $a_1 \geqslant 1$ also. □

Claim 6. $a_1 = a_2 = 1$.

Proof of Claim 6. If $a_1 \geqslant 2$ and $a_2 \geqslant 2$, then let X_{r+1} and Y_{s+1} be full strips such that $\operatorname{Supp} X_{r+1} = \mathcal{T} \setminus \mathcal{S}_1$ and $\operatorname{Supp} Y_{s+1} = \mathcal{T} \setminus \mathcal{S}_2$. Since $a_1,\ a_2 \geqslant 2$, X_{r+1} and Y_{s+1} are non-trivial strips. Set $\bar{X} = X \times X_{r+1}$ and $\bar{Y} = Y \times Y_{s+1}$. Since $T^k = XY$, we have $T^k = \bar{X}\bar{Y}$. However, the graph that corresponds to the factorisation $T^k = \bar{X}\bar{Y}$ has no vertex of valency one, which contradicts Claim 4 applied to the graph of the factorisation $T^k = \bar{X}\bar{Y}$.

Thus $\min\{a_1, a_2\} = 1$, and without loss of generality we may assume that $a_1 = 1$. If $a_2 \geqslant 2$, then we may construct Y_{s+1} and $\bar{Y}$ as in the proof of Claim 5. The graph that corresponds to the factorisation $T^k = X\bar{Y}$ has only one vertex of valency one contradicting Claim 4. □

Let us now obtain a final contradiction. It follows from Claims 4 and 6 that there is exactly one strip in X with valency 1 and exactly one such strip in Y. All other strips have valency at least 2. A forest with precisely two leaves is a path. Hence Γ is a path of the form

$$X_1 - Y_1 - \cdots - X_r - Y_r.$$

The valencies of X_1 and of Y_r are equal to one. Further, the valency of each of $Y_1, X_2, \ldots, X_r$ is equal to two. Hence $|\operatorname{Supp} X_i| = |\operatorname{Supp} Y_i| = 2$ for all i. Let

$$z = |\{(X_i, T_j) \mid T_j \in \operatorname{Supp} X_i\}|.$$

Since $|\operatorname{Supp} X_i| = 2$ for all i, we have $z = 2r$. On the other hand, by Claim 6, there exists a unique j_0 such that $T_{j_0} \notin \bigcup_i \operatorname{Supp} X_i$ and if $j \neq j_0$ then there is a unique X_i such that $T_j \in \operatorname{Supp} X_i$. Thus $2r = z = k-1$ and so $k = 2r+1$. By possibly reordering the T_i, we may assume that there exist, for $i = 1, \ldots, r$, isomorphisms $\alpha_i\colon T_{2i-1} \to T_{2i}$ and $\beta_i\colon T_{2i} \to T_{2i+1}$ such that

$$\begin{aligned} X &= \{(t_1, t_1\alpha_1, t_2, t_2\alpha_2, \ldots, t_r, t_r\alpha_r, 1) \mid t_i \in T_{2i-1}\} \text{ and} \\ Y &= \{(1, s_1, s_1\beta_1, s_2, s_2\beta_2, \ldots, s_r, s_r\beta_r) \mid s_i \in T_{2i}\}. \end{aligned}$$

Suppose that $x \in T_1$. Then there are $t_i \in T_{2i-1}$ and $s_i \in T_{2i}$ such that $(x, 1, \ldots, 1)$ equals

$$\begin{aligned} &(t_1^{-1}, t_1^{-1}\alpha_1, t_2^{-1}, t_2^{-1}\alpha_2, \ldots, t_r^{-1}, t_r^{-1}\alpha_r, 1) \\ &\qquad\qquad \cdot (1, s_1, s_1\beta_1, s_2, s_2\beta_2, \ldots, s_r, s_r\beta_r). \end{aligned}$$

Comparing the entries from the k-th to the 1-st in order, we obtain that $s_i = t_i = 1$ for all i, which is a contradiction if $x \neq 1$. Thus Theorem 4.30 is proved. □

The following example shows, for a group G that admits uniform automorphisms with some additional properties, that the direct product G^k may admit factorisations with subgroups that involve longer strips.

Example 4.31 Consider a group G. We wish to factorise G^6 as $G^6 = XY$ where X is a direct product of two strips of length three and Y is a direct product of three strips of length two. Suppose that there exist uniform automorphisms α_2, α_3 of G such that for all y_1, $y_2 \in G$ there exists t in G such that both $t^{-1}(t\alpha_2) = y_1$ and $t^{-1}(t\alpha_3) = y_2$. In other words, the map $G \to G \times G$ defined by $t \mapsto (t^{-1}(t\alpha_2), t^{-1}(t\alpha_3))$ is surjective. If G is a non-trivial finite group, then $|G \times G| > |G|$, and hence such automorphisms do not exist for finite G.

Set

$$\begin{aligned} X &= \{(t, t, t, s, s, s) \mid t,\ s \in G\}; \\ Y &= \{(t_1, t_2, t_3, t_1, t_2\alpha_2, t_3\alpha_3) \mid t_i \in G\}. \end{aligned}$$

We claim that $G^6 = XY$. Indeed, let $(x_1, \ldots, x_6) \in G^6$. Choose t in G such that

$$t(t^{-1}\alpha_2) = x_1 x_4^{-1} x_5 (x_2^{-1}\alpha_2) \quad \text{and} \quad t(t^{-1}\alpha_3) = x_1 x_4^{-1} x_6 (x_3^{-1}\alpha_3).$$

Let $t_1 \in G$ such that $tt_1 = x_1$. Then it follows by the assumptions above that

$$(t, t, t, x_4 t_1^{-1}, x_4 t_1^{-1}, x_4 t_1^{-1}) \\ \cdot (t_1, t^{-1}x_2, t^{-1}x_3, t_1, (t^{-1}x_2)\alpha_2, (t^{-1}x_3)\alpha_3) = (x_1, \ldots, x_6).$$

Therefore $G^6 = XY$.

At the time of writing, we do not know if there exists an infinite simple group admitting a pair of automorphisms as in Example 4.31.

4.9 Sets of functions as direct products

In order to simplify the notation, sometimes it will be convenient to view the k-fold direct product of a group G from a different perspective. Suppose that Δ is a set, and let $\mathsf{Func}(\Delta, G)$ denote the set of functions from Δ to G. Then the elements of $\mathsf{Func}(\Delta, G)$ can be multiplied pointwise; that is, the product of the functions f and g is the function fg that maps $\delta \mapsto (\delta f)(\delta g)$. In this way $\mathsf{Func}(\Delta, G)$ becomes a group. For $\delta \in \Delta$, set

$$G_\delta = \{f \in \mathsf{Func}(\Delta, G) \mid \delta' f = 1 \text{ for all } \delta' \in \Delta \setminus \{\delta\}\} \tag{4.11}$$

and define the coordinate projection map $\sigma_\delta \colon \mathsf{Func}(\Delta, G) \to G_\delta$ by

$$\sigma_\delta \colon f \mapsto f_\delta \text{ where } \delta' f_\delta = \begin{cases} \delta f & \text{if } \delta' = \delta; \\ 1 & \text{if } \delta' \neq \delta. \end{cases} \tag{4.12}$$

Recall that the unrestricted and restricted direct product of a collection of groups was defined in Section 3.4.

Lemma 4.32 *The set G_δ is a subgroup of $\mathsf{Func}(\Delta, G)$ and G_δ is isomorphic to G. Further, $\mathsf{Func}(\Delta, G)$ is isomorphic to the unrestricted direct product $\mathsf{Prod}_{\delta \in \Delta} G_\delta$. In particular, if Δ is finite, then the set $\{G_\delta \mid \delta \in \Delta\}$ is a direct decomposition of $\mathsf{Func}(\Delta, G)$.*

Proof By the definition of the product in $\mathsf{Func}(\Delta, G)$, if $f, g \in \mathsf{Func}(\Delta, G)$ and $\delta' \in \Delta$ such that $\delta' f = \delta' g = 1$, then $\delta'(f^{-1}g) = 1$. Therefore G_δ is a subgroup of $\mathsf{Func}(\Delta, G)$. Let us define a map

$\psi\colon G \to G_\delta$ by $g\psi = f_\delta^{(g)}$ where $f_\delta^{(g)}$ is the function in $\mathsf{Func}(\Delta, G)$ such that $\delta f_\delta^{(g)} = g$ and $\delta' f_\delta^{(g)} = 1$ for all $\delta' \in \Delta \setminus \{\delta\}$. It is routine to check, and so it is left to the reader, that ψ is an isomorphism between G and G_δ.

Let us now verify that $\mathsf{Func}(\Delta, G)$ is isomorphic to $\mathsf{Prod}_{\delta\in\Delta} G_\delta$. Let us define a map $\varphi\colon \mathsf{Func}(\Delta, G) \to \mathsf{Prod}_{\delta\in\Delta} G_\delta$ by $f\varphi = (f_\delta \mid \delta \in \Delta)$ with the f_δ as in (4.12). If $f,\ g \in \mathsf{Func}(\Delta, G)$ and $\delta \in \Delta$, then $(fg)_\delta$ is the function that maps δ to $\delta(fg) = (\delta f)(\delta g)$, and maps each $\delta' \in \Delta \setminus \{\delta\}$ to 1. Therefore $(fg)_\delta = f_\delta g_\delta$. Hence

$$(fg)\varphi = ((fg)_\delta \mid \delta \in \Delta) = (f_\delta g_\delta \mid \delta \in \Delta) = (f\varphi)(g\varphi).$$

Thus φ is a homomorphism. If $f \in \mathsf{Func}(\Delta, G)$ such that $f\varphi = 1$, then $f_\delta = 1$, for all $\delta \in \Delta$. This implies that $\delta f = 1$ for all $\delta \in \Delta$, and so $f = 1$. Thus φ is injective. Consider an arbitrary $g = (g_\delta \mid \delta \in \Delta) \in \mathsf{Prod}_{\delta\in\Delta} G_\delta$ with $g_\delta \in G_\delta$ for each δ. Define $f \in \mathsf{Func}(\Delta, G)$ by $\delta f = \delta g_\delta$ for all $\delta \in \Delta$. Then $f\varphi = (f_\delta \mid \delta \in \Delta) = (g_\delta \mid \delta \in \Delta) = g$. Therefore φ is surjective; that is, φ is an isomorphism. Thus $\mathsf{Func}(\Delta, G) \cong \mathsf{Prod}_{\delta\in\Delta} G_\delta$. The rest of the lemma follows from the definition of direct decompositions in Section 4.1. □

The unrestricted direct product is sometimes called the *cartesian product*, for instance in (Robinson 1996, Section 1.4). The restricted direct product $\prod_{\delta\in\Delta} G_\delta$ can be viewed as the set of functions in $\mathsf{Func}(\Delta, G)$ with finite support; that is

$$\prod_{\delta\in\Delta} G_\delta = \{f \in \mathsf{Func}(\Delta, G) \mid \delta f = 1 \text{ for all but finitely many } \delta\}.$$

If G and Δ are as above, then, since there is a canonical isomorphism between G_δ and G, the projection σ_δ can be viewed as a homomorphism from $\mathsf{Func}(\Delta, G)$ either onto G_δ or onto G, whichever is convenient.

5
Wreath products

For any set Ω, we show (Lemma 5.10) that the full stabiliser in $\mathsf{Sym}\,\Omega$ of a non-trivial cartesian decomposition $\mathcal{E}$ of Ω is a wreath product in product action. As usual $|\mathcal{E}|$ is assumed to be finite. Such groups play an important role in the O'Nan–Scott Theory of permutation groups and they also arise as automorphism groups of graph products (Theorem 12.3) and codes (Praeger & Schneider 2012, Theorem 1.3). This chapter is devoted to the study of wreath products in product action. For other detailed treatments, see (Dixon & Mortimer 1996, Chapter 2) or (Cameron 1999, Sections 1.10 and 4.3). Again in this chapter, groups need not be finite.

5.1 Wreath products of groups

Let G and H be groups, and let φ be a homomorphism from H to the automorphism group $\mathsf{Aut}(G)$. The *semidirect product* of G and H, denoted by $G \rtimes_\varphi H$ or simply by $G \rtimes H$, is defined as follows. The underlying set of $G \rtimes H$ is the direct product $G \times H$ of sets and the multiplication of two elements (g_1, h_1) and (g_2, h_2) is defined as

$$(g_1, h_1)(g_2, h_2) = (g_1(g_2(h_1^{-1}\varphi)), h_1 h_2).$$

An overview of semidirect products can be found in (Robinson 1996, Section 1.5) and in (Dixon & Mortimer 1996, Section 2.5). It is routine to check that the semidirect product $G \rtimes H$ is a group. It is also easy to see that $\overline{G} = \{(g, 1) \mid g \in G\}$ is a normal subgroup of $G \rtimes H$ which is isomorphic to G, and $\overline{H} = \{(1, h) \mid h \in H\}$ is a subgroup of $G \rtimes H$ which is isomorphic to H. Further, $\overline{G} \cap \overline{H} = 1$ and $\overline{G}\,\overline{H} = G \rtimes H$. Identifying G with $\overline{G}$ and H with $\overline{H}$, one may view G and H as subgroups of

$G \rtimes H$, and we will often write the element (g,h) of $G \rtimes H$ simply as gh.

For our proofs, it is most convenient to use the 'function notation' introduced in Section 4.9 for defining the wreath product and, in the next section, its product action. Let G be a group, let Δ be a set, and let H be a subgroup of $\mathsf{Sym}\,\Delta$. Since our focus will be on cartesian decompositions, which, by definition, are finite, we shall throughout assume that the set Δ is finite. Set $B = \mathsf{Func}(\Delta, G)$, the set of functions from Δ to G. As explained in Section 4.9, B is a group with respect to pointwise multiplication of its elements. It has subgroups G_δ, for $\delta \in \Delta$, defined as in (4.11), and by Lemma 4.32, each G_δ is isomorphic to G. Moreover Lemma 4.32 also proves that B is isomorphic to the direct product of these $|\Delta|$ copies of G (or the cartesian product if Δ is infinite), and the map σ_δ defined in (4.12) is the natural projection map $G \to G_\delta$.

We define a homomorphism τ from H to $\mathsf{Aut}(B)$: for $f \in B$ and $h \in H$ let $f(h\tau)$ be the function that maps

$$f(h\tau)\colon\ \delta \mapsto \delta h^{-1} f. \tag{5.1}$$

It is routine to check that τ is indeed a homomorphism. Now the *wreath product* $G \wr H$ is defined as the semidirect product $B \rtimes H$ with respect to the homomorphism τ. The subgroup B is called the *base group* of the wreath product, and H is the *top group*. As the two components of a semidirect product are considered subgroups of the semidirect product, the base group B and the top group H can also be considered subgroups of the wreath product and, in this way, B becomes a normal subgroup of $G \wr H$. Considering H as a subgroup of $G \wr H$, the conjugation action of H on B will be induced by τ in (5.1), and so we obtain that

$$(\delta h^{-1})f = \delta(f^h) \quad \text{for all} \quad h \in H,\ f \in \mathsf{Func}(\Delta, G),\ \delta \in \Delta. \tag{5.2}$$

The base group of a wreath product is usually characteristic with some explicitly described exceptions; see Theorem 5.7. The wreath product $G \wr H$ has a natural action by conjugation on the set of subgroups G_δ of its base group.

Lemma 5.1 *Let G be a group, let H be a permutation group on a finite set Δ, and set $W = G \wr H$. Then, for $\delta \in \Delta$, the subgroups G_δ are permuted by the conjugation action of W. Further, the kernel of this action is the base group B, and, considering H as a subgroup of*

W, the pair $(\delta \mapsto G_\delta, \mathrm{id}_H)$ is a permutational isomorphism between the H-actions on Δ and on $\mathcal{G} = \{G_\delta \mid \delta \in \Delta\}$. Finally

$$\mathsf{N}_W(G_\delta) = \{fh \mid f \in \mathsf{Func}(\Delta, G),\ h \in H_\delta\} = BH_\delta.$$

Proof Since the base group B is the direct product of the G_δ, each of the G_δ is a normal subgroup of B, and so B acts trivially by conjugation on the G_δ. Suppose that δ_1, $\delta_2 \in \Delta$ and $h \in H$ such that $\delta_1 h = \delta_2$. If $f \in G_{\delta_1}$ then $\delta f = 1$ for all $\delta \in \Delta \setminus \{\delta_1\}$. By (5.2), $\delta f^h = (\delta h^{-1})f$ and so $\delta f^h = 1$ for all $\delta \in \Delta \setminus \{\delta_2\}$. Hence $G_{\delta_1}^h \leqslant G_{\delta_2}$. Similar argument shows that $G_{\delta_2}^{h^{-1}} \leqslant G_{\delta_1}$, and so we obtain that $G_{\delta_1}^h = G_{\delta_2}$. This shows that H permutes the G_δ, and that $(\delta \mapsto G_\delta, \mathrm{id}_H)$ is a permutational isomorphism between the H-actions on Δ and $\mathcal{G}$. Since B acts trivially by conjugation on $\mathcal{G}$ and $W = BH$, we obtain that the conjugation action of W permutes the set $\mathcal{G}$. We have seen above that B is contained in the kernel K of the W-action on $\mathcal{G}$. If K is strictly larger than B, then $K \cap H$ is non-trivial, and so H is not faithful on $\mathcal{G}$. However, this implies that H is not faithful on Δ, which is a contradiction. Thus $K \cap H = 1$ and hence $K = B$. Now it follows that $\mathsf{N}_W(G_\delta) = B\mathsf{N}_H(G_\delta) = BH_\delta$. □

Now we prove a compendium result containing many links between subgroups of G and those of a wreath product $G \wr H$. We continue to use the notation in (4.11) and (4.12) for G_δ and the projection map $\sigma_\delta \colon \mathsf{Func}(\Delta, G) \to G_\delta$, where $\delta \in \Delta$. If $K \leqslant G$ then we interpret each function in $\mathsf{Func}(\Delta, K)$ as an element of $\mathsf{Func}(\Delta, G)$. Recall that an FCR-group is one of the form $T_1 \times \cdots \times T_k$, where the T_i are simple groups.

Lemma 5.2 *For a group G, and a permutation group H on a finite set Δ, set $W = G \wr H$ and let $B = \mathsf{Func}(\Delta, G)$ be the base group of W.*

(i) *If $K \leqslant G$, then $M := \mathsf{Func}(\Delta, K)$ is a subgroup of B normalised by H, and K is normal in G if and only if M is normal in W. Further, if $K \neq 1$, then $\mathsf{C}_W(M) = \mathsf{Func}(\Delta, \mathsf{C}_G(K))$ and, in particular, $\mathsf{C}_W(B) = \mathsf{Func}(\Delta, \mathsf{Z}(G)) = \mathsf{Z}(B)$.*
(ii) *If H is transitive on Δ and normalises a subgroup M of B, then $K := \{\delta f \mid f \in M\}$ is a subgroup of G for each $\delta \in \Delta$, and K is independent of the choice of δ. Further, $M \leqslant \mathsf{Func}(\Delta, K)$, and if M is normal in B then K is normal in G.*

(iii) *If M is a non-abelian, characteristically simple* FCR*-group which is a normal subgroup of B, then $M = \prod_{\delta\in\Delta}(M \cap G_\delta)$.*

(iv) *If H is transitive on Δ and normalises a subgroup M of B, and if in addition $M = \prod_{\delta\in\Delta}(M\cap G_\delta)$, then $M = \mathsf{Func}(\Delta, K)$, with K as in (ii). If M is a minimal normal subgroup of W then K is a minimal normal subgroup of G. Conversely, if K is a minimal normal subgroup of G then each non-trivial normal subgroup N of W contained in M is a subdirect subgroup of M, that is, $\{\delta f \mid f \in N\} = K$ for each $\delta \in \Delta$.*

(v) *For $K \leqslant G$, the set*

$$\{f \in \mathsf{Func}(\Delta, K) \mid \delta_1 f = \delta_2 f \text{ for all } \delta_1,\ \delta_2 \in \Delta\} \tag{5.3}$$

is a strip in $\mathsf{Func}(\Delta, K)$ and is centralised by H.

Proof (i) By its definition, $\mathsf{Func}(\Delta, K)$ is a subgroup of $\mathsf{Func}(\Delta, G)$. To show that M is normalised by H, let $h \in H$, $\delta \in \Delta$, and $f \in M$. Then $\delta(f^h) = (\delta h^{-1})f$ by (5.2), and so $\delta(f^h) \in K$, whence $f^h \in M$.

Since M is H-invariant, to complete the proof of (i), it is sufficient to prove that M is normal in B if and only if K is normal in G. This equivalence follows from the equality $\delta(f^g) = (\delta g)^{-1}(\delta f)(\delta g)$ for $f \in M$, $g \in B$, and $\delta \in \Delta$.

Let us now calculate $\mathsf{C}_W(M)$. First we verify that $\mathsf{Func}(\Delta, \mathsf{C}_G(K)) \leqslant \mathsf{C}_W(M)$. Indeed, if $f \in \mathsf{Func}(\Delta, \mathsf{C}_G(K))$ and $g \in M$, then, for all $\delta \in \Delta$, we have that $\delta g \in K$ and that $\delta f \in \mathsf{C}_G(K)$, and hence

$$\delta(g^f) = (\delta f^{-1})(\delta g)(\delta f) = (\delta f)^{-1}(\delta g)(\delta f) = \delta g.$$

Thus $g^f = g$ and $f \in \mathsf{C}_W(M)$. Hence $\mathsf{Func}(\Delta, \mathsf{C}_G(K)) \leqslant \mathsf{C}_W(M)$. To show the other inclusion, assume that $fh \in \mathsf{C}_W(M)$ such that $f \in \mathsf{Func}(\Delta, G)$, $h \in H$ and also assume that h is a non-trivial element of H. Then there are distinct $\delta_1,\ \delta_2 \in \Delta$ such that $\delta_1 h = \delta_2$. Let $g \in \mathsf{Func}(\Delta, K)$ such that $\delta_1 g \neq 1$, and $\delta_2 g = 1$ (such a g exists, since $K \neq 1$). Then, by (5.2),

$$\delta_2\left(g^{fh}\right) = \delta_2\left(g^f\right)^h = \left(\delta_2 h^{-1}\right) g^f = \delta_1\left(g^f\right).$$

As $\delta_1(g^f) = \delta_1(f^{-1}gf) = (\delta_1 f^{-1})(\delta_1 g)(\delta_1 f)$ and $\delta_1 g \neq 1$, we obtain that $\delta_1(g^f) \neq 1$, and hence $\delta_2(g^{fh}) \neq 1$, which shows that $g \neq g^{fh}$. Thus $fh \notin \mathsf{C}_W(M)$, and it follows that $\mathsf{C}_W(M) \leqslant B$. Suppose now that $f \in B$ such that $f \in \mathsf{C}_W(M)$. If $\delta \in \Delta$, then $\delta g = \delta(g^f) = (\delta f)^{-1}(\delta g)(\delta f)$ holds for all $g \in \mathsf{Func}(\Delta, K)$, which implies that $\delta f \in$

$\mathsf{C}_G(K)$. Thus $f \in \mathsf{Func}(\Delta, \mathsf{C}_G(K))$, and so $\mathsf{C}_W(M) \leqslant \mathsf{Func}(\Delta, \mathsf{C}_G(K))$, as claimed. Therefore $\mathsf{C}_W(M) = \mathsf{Func}(\Delta, \mathsf{C}_G(K))$. Considering the case $K = G$, we obtain that $\mathsf{C}_W(B) = \mathsf{Func}(\Delta, \mathsf{C}_G(G)) = \mathsf{Func}(\Delta, \mathsf{Z}(G)) = \mathsf{Z}(B)$.

(ii) Now suppose that H is transitive on Δ and that $M \leqslant B$ is H-invariant. Choose $\delta \in \Delta$ and set $K = \{\delta f \mid f \in M\}$. Since M is H-invariant and H is transitive on Δ, it follows from (5.2) that K is independent of the choice of δ, and hence $M \leqslant \mathsf{Func}(\Delta, K)$. Since $(\delta)(f_1 f_2^{-1}) = (\delta f_1)(\delta f_2)^{-1}$, by the definition of $\mathsf{Func}(\Delta, G)$, it follows that K is a subgroup of G. Finally suppose that M is normal in B. Each element of G is of the form δg for some $g \in G_\delta$, and we have, for $f \in M$, that $f^g \in M$ (since M is normal in B), and so K contains $\delta(f^g) = (\delta g)^{-1}(\delta f)(\delta g)$. Therefore δg normalises K and hence K is normal in G.

(iii) Suppose that $M \leqslant B$ is a non-abelian characteristically simple FCR-group. Then M is a direct product of a finite number of pairwise isomorphic non-abelian simple groups. Let T denote the common isomorphism type of the simple direct factors of M. For $\delta \in \Delta$, let σ_δ denote the coordinate projection from $B \to G_\delta$. Then the kernel of the restriction $\sigma_\delta|_M$ to M of σ_δ is a direct product of some of the direct factors of M (Theorem 4.16(iv)), and the image $M\sigma_\delta$ is the direct product of the simple direct factors of M not in $\ker \sigma_\delta|_M$. Thus $L := \prod_{\delta \in \Delta} M\sigma_\delta$ is isomorphic to T^r, for some r, and M is a subdirect subgroup of L. Moreover, since $L \leqslant B$ and M is normal in B, the group M is a normal subdirect subgroup of L. Therefore Theorem 4.16(iii)–(iv) imply that $M = \prod_{\delta \in \Delta} M\sigma_\delta$ which, in turn, gives $M = \prod_{\delta \in \Delta}(M \cap G_\delta)$.

(iv) Suppose that H is transitive on Δ, that $M \leqslant B$ is H-invariant, and in addition that $M = \prod_{\delta \in \Delta}(M \cap G_\delta)$. Then M contains the subgroup $M \cap G_\delta$, for each δ. By part (ii), the set $K = \{\delta f \mid f \in M\}$ is a subgroup of G, independent of $\delta \in \Delta$, and since $M = \prod_{\delta \in \Delta}(M \cap G_\delta)$, it follows from the definition of G_δ in (4.11) that $K = \{\delta f \mid f \in M \cap G_\delta\}$, and hence that $M = \mathsf{Func}(\Delta, K)$.

Suppose that M is a minimal normal subgroup of W, and let N_0 be a non-trivial normal subgroup of G contained in K. By part (i), $N := \mathsf{Func}(\Delta, N_0)$ is a normal subgroup of W. Moreover $N \leqslant M$ (since $N_0 \leqslant K$) and hence $N = M$ by the minimality of M. Thus $N_0 = K$ is minimal normal in G.

Conversely suppose that K is a minimal normal subgroup of G and let N be a non-trivial normal subgroup of W contained in M. Then N is normalised by H and so, by part (ii), $N \leqslant \mathsf{Func}(\Delta, N_0)$, where

$N_0 := \{\delta f \mid f \in N\}$ (independent of δ), and N_0 is a normal subgroup of G contained in K. By the minimality of K, $N_0 = K$. Thus N is a subdirect subgroup of M.

(v) Let B_1 be the set in (5.3). Then B_1 is a strip in $\mathsf{Func}(\Delta, K)$. If $f \in B_1$ then there is some $g \in K$ such that $\delta f = g$ for all $\delta \in \Delta$. If $h \in H$, then, for $\delta \in \Delta$,

$$\delta f^h = (\delta h^{-1}) f = g,$$

and hence $f^h = f$. Thus $H \leqslant \mathsf{C}_W(B_1)$, as claimed. □

Remark 5.3 Let us have a closer look at the special case $\Delta = \underline{k}$. Then, the wreath product $G \wr H$ can be described using tuples, instead of functions. For $i \in \underline{k}$, the image if can be written as f_i, and f can be given by the k-tuple $(f_1, \ldots, f_k)$; see Section 4.9. If $h \in H$ and $(g_1, \ldots, g_k) \in B$ then the conjugation action of h on $(g_1, \ldots, g_k)$ is given by

$$(g_1, \ldots, g_k)^h = (g_{1h^{-1}}, \ldots, g_{kh^{-1}}).$$

Hence H permutes the coordinates of the elements of B.

5.2 Wreath products as permutation groups

The wreath product of two permutation groups can be considered as a permutation group in two rather different actions. Both of these actions have numerous group theoretical and combinatorial applications.

5.2.1 The imprimitive action

Suppose that G, H, Δ are as above, so Δ is finite, and let us now assume, in addition, that G is a permutation group acting on a set Γ. The wreath product $G \wr H$ can be considered as a permutation group on the set $\Gamma \times \Delta$ as follows. For $f \in \mathsf{Func}(\Delta, G)$, $h \in H$, and $(\gamma, \delta) \in \Gamma \times \Delta$, define

$$(\gamma, \delta)(fh) = (\gamma(\delta f), \delta h). \tag{5.4}$$

Lemma 5.4 *Let G, H, Γ, and Δ be as above and set $W = G \wr H$. Then equation (5.4) defines a W-action on $\Gamma \times \Delta$ that satisfies the following properties.*

(i) *If* $(\gamma,\delta)\in\Gamma\times\Delta$, *then*

$$W_{(\gamma,\delta)}=\{fh\mid f\in\mathsf{Func}(\Delta,G),\ \delta f\in G_\gamma,\ h\in H_\delta\ \}.$$

(ii) *The* W*-action on* $\Gamma\times\Delta$ *is faithful.*

(iii) W *is transitive on* $\Gamma\times\Delta$ *if and only if* G *is transitive on* Γ *and* H *is transitive on* Δ.

(iv) *The stabiliser in* $\mathsf{Sym}(\Gamma\times\Delta)$ *of the partition* $\mathcal{P}=\{\Gamma\times\{\delta\}\mid\delta\in\Delta\}$ *is* $\mathsf{Sym}\,\Gamma\wr\mathsf{Sym}\,\Delta$. *In particular,* $\mathcal{P}$ *is a* W*-invariant partition of* $\Gamma\times\Delta$.

Proof First we show that (5.4) does indeed define a W-action on $\Gamma\times\Delta$. Clearly, $(\gamma,\delta)1=(\gamma,\delta)$ for all $(\gamma,\delta)\in\Gamma\times\Delta$. Suppose that $f_1,\ f_2\in\mathsf{Func}(\Delta,G)$, $h_1,\ h_2\in H$, and let $(\gamma,\delta)\in\Gamma\times\Delta$. Then, using equation (5.1),

$$\begin{aligned}&((\gamma,\delta)(f_1h_1))(f_2h_2)\\&\quad=(\gamma(\delta f_1),\delta h_1)(f_2h_2)=(\gamma(\delta f_1)((\delta h_1)f_2),\delta h_1h_2)\\&\quad=(\gamma(\delta f_1)(\delta f_2^{h_1^{-1}}),\delta h_1h_2)=(\gamma(\delta f_1f_2^{h_1^{-1}}),\delta h_1h_2),\end{aligned}$$

while

$$(\gamma,\delta)(f_1h_1f_2h_2)=(\gamma,\delta)(f_1f_2^{h_1^{-1}}h_1h_2)=(\gamma(\delta f_1f_2^{h_1^{-1}}),\delta h_1h_2).$$

Therefore $((\gamma,\delta)(f_1h_1))(f_2h_2)=(\gamma,\delta)(f_1h_1f_2h_2)$, which proves that equation (5.4) defines an action.

Let us now verify assertions (i)–(iv).

(i) The equation concerning the stabiliser $W_{(\gamma,\delta)}$ follows directly from the definition of the W-action in (5.4).

(ii) Suppose that $f\in\mathsf{Func}(\Delta,G)$ and $h\in H$ such that $fh\in W$ acts trivially on $\Gamma\times\Delta$. Then $fh\in W_{(\gamma,\delta)}$ for all $(\gamma,\delta)\in\Gamma\times\Delta$. This implies that h acts trivially on Δ, and δf acts trivially on Γ for each δ; that is, fh is the identity element of W. Thus W is faithful on $\Gamma\times\Delta$.

(iii) If $\Gamma_1\subset\Gamma$ is a G-invariant subset, then $\Gamma_1\times\Delta$ is W-invariant. Similarly, if $\Delta_1\subset\Delta$ is H-invariant, then $\Gamma\times\Delta_1$ is W-invariant. Thus, if either G or H is intransitive, then so is W.

Suppose now that both G and H are transitive. Choose two pairs (γ_1,δ_1), $(\gamma_2,\delta_2)\in\Gamma\times\Delta$. Suppose that $h\in H$ such that $\delta_1h=\delta_2$ and let $f\in\mathsf{Func}(\Delta,G)$ such that $\gamma_1(\delta_1f)=\gamma_2$. Such h and such f exist because of the transitivity of G and H. Then

$$(\gamma_1,\delta_1)fh=(\gamma_1(\delta_1f),\delta_1h)=(\gamma_2,\delta_2).$$

Thus W is transitive on $\Gamma \times \Delta$.

(iv) Set $X = \mathsf{Sym}(\Gamma) \wr \mathsf{Sym}(\Delta)$ and consider X as a subgroup of $\mathsf{Sym}(\Gamma \times \Delta)$. Let us first show that X stabilises $\mathcal{P}$. Let $\delta \in \Delta$, let $f \in \mathsf{Func}(\Delta, \mathsf{Sym}\,\Gamma)$, and let $h \in \mathsf{Sym}\,\Delta$. If $\delta h = \delta'$ for some $\delta' \in \Delta$, then $(\Gamma \times \{\delta\})fh = \Gamma \times \{\delta'\}$, and so the partition $\{\Gamma \times \{\delta\} \mid \delta \in \Delta\}$ is X-invariant, as claimed.

Let K be the stabiliser in $\mathsf{Sym}(\Gamma \times \Delta)$ of the partition $\mathcal{P}$. By the argument in the previous paragraph, $X \leqslant K$, and hence we are only required to show here that $K \leqslant X$. Let $x \in K$. Then x induces a permutation, say h, of the parts of $\mathcal{P}$. We have just shown that the wreath product preserves $\mathcal{P}$ so, identifying h with this element of the top group of X, we see that h preserves $\mathcal{P}$. Thus xh^{-1} preserves $\mathcal{P}$ and induces the identity permutation on the parts of $\mathcal{P}$; that is to say, xh^{-1} fixes each part of $\mathcal{P}$ setwise. For each $\delta \in \Delta$, let g_δ be the permutation in $\mathsf{Sym}\,\Gamma$ induced by xh^{-1} on the part $\Gamma \times \{\delta\}$, and let $f \in \mathsf{Func}(\Delta, \mathsf{Sym}\,\Gamma)$ defined by $\delta f = g_\delta$ (for $\delta \in \Delta$). Then $xh^{-1} = f$, and hence $x = fh \in X$. □

This action of $G \wr H$ on $\Gamma \times \Delta$ is called the *imprimitive action* of $G \wr H$; see (Dixon & Mortimer 1996, Section 2.6) for more details. By Lemma 5.4, if W is transitive and $|\Gamma|$, $|\Delta| \geqslant 2$, then W is an imprimitive permutation group preserving the non-trivial partition of $\Gamma \times \Delta$ given in Lemma 5.4(iv). Hence, imprimitive actions, as the name also suggests, are linked to systems of imprimitivity. The following theorem can be seen as a converse to Lemma 5.4(iv) and states that imprimitive permutation groups can be embedded into wreath products in imprimitive action. Recall, from Section 2.1, that if G is a group acting on a set Ω and $\Delta \subseteq \Omega$, then G^{Δ} denotes the permutation group induced on Δ by the setwise stabiliser G_{Δ}. As introduced after Lemma 2.13, for a G-invariant partition $\mathcal{P}$, $G^{\mathcal{P}}$ denotes the permutation group induced by G on $\mathcal{P}$.

Theorem 5.5 (Imprimitive Wreath Embedding Theorem) *Let G be a permutation group on a set Ω, let $\mathcal{P}$ be a G-invariant partition of Ω with finitely many blocks, and let $\Delta_0 \in \mathcal{P}$. Assume that G induces a transitive permutation group on $\mathcal{P}$. Then G is permutationally isomorphic to a subgroup of $G^{\Delta_0} \wr G^{\mathcal{P}}$ acting on $\Delta_0 \times \mathcal{P}$.*

Proof We construct a permutational embedding (ϑ, χ) where $\vartheta\colon \Omega \to \Delta_0 \times \mathcal{P}$ is a bijection and $\chi\colon G \to G^{\Delta_0} \wr G^{\mathcal{P}}$ is an injective homomorphism. By assumption, $G^{\mathcal{P}}$ is a transitive subgroup of $\mathsf{Sym}\,\mathcal{P}$. Hence for all $\Delta \in \mathcal{P}$ we choose and fix an element $g_\Delta \in G$ such that $\Delta g_\Delta = \Delta_0$. Note that the restriction $g_\Delta|_\Delta$ is a bijection from Δ to Δ_0.

Definition of ϑ*:* Let $\omega \in \Omega$. Then there is a unique $\Delta \in \mathcal{P}$ such that $\omega \in \Delta$. Set $\omega\vartheta = (\omega g_\Delta, \Delta)$. This rule defines a map $\vartheta\colon \Omega \to \Delta_0 \times \mathcal{P}$. We claim that ϑ is a bijection. If $\omega_1\vartheta = \omega_2\vartheta = (\delta, \Delta)$ for some ω_1, $\omega_2 \in \Omega$, $\delta \in \Delta_0$, and $\Delta \in \mathcal{P}$, then ω_1, $\omega_2 \in \Delta$ and $\omega_1 g_\Delta = \omega_2 g_\Delta = \delta$. Since g_Δ is injective, it follows that $\omega_1 = \omega_2$, and hence ϑ is injective. We claim that ϑ is surjective. Suppose that $(\delta, \Delta) \in \Delta_0 \times \mathcal{P}$. Let ω be the unique element of Δ such that $\delta g_\Delta^{-1} = \omega$. Then, by the definition of ϑ, we have that $\omega\vartheta = (\delta, \Delta)$, and so ϑ is surjective.

Definition of χ*:* Let $x \in G$. Then x induces a permutation on the partition $\mathcal{P}$ and we denote this permutation by π_x. For $\Delta \in \mathcal{P}$, let $\sigma_{x,\Delta}$ denote the composition $g_\Delta^{-1}(x|_\Delta)g_{\Delta\pi_x}$. Note that $\sigma_{x,\Delta} \in \mathsf{Sym}\,\Delta_0$ and let f_x denote the function in $\mathsf{Func}(\mathcal{P}, \mathsf{Sym}\,\Delta_0)$ that maps $\Delta \mapsto \sigma_{x,\Delta}$ for all $\Delta \in \mathcal{P}$. Set $x\chi = f_x\pi_x$. Then, clearly, χ is a map $G \to \mathsf{Sym}\,\Delta_0 \wr \mathsf{Sym}\,\mathcal{P}$. Further, since π_x is the permutation of $\mathcal{P}$ induced by $x \in G$, we have that $\pi_x \in G^{\mathcal{P}}$. Since $\sigma_{x,\Delta} = g_\Delta^{-1}(x|_\Delta)g_{\Delta\pi_x}$, the permutation $\sigma_{x,\Delta}$ on Δ_0 is induced by the element $g_\Delta^{-1}xg_{\Delta\pi_x}$ of G. Since $g_\Delta^{-1}xg_{\Delta\pi_x} \in G_{\Delta_0}$, we have that $\sigma_{x,\Delta} \in G^{\Delta_0}$. Therefore, we have in fact that $x\chi \in G^{\Delta_0} \wr G^{\mathcal{P}}$.

We claim that χ is an injective homomorphism. First we show that χ is a homomorphism. Let x, $y \in G$ and let $(\delta, \Delta) \in \Delta_0 \times \mathcal{P}$. Then

$$\begin{aligned}(\delta,\Delta)(x\chi)(y\chi) &= (\delta,\Delta)f_x\pi_x f_y\pi_y = (\delta,\Delta)f_x f_y^{\pi_x^{-1}}\pi_x\pi_y\\ &= (\delta(\Delta f_x f_y^{\pi_x^{-1}}), \Delta\pi_x\pi_y) = (\delta(\Delta f_x)(\Delta\pi_x)f_y, \Delta\pi_x\pi_y),\end{aligned}$$

while

$$\begin{aligned}(\delta,\Delta)((xy)\chi) &= (\delta g_\Delta^{-1}(xy)|_\Delta g_{\Delta\pi_{xy}}, \Delta\pi_{xy})\\ &= (\delta g_\Delta^{-1}(x|_\Delta)(y|_{\Delta\pi_x})g_{\Delta\pi_x\pi_y}, \Delta\pi_x\pi_y)\\ &= (\delta g_\Delta^{-1}(x|_\Delta)g_{\Delta\pi_x}g_{\Delta\pi_x}^{-1}(y|_{\Delta\pi_x})g_{\Delta\pi_x\pi_y}, \Delta\pi_x\pi_y)\\ &= (\delta(\Delta f_x)(\Delta\pi_x)f_y, \Delta\pi_x\pi_y).\end{aligned}$$

Therefore $(x\chi)(y\chi) = (xy)\chi$ and so χ is a homomorphism.

Let us now show that $\ker\chi = 1$. Suppose that $x \in G$ is such that $x\chi = 1$. Then x preserves all blocks $\Delta \in \mathcal{P}$. Further, $\sigma_{x,\Delta} = 1$ for all Δ, which means that x induces the identity permutation on each Δ. Hence $x = 1$.

Let us finally show that (ϑ,χ) is a permutational embedding. Let $\omega\in\Omega$ such that $\omega\in\Delta$ and let $x\in G$. Then

$$\begin{aligned}(\omega\vartheta)(x\chi) &= (\omega g_\Delta,\Delta)(x\chi)=(\omega g_\Delta\sigma_{x,\Delta},\Delta\pi_x)\\ &= (\omega g_\Delta g_\Delta^{-1}(x|_\Delta)g_{\Delta\pi_x},\Delta\pi_x)=(\omega x g_{\Delta\pi_x},\Delta\pi_x)=(\omega x)\vartheta.\end{aligned}$$

□

Applying Theorem 5.5 to the right regular action (see Section 2.3) of an abstract group G, we obtain the following immediate corollary.

Corollary 5.6 *Let G be a group and let K be a subgroup of G with finite index. Suppose that H is the permutation group induced by G on the right coset space $[G\colon K]$. Then G can be embedded into $K \wr H$.*

Using the imprimitive action of the wreath product we can explicitly determine the cases when the base group of a wreath product is not characteristic. The exceptions were determined by (Neumann 1964) for the case when H is regular and by (Gross 1992) and, in the context of a more general study, by (Brewster, Passman & Wilcox 2011) for the case when G is finite. A group X is said to be *special dihedral* if it is a semidirect product $X_0\rtimes_\varphi\langle y\rangle$ of an abelian group X_0 and a cyclic group $\langle y\rangle$ of order 2 such that $x(y\varphi)=x^{-1}$ for all $x\in X_0$ and each element of X_0 has a unique square root in X_0. Note that a finite group X is special dihedral if and only if $X=X_0\rtimes_\varphi\langle y\rangle$ as above, and X_0 is an abelian group of odd order.

Theorem 5.7 *Let G be a group, let H be a permutation group on a finite set Δ of size n, set $W=G\wr H$, and let B be the base group of W.*

(i) *If H is regular on Δ, then the following are equivalent:*
   (a) *B is not a characteristic subgroup of W;*
   (b) *$H\cong C_2$ and G is a special dihedral group.*

(ii) *If G is finite and H acts faithfully on its orbits in Δ, then the following are equivalent:*
   (a) *B is not a characteristic subgroup of W;*
   (b) *G is a finite special dihedral group, n is even, and H is permutationally isomorphic to $\mathsf{S}_2\wr Y$ where $Y\leqslant \mathsf{S}_{n/2}$ and $\mathsf{S}_2\wr Y$ is considered as a permutation group on $\underline{2}\times\underline{n/2}$ in its imprimitive action.*

Remark 5.8 Part (i) of Theorem 5.7 was first proved in (Neumann 1964, Theorem 9.12) without assuming that Δ is finite; a proof is also presented in (Meldrum 1995, Section 5.1). Part (ii) is proved in the unpublished manuscript (Brewster *et al.* 2011, Theorem 1.3). A variation of part (ii) with the stronger condition that H is transitive appeared in (Gross 1992, 5.1 Theorem) and in (Wilcox 2010, Theorems 2.2.12 and 2.2.22). Another version of part (ii), for not necessarily finite groups, appears also in (Bodnarchuk 1984, Theorem, page 1); however that result contains an unfortunate error. Gross pointed out the error in (Gross 1992, Introduction) and presented some counterexamples to the statement in (Gross 1992, Section 6). Gross also discussed how Theorem 5.7(ii) (for finite groups) could be deduced from Bodnarchuk's arguments.

5.2.2 *The product action*

A particular focus of this book is the investigation of another action: the *product action* of $G \wr H$ on $\Pi = \mathsf{Func}(\Delta, \Gamma)$ is defined as follows. For $G \leqslant \mathsf{Sym}\,\Gamma$ and $H \leqslant \mathsf{Sym}\,\Delta$, let $f \in \mathsf{Func}(\Delta, G)$, $h \in H$ and set $g = fh$. For $\varphi \in \Pi$ we define φg as the function that maps $\delta \in \Delta$ to

$$\delta(\varphi g) = (\delta h^{-1}\varphi)(\delta h^{-1} f). \tag{5.5}$$

Note that $\delta h^{-1}\varphi \in \Gamma$, and $\delta h^{-1} f \in \mathsf{Sym}\,\Gamma$. Thus $(\delta h^{-1}\varphi)(\delta h^{-1} f) \in \Gamma$, and so $\varphi g \in \mathsf{Func}(\Delta, \Gamma) = \Pi$, as required.

Lemma 5.9 *Equation* (5.5) *defines a* $(G \wr H)$*-action on* $\mathsf{Func}(\Delta, \Gamma)$*. Further, if* $\gamma \in \Gamma$ *and* $\varphi\colon \Delta \to \Gamma$ *is the constant function that maps* $\delta \mapsto \gamma$ *for all* δ*, then* $(G \wr H)_\varphi = G_\gamma \wr H$*. In particular, the* $(G \wr H)$*-action on* $\mathsf{Func}(\Delta, \Gamma)$ *is faithful.*

Proof Set $W = G \wr H$ and $\Pi = \mathsf{Func}(\Delta, \Gamma)$. The claim that (5.5) defines a W-action on Π can be proved similarly to the claim concerning (5.4) in Lemma 5.4. Let us compute the stabiliser W_φ of φ. Clearly $H \leqslant (G \wr H)_\varphi$ since, if $h \in H$, then

$$\delta(\varphi h) = \delta h^{-1}\varphi = \gamma.$$

Therefore $(G \wr H)_\varphi = B_\varphi H$. Suppose that $f \in B$. Then the image of $\delta \in \Delta$ under φf is

$$(\delta\varphi)(\delta f) = \gamma(\delta f).$$

Hence $f \in B_\varphi$ if and only if $\delta f \in G_\gamma$ for all $\delta \in \Delta$. Thus

$$(G \wr H)_\varphi = \mathsf{Func}(\Delta, G_\gamma)H = G_\gamma \wr H \tag{5.6}$$

as claimed.

Suppose that $f \in \mathsf{Func}(\Delta, G)$ and $h \in H$ such that $fh \in W$ acts trivially on Π. Then fh stabilises all the constant functions, and so the expression for the stabiliser of φ implies that $f = 1$. Let $\delta_0 \in \Delta$, let γ, γ' be distinct elements of Γ, and let $\psi\colon \Delta \to \Gamma$ be the function that maps $\delta_0\psi = \gamma'$ and $\delta\psi = \gamma$ if $\delta \neq \delta_0$. Then $\psi h = \psi$, and hence

$$\gamma' = \delta_0\psi = \delta_0(\psi h) = (\delta_0 h^{-1})\psi.$$

Thus $\delta_0 h^{-1} = \delta_0$. Since this must be true for all $\delta_0 \in \Delta$, h acts trivially on Δ, and so $h = 1$. Thus the identity is the unique element of W that acts trivially on Π; in other words W acts faithfully on Π. □

More details about the product action can be found in (Dixon & Mortimer 1996, Section 2.7). Since Δ is finite, it is worth expressing the product action of the wreath product in coordinate notation. Suppose that $\Delta = \underline{k}$, and view $\mathsf{Func}(\Delta, G)$ and $\mathsf{Func}(\Delta, \Gamma)$ as G^k and Γ^k, respectively, as in Remark 5.3. Then for $(\gamma_1, \ldots, \gamma_k) \in \Gamma^k$, $(g_1, \ldots, g_k)h \in G \wr H$ we have

$$(\gamma_1, \ldots, \gamma_k)((g_1, \ldots, g_k)h) = (\gamma_{1h^{-1}} g_{1h^{-1}}, \ldots, \gamma_{kh^{-1}} g_{kh^{-1}}). \tag{5.7}$$

In order to facilitate our discussion of subgroups of wreath products we invoke the language of cartesian decompositions (which will be used to describe innately transitive subgroups of wreath products in product action). Consider the set $\Pi = \mathsf{Func}(\Delta, \Gamma)$, and define, for each $\delta \in \Delta$, a partition Γ_δ of Π as follows. Set

$$\Gamma_\delta = \{\gamma_\delta \mid \gamma \in \Gamma\}, \text{ where } \gamma_\delta := \{\psi \in \Pi \mid \delta\psi = \gamma\}. \tag{5.8}$$

It is routine to check that Γ_δ is indeed a partition of Π. Our notation reflects two important facts. Firstly, the map $\delta \mapsto \Gamma_\delta$ is a bijection between Δ and $\{\Gamma_\delta \mid \delta \in \Delta\}$. Secondly, for a fixed $\delta \in \Delta$, the map $\gamma \mapsto \gamma_\delta$ is a bijection between Γ and Γ_δ. For $\gamma \in \Gamma$ and $\delta \in \Delta$, the element $\gamma_\delta \in \Gamma_\delta$ can be considered as 'the copy' of γ in Γ_δ, and is usually called the *γ-part of* Γ_δ.

The cartesian product $\prod_{\delta \in \Delta} \Gamma_\delta$ can be bijectively identified with the original set Π: choosing $\gamma_\delta \in \Gamma_\delta$, one for each $\delta \in \Delta$, the intersection $\bigcap_{\delta \in \Delta} \gamma_\delta$ consists of a single element of Π, namely the map that sends each δ to the element $\gamma \in \Gamma$ that corresponds to γ_δ. This gives rise to a bijection from the cartesian product $\prod_{\delta\in\Delta}\Gamma_\delta$ to Π. Therefore the set

$$\mathcal{E} = \{\Gamma_\delta \mid \delta \in \Delta\} \tag{5.9}$$

is a cartesian decomposition of Π. In fact, this set of partitions is viewed as the *natural cartesian decomposition* of Π. As $\mathsf{Sym}\,\Gamma \wr \mathsf{Sym}\,\Delta$ is a permutation group acting on Π, the action of $\mathsf{Sym}\,\Gamma \wr \mathsf{Sym}\,\Delta$ can be extended to subsets of Π, subsets of subsets, etc. In particular, one can consider the action of $\mathsf{Sym}\,\Gamma \wr \mathsf{Sym}\,\Delta$ on the set of partitions of Π. We will see that $\{\Gamma_\delta \mid \delta \in \Delta\}$ is invariant under this action. The natural product action of $\mathsf{Sym}\,\Gamma \wr \mathsf{Sym}\,\Delta$ on $\prod_{\delta\in\Delta}\Gamma_\delta$ is permutationally isomorphic to its action on Π, and indeed the stabiliser in $\mathsf{Sym}\,\Pi$ of this cartesian decomposition is the wreath product $\mathsf{Sym}\,\Gamma \wr \mathsf{Sym}\,\Delta$.

Lemma 5.10 *Using the notation above, the stabiliser in* $\mathsf{Sym}\,\Pi$ *of the natural cartesian decomposition* $\mathcal{E}$ *of* Π *is* $W = \mathsf{Sym}\,\Gamma \wr \mathsf{Sym}\,\Delta$. *Further, the bijection* $\vartheta\colon \delta \mapsto \Gamma_\delta$ *results in a permutational isomorphism* $(\vartheta, \mathrm{id}_W)$ *between the* W*-action on* Δ *defined in* (5.1) *and the* W*-action on* $\mathcal{E}$. *In particular, if* $G \leqslant \mathsf{Sym}\,\Gamma$ *and* $H \leqslant \mathsf{Sym}\,\Delta$, *then* $\mathcal{E}$ *is a* $(G \wr H)$*-invariant cartesian decomposition of* Π.

Proof Set $W = \mathsf{Sym}\,\Gamma \wr \mathsf{Sym}\,\Delta$. We show first that $\mathcal{E}$ is W-invariant. Let $\delta \in \Delta$, $g = fh \in W$ and let us compute the image of Γ_δ under the action of g. Let $\gamma \in \Gamma$ and $\delta \in \Delta$, and consider the γ-part γ_δ of Γ_δ as defined in (5.8). Let $\varphi \in \gamma_\delta$ so that $\delta\varphi = \gamma$. Then, by (5.5),

$$(\delta h)(\varphi g) = (\delta\varphi)(\delta f) = \gamma(\delta f).$$

Thus $\gamma_\delta g \subseteq (\gamma(\delta f))_{\delta h}$. Conversely let $\varphi \in (\gamma(\delta f))_{\delta h}$, that is to say, by (5.8), $(\delta h)\varphi = \gamma(\delta f)$. First we apply h^{-1} and obtain $\varphi' := \varphi h^{-1}$. By (5.5) we have $\delta\varphi' = \delta(\varphi h^{-1}) = \delta h \varphi = \gamma(\delta f)$, and hence $\varphi' = (\gamma(\delta f))_\delta$ by (5.8). Next we apply f^{-1} to φ' and obtain $\varphi'' = \varphi' f^{-1}$. Again by (5.5) we have $\delta\varphi'' = \delta(\varphi' f^{-1}) = (\delta\varphi')(\delta f^{-1}) = (\gamma(\delta f))(\delta f^{-1}) = \gamma\delta(ff^{-1}) = \gamma$, so that $\varphi'' \in \gamma_\delta$. Thus $\varphi g^{-1} = \varphi'' \in \gamma_\delta$, so $\varphi \in \gamma_\delta g$ and equality $\gamma_\delta g = (\gamma(\delta f))_{\delta h}$ holds.

Therefore g maps each part of Γ_δ to a part of $\Gamma_{\delta h}$, so $\Gamma_\delta g \subseteq \Gamma_{\delta h}$. The same argument with g, δ replaced by $g^{-1}, \delta h$ shows that g^{-1} maps each part of $\Gamma_{\delta h}$ to a part of Γ_δ, and hence $(\Gamma_{\delta h})g^{-1} \subseteq \Gamma_\delta$ and it follows that $\Gamma_\delta g = \Gamma_{\delta h}$. Thus the elements of W permute the elements of $\mathcal{E}$, and so $\mathcal{E}$ is W-invariant, and it follows moreover that $(\vartheta, \mathrm{id}_W)$ defines a permutational isomorphism between the W-actions on Δ and $\mathcal{E}$.

Let S denote the stabiliser of $\mathcal{E}$ in $\mathsf{Sym}\,\Pi$. We have shown that $W \leqslant S$, and that W induces the full symmetric group $\mathsf{Sym}\,\mathcal{E}$ in its induced action on $\mathcal{E}$. Let K denote the kernel of the action of S on $\mathcal{E}$. Then we have $S^{\mathcal{E}} = \mathsf{Sym}\,\mathcal{E}$. Further, as both S and WK contain the normal subgroup K and both S/K and $WK/K = W/(K \cap W) = W/B$ (where B is the base group) are isomorphic to the finite group $\mathsf{Sym}\,\mathcal{E}$, we have $S = WK$. Now K contains the base group $B = \mathsf{Func}(\Delta, \mathsf{Sym}\,\Gamma) = \prod_{\delta\in\Delta} \mathsf{Sym}\,\Gamma_\delta$. Moreover K fixes setwise each of the elements of $\mathcal{E}$, and acts faithfully as a permutation group on $\bigcup_{\delta\in\Delta} \Gamma_\delta$ preserving each of the Γ_δ. Thus, by Lemma 4.1, K can be viewed as a subgroup of $\prod_{\delta\in\Delta} \mathsf{Sym}\,\Gamma_\delta$, and since K contains B it follows that $K = B$ and hence that $S = WK = W$.

If $G \leqslant \mathsf{Sym}\,\Gamma$ and $H \leqslant \mathsf{Sym}\,\Delta$ then $G \wr H \leqslant \mathsf{Sym}\,\Gamma \wr \mathsf{Sym}\,\Delta$. Since $\mathsf{Sym}\,\Gamma \wr \mathsf{Sym}\,\Delta$ preserves the $\mathcal{E}$, so does $G \wr H$. □

We finish this section by stating a result about the maximality of finite wreath products in product action in their parent symmetric or alternating group. The proof can be found in (Liebeck, Praeger & Saxl 1987, Theorem, page 366).

Theorem 5.11 *Suppose that Γ and Δ are finite sets such that $|\Gamma| \geqslant 5$ and that $|\Delta| \geqslant 2$. Then $\mathsf{Sym}\,\Gamma \wr \mathsf{Sym}\,\Delta$ is a maximal subgroup of $\mathsf{Sym}\,\mathsf{Func}(\Delta, \Gamma)$ or of $\mathsf{Alt}\,\mathsf{Func}(\Delta, \Gamma)$.*

Remark 5.12 (1) The requirement that $|\Gamma| \geqslant 5$ in Theorem 5.11 is necessary, since the wreath products for smaller values of $|\Gamma|$ are properly contained in affine groups.

(2) Note that $\mathsf{Sym}\,\Gamma \wr \mathsf{Sym}\,\Delta$ can be contained in $\mathsf{Alt}\,\mathsf{Func}(\Delta, \Gamma)$. We leave it to the reader to verify that $\mathsf{Sym}\,\Gamma \wr \mathsf{Sym}\,\Delta \leqslant \mathsf{Alt}\,\mathsf{Func}(\Delta, \Gamma)$ if and only if either $|\Delta| = 2$ and $4 \mid |\Gamma|$, or $|\Delta| \geqslant 3$ and $2 \mid |\Gamma|$.

(3) The intersection $(\mathsf{Sym}\,\Gamma \wr \mathsf{Sym}\,\Delta) \cap \mathsf{Alt}\,\mathsf{Func}(\Delta, \Gamma)$ may be imprimitive. This occurs if and only if $|\Delta| = 2$ and $|\Gamma| \equiv 2 \pmod 4$; see (Liebeck *et al.* 1987, Remark 2, page 369).

5.3 Cartesian decompositions and embedding theorems

Subgroups of wreath products in product action arise in a number of different contexts. Their importance for actions of finite groups is primarily due to the fact that such subgroups give rise to several of the 'O'Nan–Scott types' of finite primitive and quasiprimitive permutation groups which we discuss in Chapter 7 (see also (Dixon & Mortimer 1996, Chapter 4), (Cameron 1999, Chapter 4), and (Praeger 1993)). They also arise in similar analyses of infinite primitive permutation groups (Macpherson & Praeger 1994, Smith 2015). Subgroups of wreath products have received special attention in other contexts, such as the work of Aschbacher and Shareshian (Aschbacher 2009a, Aschbacher 2009b, Aschbacher & Shareshian 2009) on intervals in subgroup lattices, and they are of critical importance for studying invariant cartesian decompositions. Moreover, if X is a subgroup of $\mathsf{Sym}\,\Gamma \wr \mathsf{Sym}\,\Delta$ in its product action on $\Pi = \mathsf{Func}(\Delta, \Gamma)$, and $\Delta = \{1, \ldots, m\}$, then we identify Π with the set Γ^m of ordered m-tuples of elements of Γ, and in this case subgroups of $\mathsf{Sym}\,\Gamma \wr \mathsf{Sym}\,\Delta$ arise as automorphism groups of various kinds of graph products (Theorem 12.3), as automorphism groups of codes of length m over the alphabet Γ, regarded as subsets of Γ^m (Praeger & Schneider 2012, Theorem 1.3), and as automorphism groups of a special class of chamber systems in the sense of Tits; see Section 12.3 for a more detailed comparison.

Our first embedding theorem, Theorem 5.13, gives a hint as to why subgroups of wreath products in product action are so important for studying symmetric structures. Recall from Proposition 4.5 and Remark 4.6 that the action of a permutation group stabilising an inhomogeneous cartesian decomposition of a set Ω is permutationally isomorphic to a subgroup of a direct product $W_1 \times \cdots \times W_s$ in its product action on $\Omega_1 \times \cdots \times \Omega_s$ where each factor W_i is the stabiliser in $\mathsf{Sym}\,\Omega_i$ of a possibly trivial homogeneous cartesian decomposition of Ω_i. Theorem 5.13 shows the relevance of wreath products for addressing the homogeneous case. It shows that, for each permutation group X on a set Ω that preserves a homogeneous cartesian decomposition $\mathcal{E}_0$ of Ω, there is a permutational embedding of X into a wreath product in product action that maps $\mathcal{E}_0$ to the natural cartesian decomposition of the underlying set of functions on which the wreath product acts (as defined in (5.9) before Lemma 5.10).

Let X be a permutation group on a set Ω and suppose that X preserves a homogeneous cartesian decomposition $\mathcal{E}_0$ of Ω. Let Δ be a finite set labelling the partitions in $\mathcal{E}_0$ so that $\mathcal{E}_0 = \{\Gamma_\delta \mid \delta \in \Delta\}$. We construct a permutational embedding of X into $\mathsf{Sym}\,\Gamma \wr \mathsf{Sym}\,\Delta$ in its product action on $\mathsf{Func}(\Delta, \Gamma)$ for a chosen $\Gamma \in \mathcal{E}_0$.

Theorem 5.13 (Wreath Embedding Theorem) *Let X be a permutation group on a set Ω preserving a homogeneous cartesian decom-*

position $\mathcal{E}_0 = \{\Gamma_\delta \mid \delta \in \Delta\}$ *of* Ω, *and let* $\Gamma \in \mathcal{E}_0$. *Then there is a permutational isomorphism that maps* X *to a subgroup of* $\mathsf{Sym}\,\Gamma \wr \mathsf{Sym}\,\Delta$ *in its product action on* $\mathsf{Func}(\Delta, \Gamma)$, *and maps* $\mathcal{E}_0$ *to the natural cartesian decomposition of* $\mathsf{Func}(\Delta, \Gamma)$ *defined in* (5.9).

Proof We construct a permutational isomorphism (ϑ, χ), as defined in Section 2.2, from X acting on Ω to a subgroup of $\mathsf{Sym}\,\Gamma \wr \mathsf{Sym}\,\Delta$ in its product action on $\mathsf{Func}(\Delta, \Gamma)$. Recall that $\mathsf{Sym}\,\Gamma \wr \mathsf{Sym}\,\Delta = \mathsf{Func}(\Delta, \mathsf{Sym}\,\Gamma)\,\mathsf{Sym}\,\Delta$.

Definition of ϑ: The mapping $\delta \mapsto \Gamma_\delta$ is a natural bijection between Δ and $\mathcal{E}_0$. (As $\mathcal{E}_0$ is finite, so is Δ.) As X acts on $\mathcal{E}_0$, this bijection gives rise to an X-action on Δ, namely $g \mapsto g^\Delta$ where $g^\Delta \in \mathsf{Sym}\,\Delta$ satisfies $\delta g^\Delta = \delta'$ if and only if $\Gamma_\delta g = \Gamma_{\delta'}$. Let $\Delta_1, \ldots, \Delta_r$ be the X-orbits in Δ under this action. For $1 \leqslant i \leqslant r$ choose $\delta_i \in \Delta_i$, set $\Gamma := \Gamma_{\delta_1}$, and choose a bijection $a_i \colon \Gamma \to \Gamma_{\delta_i}$, such that a_1 is the identity map. Note that, since $\mathcal{E}_0$ is homogeneous, such bijections a_i exist, and $r = 1$ if X is transitive on $\mathcal{E}_0$. Further, for each i and each $\delta \in \Delta_i$, choose an element $t_\delta \in X$ such that $\delta_i t_\delta = \delta$, or equivalently, $\Gamma_{\delta_i} t_\delta = \Gamma_\delta$.

By the definition of a cartesian decomposition (Definition 1.1), for each $\omega \in \Omega$, and each $\delta \in \Delta$, there is a unique $\beta_{\omega,\delta} \in \Gamma_\delta$ containing ω, and $\{\omega\} = \bigcap_{\delta\in\Delta} \beta_{\omega,\delta}$. Define $\vartheta \colon \Omega \to \mathsf{Func}(\Delta, \Gamma)$ as follows. If $\{\omega\} = \bigcap_{\delta\in\Delta} \beta_{\omega,\delta}$, then

$$\vartheta \colon \omega \mapsto f_\omega, \text{where } \delta f_\omega = \beta_{\omega,\delta} t_\delta^{-1} a_i^{-1} \text{ for } \delta \in \Delta_i. \tag{5.10}$$

By the definition of the t_δ and a_i, it follows that $\beta_{\omega,\delta} t_\delta^{-1} a_i^{-1} \in \Gamma$, and hence $f_\omega \in \mathsf{Func}(\Delta, \Gamma)$. The existence and uniqueness of the $\beta_{\omega,\delta}$, for a given ω, imply that ϑ is well defined, and these properties combined with the fact that each of the $a_i t_\delta$ is a bijection imply that ϑ is a bijection.

For a fixed part $\beta_{\omega,\delta}$ of the partition Γ_δ, where $\delta \in \Delta_i$, the set of points ω' in $\beta_{\omega,\delta}$ is mapped by ϑ to the set of functions $\psi \in \mathsf{Func}(\Delta, \Gamma)$ with the property $\delta\psi = \beta_{\omega,\delta} t_\delta^{-1} a_i^{-1} = \gamma$, say, that is to say, $\beta_{\omega,\delta}\vartheta$ is the γ-part (defined in Section 5.2.2) of the partition $\Gamma_\delta^{\mathcal{E}}$ labelled by δ in the natural cartesian decomposition $\mathcal{E}$ of $\mathsf{Func}(\Delta, \Gamma)$ defined in (5.9). It follows that $\Gamma_\delta\vartheta$ is the partition $\Gamma_\delta^{\mathcal{E}}$ of $\mathcal{E}$ labelled by δ, and hence that $\mathcal{E}_0$ is mapped by ϑ to $\mathcal{E}$.

Definition of χ: For each $i \in \underline{r}$, let X_i be the setwise stabiliser of Γ_{δ_i} in X. Then X_i induces a natural action on Γ_{δ_i}, and hence the map $\varphi_i \colon X_i \to \mathsf{Sym}\,\Gamma$ defined by $\gamma(x\varphi_i) := \gamma a_i x a_i^{-1}$ is a well defined action of X_i on Γ. Let $g \in X$ and $\delta \in \Delta_i$. Then $t_\delta g t_{\delta g}^{-1}$ leaves Γ_{δ_i} invariant and so lies in X_i. Thus we may define an element $f_g \in \mathsf{Func}(\Delta, \mathsf{Sym}\,\Gamma)$ by

$$\delta f_g := (t_\delta g t_{\delta g}^{-1})\varphi_i \quad \text{for} \quad \delta \in \Delta_i.$$

Recalling that g^Δ denotes the permutation of Δ induced by g, we define $\chi\colon X \to \mathsf{Sym}\,\Gamma \wr \mathsf{Sym}\,\Delta$ by $g\chi := f_g g^\Delta$. To show that χ is a homomorphism, let $g, h \in X$. We claim that $f_{gh} = f_g(g^\Delta f_h (g^\Delta)^{-1})$. Note that, for $\delta \in \Delta_i$, we also have $\delta g^\Delta = \delta g \in \Delta_i$, and hence

$$\begin{aligned} \delta(f_g(g^\Delta f_h(g^\Delta)^{-1})) &= (\delta f_g)((\delta g^\Delta) f_h) = (t_\delta g t_{\delta g}^{-1})\varphi_i (t_{\delta g} h t_{\delta g h}^{-1})\varphi_i \\ &= (t_\delta g h t_{\delta g h}^{-1})\varphi_i = \delta f_{gh}. \end{aligned}$$

Thus the claim is proved. It follows that $(gh)\chi = f_{gh}(gh)^\Delta = f_g(g^\Delta f_h(g^\Delta)^{-1}) g^\Delta h^\Delta = f_g g^\Delta f_h h^\Delta = (g\chi)(h\chi)$. Thus χ is a homomorphism. If $g \in G$ such that $g\chi$ is the identity element of $\mathsf{Sym}\,\Gamma \wr \mathsf{Sym}\,\Delta$, then $f_g = 1$ and $g^\Delta = 1$. Hence g stabilises $\mathcal{E}_0$ pointwise, and the definition of f_g implies that g must act trivially on each Γ_δ. As, for each $\omega \in \Omega$, the singleton $\{\omega\}$ can be written in the form $\bigcap_{\delta\in\Delta} \beta_{\omega,\delta}$ with $\beta_{\omega,\delta} \in \Gamma_\delta$, we obtain that g is the trivial permutation on Ω. Hence $g = 1$ and χ is injective.

Finally, to prove that (ϑ, χ) is a permutational embedding, let $\omega \in \Omega$ and $g \in X$. We must show that $(\omega g)\vartheta = (\omega\vartheta)(g\chi)$. Now $\{\omega\} = \bigcap_{\delta\in\Delta} \beta_{\omega,\delta}$, where $\beta_{\omega,\delta}$ is the part of Γ_δ containing ω, for each $\delta \in \Delta$. Thus $\{(\omega g)\vartheta\} = \bigcap_{\delta\in\Delta} (\beta_{\omega,\delta} g)\vartheta$. For $\delta \in \Delta_i$ we have also $\delta g \in \Delta_i$, and we showed above that $(\beta_{\omega,\delta} g)\vartheta$ is the $(\beta_{\omega,\delta} g t_{\delta g}^{-1} a_i^{-1})$-part of $\Gamma_{\delta g}^{\mathcal{E}}$. On the other hand $\{(\omega\vartheta)(g\chi)\} = \bigcap_{\delta\in\Delta} (\beta_{\omega,\delta}\vartheta(f_g g^\Delta))$. Now for $\delta \in \Delta_i$, as we showed above, $\beta_{\omega,\delta}\vartheta$ is the $(\beta_{\omega,\delta} t_\delta^{-1} a_i^{-1})$-part of $\Gamma_\delta^{\mathcal{E}}$, and this is mapped by f_g to the β'-part of $\Gamma_\delta^{\mathcal{E}}$, where

$$\begin{aligned} \beta' &= (\beta_{\omega,\delta} t_\delta^{-1} a_i^{-1})(t_\delta g t_{\delta g}^{-1})\varphi_i \\ &= (\beta_{\omega,\delta} t_\delta^{-1} a_i^{-1})(a_i t_\delta g t_{\delta g}^{-1} a_i^{-1}) = \beta_{\omega,\delta} g t_{\delta g}^{-1} a_i^{-1}. \end{aligned}$$

Moreover, the β'-part of $\Gamma_\delta^{\mathcal{E}}$ is mapped by g^Δ to the β'-part of $\Gamma_{\delta g}^{\mathcal{E}}$. This proves that $\{(\omega\vartheta)(g\chi)\}$ is equal to the intersection of the same collection of subsets as the intersection that defines $\{(\omega g)\vartheta\}$. Hence $(\omega g)\vartheta = (\omega\vartheta)(g\chi)$, and we conclude that (ϑ, χ) defines a permutational isomorphism from X to a subgroup of $\mathsf{Sym}\,\Gamma \wr \mathsf{Sym}\,\Delta$. □

For a permutation group X on Ω preserving a cartesian decomposition $\mathcal{E} = \{\Gamma_\delta \mid \delta \in \Delta\}$, X permutes the partitions Γ_δ and we denote the stabiliser $\{x \in X \mid \Gamma_\delta x = \Gamma_\delta\}$ in X of Γ_δ by X_{Γ_δ}. Now X_{Γ_δ} induces a permutation group X^{Γ_δ} on Γ_δ called the *δ-component* of X.

In the special case where $\mathcal{E}$ is homogeneous we may, by the Wreath Embedding Theorem 5.13, assume that $X \leqslant \mathsf{Sym}\,\Gamma \wr \mathsf{Sym}\,\Delta$ acting on $\Pi = \mathsf{Func}(\Delta, \Gamma)$, and that $\mathcal{E}$ is the natural cartesian decomposition $\{\Gamma_\delta \mid \delta \in \Delta\}$ defined in (5.9). From now on we make these assumptions. This allows us to define the δ-component of X as a subgroup of $\mathsf{Sym}\,\Gamma$ more directly. Recall that each element of X is of the form fh, where $f \in \mathsf{Func}(\Delta, \mathsf{Sym}\,\Gamma)$ and $h \in \mathsf{Sym}\,\Delta$. Recall also the definition of Γ_δ in (5.8). Then the stabiliser X_{Γ_δ} of Γ_δ is $X_{\Gamma_\delta} = \{fh \in X \mid \delta h = \delta\}$, and the δ-component X^{Γ_δ} of X is the image of X_{Γ_δ} in $\mathsf{Sym}\,\Gamma$ under the map $fh \mapsto \delta f$, namely

$$X^{\Gamma_\delta} := \{\delta f \mid f \in \mathsf{Func}(\Delta, \mathsf{Sym}\,\Gamma) \text{ such that there exists } h \in \mathsf{Sym}\,\Delta \text{ with } fh \in X_{\Gamma_\delta}\}. \quad (5.11)$$

Recall from (5.9) that $\Gamma_\delta \in \mathcal{E}$ consists of its γ-parts γ_δ for $\gamma \in \Gamma$. The bijection $\gamma_\delta \mapsto \gamma$ is equivariant with respect to the actions of X_{Γ_δ} on Γ_δ and X^{Γ_δ} on Γ in the sense that, for $x \in X_{\Gamma_\delta}$ and $\gamma \in \Gamma$, we have $(\gamma_\delta)x = (\gamma x)_\delta$. Moreover, for the W-actions on Δ defined in (5.1) and on $\mathcal{E}$, it follows from Lemma 5.10 that $\Gamma_\delta x = \Gamma_{\delta x}$ for $x \in X$.

We are interested in X up to permutational isomorphism, and wish to replace X by some conjugate in W which gives a simple form with respect to the product action, both for X and the structures on which it acts. This has been done in detail by Kovács (Kovács 1989a) in the cases where X is primitive on Π and where X induces a transitive group on Δ. We give a uniform treatment for arbitrary subgroups X. One way to handle the general case is to proceed indirectly by appealing to the Imprimitive Wreath Embedding Theorem (Theorem 5.5) for subgroups of W acting on $\Gamma \times \Delta$ (see for example (Bhattacharjee, Macpherson, Möller & Neumann 1997, Theorem 8.5)). However this indirect method does not allow us to keep track of important properties of the underlying product structure. For example, if X is an automorphism group of a code $C \subset \Gamma^m$ then we may wish to maintain the property that C contains a specified codeword, say $(\gamma, \ldots, \gamma)$ for a fixed $\gamma \in \Gamma$, as well as to obtain a simple form for the group X. Thus a direct approach is highly desirable.

Our next result shows how to choose a form for X so that the δ-component depends only on the X-orbit in Δ containing δ. The easiest way to prove this is by showing that each subgroup of $\mathsf{Sym}\,\Gamma \wr \mathsf{Sym}\,\Delta$ is conjugate by an element of the base group of the wreath product to a subgroup with this property, since the permutational embedding homomorphism χ in Theorem 5.13 may be composed with the conjugating element of $\mathsf{Sym}\,\Gamma \wr \mathsf{Sym}\,\Delta$ to give a permutational embedding to a group Y with components constant on the Y-orbits in Δ. We construct such a conjugating element of the base group in Theorem 5.14.

Theorem 5.14 *Suppose that* $W = \mathsf{Sym}\,\Gamma \wr \mathsf{Sym}\,\Delta$ *acts in product action on* $\mathsf{Func}(\Delta, \Gamma)$ *with base group* $B = \mathsf{Func}(\Delta, \mathsf{Sym}\,\Gamma)$ *(where* Δ *is finite but* Γ *has arbitrary cardinality). Let* $X \leqslant W$, $\varphi \in \mathsf{Func}(\Delta, \Gamma)$ *and* $\delta_1 \in \Delta$. *Then the following hold.*

(i) *There is an element* $x \in B$ *such that the components of* $x^{-1}Xx$, *as defined in* (5.11), *are constant on each* X*-orbit in* Δ. *Moreover, if the* δ*-component of* X *is transitive on* Γ *for each* $\delta \in \Delta$, *then the element* x *can be chosen to fix* φ.

(ii) *If the group* H *induced by* X *on* Δ *is transitive, and if* G *is the* δ_1*-component of* X, *then the element* x *may be chosen in* $\mathsf{Func}(\Delta, \mathsf{Sym}\,\Gamma)$ *such that* $x^{-1}Xx \leqslant G \wr H$, *(and also such that* $\varphi x = \varphi$ *if* G *is transitive on* Γ*).*

Note that, in part (ii), $G \wr H$ denotes a particular subgroup of W (as discussed in Section 5.2.2) and not just an isomorphism class of groups. If the subgroup X is transitive on $\mathsf{Func}(\Delta, \Gamma)$ then we will prove in Theorem 5.15 that all of its components are transitive, so the additional condition on the element x in Theorem 5.14 to fix a given point is possible.

Theorem 5.15 *Let* X *be a transitive permutation group on a set* Ω *and let* $\mathcal{E}$ *be an* X*-invariant cartesian decomposition of* Ω. *Then each component of* X *is transitive. Moreover, if* X *is transitive on* $\mathcal{E}$, *then the kernel* $X_{(\mathcal{E})}$ *of the action of* X *on* $\mathcal{E}$ *also has transitive components.*

First we prove Theorem 5.14.

Proof of Theorem 5.14. Note that B is the kernel of the induced action of W on Δ, so if $x \in B$, then the X-orbits in Δ coincide with the $x^{-1}Xx$-orbits in Δ. For the computations in the proof we often use the properties given in (5.2) and (5.5), and the equality $\delta(ff') = (\delta f)(\delta f')$, for $f, f' \in B$ and $\delta \in \Delta$.

Let $\Delta_1, \ldots, \Delta_r$ be the X-orbits in Δ under the action induced by X on Δ. For $1 \leqslant i \leqslant r$ choose $\delta_i \in \Delta_i$, and for each $\delta \in \Delta_i$, choose

$t_\delta \in X$ such that $\Gamma_{\delta_i} t_\delta = \Gamma_\delta$, and in particular take $t_{\delta_i} = 1$ for all i. Then $t_\delta = f_\delta h_\delta$ with $f_\delta \in B$ and $h_\delta \in \mathsf{Sym}\,\Delta$ such that $\delta_i h_\delta = \delta$. Also $X_{\Gamma_\delta} = (X_{\Gamma_{\delta_i}})^{t_\delta}$.

Claim 1. If the δ_i-component is transitive on Γ, then the t_δ can be chosen so that, for each $\delta \in \Delta_i$, $\delta_i f_\delta$ fixes the point $\delta\varphi$ of Γ.

Proof of Claim 1. Since we have $t_{\delta_i} = 1$, the element $\delta_i f_{\delta_i}$ is the identity of $\mathsf{Sym}\,\Gamma$ and hence fixes $\delta\varphi$. Let $\delta \in \Delta_i \setminus \{\delta_i\}$ and consider $s_\delta = fh \in X_{\Gamma_{\delta_i}}$ with $f \in B$ and $h \in \mathsf{Sym}\,\Delta$. Then $\delta_i h = \delta_i$, and the element $s_\delta t_\delta$ is equal to $f'_\delta h'_\delta$, with $f'_\delta = f f_\delta^{h^{-1}}$ and $h'_\delta = h h_\delta$, and satisfies $\Gamma_{\delta_i} s_\delta t_\delta = \Gamma_\delta$. Moreover $\delta_i f'_\delta = (\delta_i f)((\delta_i h) f_\delta) = (\delta_i f)(\delta_i f_\delta)$, and we note that $\delta_i f \in \mathsf{Sym}\,\Gamma$ lies in the δ_i-component of X, see (5.11). If the δ_i-component is transitive on Γ, then we may choose s_δ in $X_{\Gamma_{\delta_i}}$ such that the element $(\delta_i f)(\delta_i f_\delta)$ fixes $\delta\varphi$. Replacing t_δ by $s_\delta t_\delta$ gives the required properties. □

Claim 2. For $\delta \in \Delta_i$, the δ-component X^{Γ_δ} equals $(X^{\Gamma_{\delta_i}})^{\delta_i f_\delta}$.

Proof of Claim 2. Let $\delta_i f \in X^{\Gamma_{\delta_i}}$. By (5.11) there exists $h \in \mathsf{Sym}\,\Delta$ such that $\delta_i h = \delta_i$ and $fh \in X_{\Gamma_{\delta_i}}$. Therefore X_{Γ_δ} contains

$$(fh)^{t_\delta} = f^{t_\delta} h^{f_\delta h_\delta} = f^{t_\delta}(f_\delta^{-1} f_\delta^{h^{-1}} h)^{h_\delta} = f^{t_\delta}(f_\delta^{-1})^{h_\delta} f_\delta^{h^{-1}h_\delta} h^{h_\delta}.$$

Since $\delta h^{h_\delta} = \delta$, this implies, using (5.2), that the δ-component X^{Γ_δ} contains

$$\delta(f^{t_\delta}(f_\delta^{-1})^{h_\delta} f_\delta^{h^{-1}h_\delta}) = ((\delta h_\delta^{-1}) f^{f_\delta})((\delta h_\delta^{-1}) f_\delta^{-1})((\delta h_\delta^{-1} h) f_\delta)$$

and using the facts that $\delta_i h_\delta = \delta$ and $\delta_i h = \delta_i$, this is equal to

$$(\delta_i f^{f_\delta})(\delta_i f_\delta^{-1})(\delta_i f_\delta) = \delta_i (f^{f_\delta} f_\delta^{-1} f_\delta) = \delta_i f^{f_\delta} = (\delta_i f)^{\delta_i f_\delta}.$$

Thus X^{Γ_δ} contains $(X^{\Gamma_i})^{\delta_i f_\delta}$, and a similar argument proves the reverse inclusion. Hence equality holds and the claim is proved. □

Definition of x: Define $x \in B = \mathsf{Func}(\Delta, \mathsf{Sym}\,\Gamma)$ as the function satisfying, for each i and each $\delta \in \Delta_i$, $\delta x = \delta_i f_\delta^{-1}$. If all components of X are transitive on Γ then we assume (as we may by Claim 1) that, for each i and $\delta \in \Delta_i$, $\delta_i f_\delta$ fixes the point $\delta\varphi$, and hence $\delta x = (\delta_i f_\delta)^{-1}$ fixes $\delta\varphi$. Thus in this case x fixes φ.

Claim 3. The components of $x^{-1}Xx$ are constant on each of the Δ_i.

Proof of Claim 3. Since x acts trivially on Δ, we have that $(X^x)_{\Gamma_\delta} =$

$(X_{\Gamma_\delta})^x$ for each $\delta \in \Delta$. Thus δf lies in the δ-component X^{Γ_δ} if and only if there exists $h \in \mathsf{Sym}\,\Delta$ such that $fh \in X_{\Gamma_\delta}$ or equivalently, $(fh)^x = f^x x^{-1} x^{h^{-1}} h \in (X^x)_{\Gamma_\delta}$. This implies that the δ-component of X^x contains

$$\delta(f^x x^{-1} x^{h^{-1}}) = \delta(x^{-1} f x^{h^{-1}}) = (\delta x^{-1})(\delta f)((\delta h)x) = (\delta f)^{\delta x}$$

since $\delta h = \delta$. Thus the δ-component of X^x contains $(X^{\Gamma_\delta})^{\delta x}$ and a similar argument proves the reverse inclusion, so equality holds. Now $\delta x = \delta_i f_\delta^{-1} = (\delta_i f_\delta)^{-1}$, which by Claim 2 conjugates X^{Γ_δ} to $X^{\Gamma_{\delta_i}}$. Thus

$$(X^x)^{\Gamma_\delta} = (X^{\Gamma_\delta})^{\delta x} = (X^{\Gamma_\delta})^{\delta_i f_\delta^{-1}} = X^{\Gamma_{\delta_i}}$$

for all $\delta \in \Delta_i$. This completes the proof of Claim 3, and part (i) also follows. □

Let us now prove part (ii). From now on we assume that the group H induced by X on Δ is transitive, and let G be the δ_1-component of X. From what we have just proved, each component of X^x is equal to G. Let g' be an arbitrary element of X^x. Then $g' = x^{-1}gx$ for some $g \in X$, and we have $g = fh$ with $f \in B$ and $h \in \mathsf{Sym}\,\Delta$. By the definition of H we have $h \in H$. Also

$$g' = x^{-1} fhx = (x^{-1} f x^{h^{-1}})h = f'h, \text{ say.}$$

Thus, in order to prove that $g' \in G \wr H$, it is sufficient to prove that, for each $\delta \in \Delta$, $\delta f' \in G$.

Choose $\delta \in \Delta$, and let $\delta' := \delta h$. Then $hh_{\delta'}^{-1} h_\delta$ fixes δ, and so X_{Γ_δ} contains

$$gt_{\delta'}^{-1} t_\delta = fhh_{\delta'}^{-1} f_{\delta'}^{-1} f_\delta h_\delta = f(f_{\delta'}^{-1} f_\delta)^{h_{\delta'} h^{-1}} hh_{\delta'}^{-1} h_\delta.$$

Hence $(X^x)_{\Gamma_\delta} = (X_{\Gamma_\delta})^x$ contains

$$x^{-1} gt_{\delta'}^{-1} t_\delta x = (x^{-1} f (f_{\delta'}^{-1} f_\delta)^{h_{\delta'} h^{-1}} x^{h_\delta^{-1} h_{\delta'} h^{-1}}) hh_{\delta'}^{-1} h_\delta$$

which equals $f'' hh_{\delta'}^{-1} h_\delta$ say. This means that the δ-component G of X^x contains

$$\begin{aligned} \delta f'' &= (\delta x^{-1} f)((\delta hh_{\delta'}^{-1})(f_{\delta'}^{-1} f_\delta))((\delta hh_{\delta'}^{-1} h_\delta)x) \\ &= (\delta x^{-1} f)(\delta_1 (f_{\delta'}^{-1} f_\delta))(\delta x). \end{aligned}$$

By the definition of x, $\delta_1(f_{\delta'}^{-1} f_\delta) = (\delta_1 f_{\delta'}^{-1})(\delta_1 f_\delta) = (\delta' x)(\delta x)^{-1}$. It follows that

$$\delta f'' = (\delta x^{-1} f)(\delta' x) = (\delta x^{-1} f)(\delta x^{h^{-1}}) = \delta(x^{-1} f x^{h^{-1}}) = \delta f'.$$

Therefore $\delta f' \in G$, as required. This completes the proof. □

Now we proceed to prove Theorem 5.15, but first we state a simple number theoretic lemma; see for instance (Jones & Jones 1998, Exercise 2.20).

Lemma 5.16 *If $n \geqslant 1$ and p is a prime, then $p^n \nmid n!$.*

Proof of Theorem 5.15. By Proposition 4.5 and Remark 4.6, if $\mathcal{E}$ is an intransitive cartesian decomposition, then X can be embedded into a direct product $W_1 \times \cdots \times W_s$ acting on $\Omega_1 \times \cdots \times \Omega_s$ where each factor W_i is the stabiliser in $\mathsf{Sym}\,\Omega_i$ of a possibly trivial transitive cartesian decomposition of Ω_i. Further, for $\Gamma \in \mathcal{E}$ the component X^Γ coincides with the component $(X^{\Omega_i})^{\bar{\Gamma}}$ where Ω_i is the appropriate factor in the direct product $\Omega_1 \times \cdots \times \Omega_s$ and $\bar{\Gamma}$ is the partition of Ω_i constructed as in (4.1). Thus we may assume that $\mathcal{E}$ is a transitive cartesian decomposition. By Theorem 5.13, we may assume without loss of generality that X is a subgroup of $W = \mathsf{Sym}\,\Gamma \wr \mathsf{Sym}\,\Delta$ acting in product action on $\Pi = \mathsf{Func}(\Delta, \Gamma)$ with base group $B = \mathsf{Func}(\Delta, \mathsf{Sym}\,\Gamma)$, where Δ is finite and Γ has arbitrary cardinality. We also assume that $\mathcal{E}$ is the natural cartesian decomposition of Π. Suppose that X is a transitive subgroup of W, and let $K := X \cap B$. Note that K has finite index in X since Δ is finite.

So assume that X is transitive on Δ. Let $r := |\Delta|$. Let $\delta \in \Delta$, let $G = X^{\Gamma_\delta}$ be the δ-component of X, and let H be the transitive permutation group of Δ induced by X. Then by Theorem 5.14, we may assume that $X \leqslant G \wr H$. Now we have, for each $\delta \in \Delta$, that K is a normal subgroup of the stabiliser X_{Γ_δ}, and hence that, by (5.11), the δ-component K^{Γ_δ} of K is a normal subgroup of $X^{\Gamma_\delta} = G$. Hence X^{Γ_δ} permutes the set Ξ_δ of K^{Γ_δ}-orbits in Γ. Let $\Phi \in \prod_{\delta \in \Delta} \Xi_\delta$, that is to say, Φ is an r-tuple with δ-entry, denoted Φ_δ, lying in Ξ_δ for each $\delta \in \Delta$. Define $\Pi(\Phi) := \{\varphi \in \Pi \mid \delta\varphi \in \Phi_\delta,\ \text{for each } \delta \in \Delta\}$. Then the set $\hat{\Pi} := \{\Pi(\Phi) \mid \Phi \in \prod_{\delta \in \Delta} \Xi_\delta\}$ is a partition of Π, and we claim that $\hat{\Pi}$ is X-invariant. Indeed, if $x = fh \in X$ with $f \in \mathsf{Func}(\Delta, G)$ and $h \in H$, then for $\varphi \in \Pi(\Phi)$, we have, for each $\delta \in \Delta$, that $\delta(\varphi x) = (\delta h^{-1}\varphi)(\delta h^{-1} f)$ (by (5.5)) which lies in the image of $\Phi_{\delta h^{-1}}$ under $\delta h^{-1} f$. Now $\delta h^{-1} f$ lies in the (δh^{-1})-component of X and so $\delta h^{-1} f$ permutes $\Xi_{\delta h^{-1}}$. This implies that $\Phi_{\delta h^{-1}}(\delta h^{-1} f) \in \Xi_{\delta h^{-1}}$. Since this holds for all $\delta \in \Delta$ and for all $\varphi \in \Pi(\Phi)$, it follows that $\Pi(\Phi)x \subseteq \Pi(\Phi')$, where $\Phi'_\delta = \Phi_{\delta h^{-1}}(\delta h^{-1} f)$ for all $\delta \in \Delta$. A similar argument proves the other inclusion, and hence the equality $\Pi(\Phi)x = \Pi(\Phi')$. Hence $\hat{\Pi}$ is X-invariant.

Since X is transitive on Π, it is also transitive on $\hat{\Pi}$, and the subgroup K is contained in the kernel of the X-action on $\hat{\Pi}$. Thus $|\hat{\Pi}|$ divides $|X\colon K|$. Moreover $X/K \cong H \leqslant \mathsf{Sym}\,\Delta$, and so $|X\colon K|$ divides $r!$. In particular, $s := |\Xi_\delta|$ is finite, and as X is transitive on Δ, s is independent of $\delta \in \Delta$. Thus $|\hat{\Pi}| = s^r$. If $s > 1$ and p is a prime dividing s, then p^r divides s^r, which divides $r!$, which is a contradiction, by Lemma 5.16. This proves both assertions of Theorem 5.15. □

The following corollary for transitive subgroups of $\mathsf{Sym}\,\Gamma \wr \mathsf{Sym}\,\Delta$ follows immediately from Theorem 5.15.

Corollary 5.17 *Let $W = \mathsf{Sym}\,\Gamma \wr \mathsf{Sym}\,\Delta$ act in product action on $\mathsf{Func}(\Delta,\Gamma)$ with base group $B = \mathsf{Func}(\Delta, \mathsf{Sym}\,\Gamma)$, where Δ is a finite set and Γ has arbitrary cardinality, and let X be a transitive subgroup of W. Then each component of X is transitive on Γ. Moreover, if X acts transitively on Δ then each component of the intersection $X \cap B$ is transitive on Γ.*

5.4 Primitivity-type properties of wreath products

Wreath products in product action form important classes of primitive, quasiprimitive, and innately transitive permutation groups. The following result provides a necessary and sufficient condition for verifying whether a wreath product in product action has one of these properties. We deal with innately transitive groups in the case where the plinth is an FCR-group (see Section 4.7).

Theorem 5.18 *Suppose that Γ and Δ are sets with Δ finite and Γ of arbitrary cardinality, and $|\Delta| > 1$, $|\Gamma| > 1$. Let G and H be permutation groups on Γ and Δ, respectively. Set $\Pi = \mathsf{Func}(\Delta,\Gamma)$, $W = G \wr H$, and let $\mathcal{Q}$ be one of the properties 'primitive', 'quasiprimitive', or 'innately transitive with an FCR-plinth'. Then the following are equivalent.*

(i) *the group W has property $\mathcal{Q}$ on Π;*
(ii) *the group G has property $\mathcal{Q}$ on Γ, and also G is not cyclic of prime order and H is transitive on Δ.*

Proof We use the notation σ_δ, G_δ introduced in (4.11) and (4.12).

In particular, the base group B is isomorphic to $\prod_{\delta\in\Delta} G_\delta$, and in the following we will often identify B with this direct product. For later use in the proof, let γ be a fixed element of Γ, and let φ be the constant map of $\mathsf{Func}(\Delta,\Gamma)$ that maps $\delta \mapsto \gamma$ for each $\delta\in\Delta$. Then, by (5.6),

$$W_\varphi = \{fh \mid \delta f \in L \text{ for all } \delta\in\Delta \text{ and } h\in H\} = \mathsf{Func}(\Delta, L)H, \quad (5.12)$$

where $L = G_\gamma$, the stabiliser of γ in G. Note that this implies $B_\varphi = B\cap W_\varphi = \mathsf{Func}(\Delta, L)$ and W_φ contains H. For the first part of the proof we assume that part (i) holds, and prove part (ii) in the following four claims.

Claim 1. If W is innately transitive with FCR-plinth M, then $M \leqslant B$ and $M\sigma_\delta$ is transitive on Γ for each $\delta\in\Delta$.

Proof of Claim 1. In this case, M is a minimal normal subgroup that is transitive on Π. Since the base group B is a non-trivial normal subgroup of W, it follows from Lemma 3.4 that either $M \leqslant B$ or $M\cap B = 1$. In the latter case $M \leqslant \mathsf{C}_W(B)$, but this contradicts the fact that $\mathsf{C}_W(B) = \mathsf{Z}(B) \leqslant B$, proved in Lemma 5.2(i). Thus B contains M. Since M is transitive on Π it follows from Lemma 4.2 that $M\sigma_\delta$ is transitive on Γ, for each $\delta\in\Delta$. □

Claim 2. If part (i) holds, then H is transitive on Δ.

Proof of Claim 2. Suppose to the contrary that Δ_1 is an H-orbit in Δ, and $\Delta_2 := \Delta\setminus\Delta_1 \neq \emptyset$. We claim that, for $i = 1, 2$, the subgroup

$$B_i = \{f\in B \mid \delta f = 1 \text{ for } \delta\in\Delta_{3-i}\}$$

is normal in W and intransitive on Π. It is sufficient to give the proof for $i = 1$. Let $f\in B_1$, $h\in H$, and $\delta\in\Delta_2$. Then $\delta h^{-1}\in\Delta_2$ since Δ_2 is H-invariant, and therefore by (5.2),

$$\delta\left(f^h\right) = (\delta h^{-1})f = 1.$$

Thus $f^h\in B_1$ and so B_1 is normalised by H. Also B_1 is normal in B since it is the direct product $\prod_{\delta\in\Delta_1} G_\delta$. Thus B_1 is normalised by $BH = W$. Recall that the element φ defined above lies in the γ-part γ_δ for each $\delta\in\Delta$. Let $f\in B_1$ and $\delta\in\Delta_2$. Then $\delta f = 1$ (definition of B_1) and hence the point φf lies in the γ-part γ_δ of Γ_δ. It follows, since $|\Gamma| > 1$, that the B_1-orbit $\varphi B_1 \neq \Pi$, so B_1 is intransitive. This proves the claim about the B_i, and we note that $B = B_1\times B_2$.

Thus W has an intransitive normal subgroup B_1, and hence is neither quasiprimitive nor primitive (Lemma 2.20). Therefore W is innately transitive with an FCR-plinth M, and by Claim 1, $M \leqslant B$. Since M is transitive but the B_i are intransitive, we have $M \not\leqslant B_i$, and since M is a minimal normal subgroup of W, this implies that $M \cap B_i = 1$, for $i = 1, 2$. Therefore $M \leqslant \mathsf{C}_W(B_1) \cap \mathsf{C}_W(B_2) = \mathsf{C}_W(B)$ which is contained in $\mathsf{Z}(B)$ by Lemma 5.2(i). In particular, for $\delta \in \Delta$, $M\sigma_\delta$ centralises $B\sigma_\delta = G_\delta$, and by Claim 1, $M\sigma_\delta$ is transitive on Γ_δ. It follows from Theorem 3.2(iv) that $M\sigma_\delta = G_\delta$ and is abelian and regular on Γ_δ. Hence $B = \mathsf{Func}(\Delta, G)$ is regular on Π, and since $M \leqslant B$ and M is transitive on Π it follows that $M = B$, and in particular M contains B_1, which is a contradiction. Thus H is transitive on Δ. □

Claim 3. If part (i) holds, then G is not cyclic of prime order.

Proof of Claim 3. Suppose that $G = C_p$ for some prime p, so that $|\Gamma| = p$ also and G is regular on Γ. In this case $B = \prod_{\delta\in\Delta} G_\delta$ is abelian and regular on Π, and $|\Pi| = p^{|\Delta|}$ is finite and of size at least p^2 since $|\Delta| > 1$. By Theorem 3.12, it follows that B is a minimal normal subgroup of W. Consider the subgroup

$$B_3 := \{f \in B \mid \delta_1 f = \delta_2 f \text{ for all } \delta_1,\ \delta_2 \in \Delta\}$$

of B formed by the constant mappings from Δ to G. Since B is abelian, B normalises B_3, and also the subgroup H normalises B_3. Thus B_3 is normal in $W = BH$. However $|B_3| = |G_\delta| = p < |B|$ and so B_3 is properly contained in the minimal normal subgroup B, which is a contradiction. Thus G is not cyclic of prime order. □

Claim 4. If part (i) holds, then G has property $\mathcal{Q}$ on Γ.

Proof of Claim 4. Suppose first that W is primitive on Π. Suppose that G is not primitive on Γ, so that, by Lemma 2.14, there exists a subgroup R such that $L = G_\gamma < R < G$. Then $\mathsf{Func}(\Delta, R)$ is normalised by H, and we have the subgroup chain, $W_\varphi = \mathsf{Func}(\Delta, L)H < \mathsf{Func}(\Delta, R)H < \mathsf{Func}(\Delta, G)H = W$. This implies, by Lemma 2.14, that W is imprimitive on Π, which is a contradiction. Hence G is primitive on Γ.

Next suppose that W is quasiprimitive on Π, and suppose that G is not quasiprimitive on Γ. Then there exists a non-trivial intransitive normal subgroup K of G. This implies that the subgroup $N := \mathsf{Func}(\Delta, K)$ is intransitive on Π, and N is normalised by $B = \mathsf{Func}(\Delta, G)$ and by H, and hence by $W = BH$. Hence W is not quasiprimitive on Π, which is a contradiction, so G is quasiprimitive on Γ.

Finally suppose that W is innately transitive on Π with FCR-plinth M. By Claim 1, $M \leqslant B$ and $M\sigma_\delta$ is transitive on Γ_δ for each δ.

By definition of an FCR-group, M is a direct product $T_1 \times \cdots \times T_k$ of finitely many simple groups T_i, and since M is minimal normal, M is characteristically simple and the T_i are pairwise isomorphic, by Corollary 4.17. If the T_i are cyclic of prime order, then M, and hence also Π, is finite. In this case, M is abelian and so W is primitive by Corollary 3.13. Therefore, as we showed above, G is primitive on the finite set Γ, and hence a minimal normal subgroup of G is transitive on Γ by Corollary 2.21 (and being finite, is an FCR-group). That is, G is innately transitive on Γ with FCR-plinth.

Thus we may assume that the T_i are non-abelian simple groups. Then by Lemma 5.2(iii), $M = \prod_{\delta \in \Delta}(M \cap G_\delta)$. For a chosen $\delta \in \Delta$, define $K := \{\delta f \mid f \in M\}$. Then by Lemma 5.2 parts (ii) and (iv), K is a minimal normal subgroup of G, K is independent of δ, and $M = \mathsf{Func}(\Delta, K)$. Also, by Lemma 4.2, K is transitive on Γ, so G is innately transitive with plinth K. Finally, since $M = \mathsf{Func}(\Delta, K) = \prod_{\delta \in \Delta}(M \cap G_\delta)$, it follows that $K \cong M \cap G_\delta$, so K is isomorphic to a direct product of some of the T_i by Theorem 4.16(iv), and hence K is an FCR-group. Thus G is innately transitive on Γ with FCR-plinth. □

Thus, by Claims 2–4, part (i) implies part (ii). Suppose conversely that part (ii) holds, so G has property $\mathcal{Q}$, H is transitive on Δ, and G is not cyclic of prime order. We must prove that W has property $\mathcal{Q}$ on Π.

Claim 5. If (ii) holds and G is primitive on Γ, then W is primitive on Π.

Proof of Claim 5. As the stabiliser L in G of a point $\gamma \in \Gamma$ is a maximal subgroup of G, and G is not cyclic of prime order, we find that $L \neq 1$, and so G is not regular on Γ. In particular, L is not normal in G. We are required to show that $W_\varphi = \mathsf{Func}(\Delta, L)H$, as given in (5.12), is a maximal subgroup of W. Note that $B_\varphi = W_\varphi \cap B = \mathsf{Func}(\Delta, L)$. Let W_1 be a subgroup of W such that $W_\varphi < W_1$. We claim that $W_1 = W$. As $H \leqslant W_\varphi$, we obtain that $W_1 = B_1 H$, where $B_1 = W_1 \cap B$, and that B_φ is a proper subgroup of B_1. Thus there exists a map $f \in B_1 \setminus B_\varphi$. By (5.12), $\delta f \notin L$ for some $\delta \in \Delta$. Thus under the projection σ_δ, the image $B_1\sigma_\delta$ is a subgroup of G_δ properly containing $B_\varphi\sigma_\delta \cong L$. As L is maximal in G, this implies that $B_1\sigma_\delta = G_\delta$. By Theorem 4.8(i), $B_1 \cap G_\delta$ is a normal subgroup of $B_1\sigma_\delta$, and as $B_\varphi = \mathsf{Func}(\Delta, L) \leqslant B_1$, we find that the maximal (and non-normal) subgroup $B_\varphi\sigma_\delta \cong L$ of G_δ is contained in $B_1 \cap G_\delta$. Hence $B_1 \cap G_\delta = G_\delta$, and so $G_\delta \leqslant B_1$. Since

H is transitive on Δ, we find that $G_\delta \leqslant B_1$ for all $\delta \in \Delta$, that is, $B \leqslant B_1$. Thus $W_1 = B_1 H = BH = W$, and we conclude that W_φ is a maximal subgroup of W, so W is primitive on Π. □

Claim 6. If (ii) holds and G is quasiprimitive on Γ, then W is quasiprimitive on Π.

Proof of Claim 6. If G is primitive on Γ, then W is primitive on Π by Claim 5, and hence W is quasiprimitive on Π, by Corollary 2.21. Thus we may assume that G is imprimitive and quasiprimitive on Γ. Then each non-trivial normal subgroup of G is non-abelian by Corollary 3.13, and in particular $\mathsf{Z}(G) = 1$. Let M be a non-trivial normal subgroup of W. If $B \leqslant M$ then M is transitive on Π since B is transitive on Π. So we may assume that M does not contain B. If $M \cap B = 1$ then $M \leqslant \mathsf{C}_W(B) = \mathsf{Z}(B)$ by Lemma 5.2(i), which is a contradiction. Thus $M \cap B \neq 1$.

We claim that $M \cap G_\delta \neq 1$, for each $\delta \in \Delta$. Since H is transitive on Δ and normalises $M \cap B$, it is sufficient to prove this for a single δ. For $f \in M \cap B$, define $S(f) := \{\delta \in \Delta \mid \delta f \neq 1\}$. Now let $f \in (M \cap B) \setminus \{1\}$ be such that $S(f)$ is of minimal size (at least 1 since $f \neq 1$). Let $\delta \in S(f)$. Then $\delta f \neq 1$, and since $\mathsf{Z}(G) = 1$ there exists $x \in G_\delta \setminus \mathsf{C}_{G_\delta}(\delta f)$. Now $[f, x] = f^{-1} f^x$ lies in $M \cap B$ since $M \cap B$ is normal in W, and $[f, x] \neq 1$ since $f^x \neq f$. Further $S([f, x]) = \{\delta\}$, since $\delta'[f, x] = \delta'(f^{-1} f) = 1$ for $\delta' \neq \delta$. It follows from the minimality of $S(f)$ that $S(f) = \{\delta\}$, and hence $f \in M \cap G_\delta$, proving the claim. Since H is transitive on Δ and normalises $M \cap B$, it follows that $M_0 := \prod_{\delta \in \Delta}(M \cap G_\delta)$ is a non-trivial normal subgroup of W contained in $M \cap B$. By Lemma 5.2 parts (ii) and (iv), $K = \{\delta f \mid f \in M_0\}$ is a subgroup of G independent of δ, and $M_0 = \mathsf{Func}(\Delta, K)$. Moreover, since $M_0 := \prod_{\delta \in \Delta}(M \cap G_\delta)$, the subgroup $K = \delta(M \cap G_\delta)$ for all δ. Also, by Lemma 5.2(ii), K is a non-trivial normal subgroup of G. As G is quasiprimitive, the subgroup K is transitive on Γ, and hence $M_0 = \mathsf{Func}(\Delta, K)$ is transitive on Π. Thus M is transitive on Π, and so W is quasiprimitive on Π. □

Claim 7. If (ii) holds and G is innately transitive on Γ with FCR-plinth, then W is innately transitive on Π with FCR-plinth.

Proof of Claim 7. Let K be an FCR-plinth for G on Γ. By definition of an FCR-group, K is a direct product $T_1 \times \cdots \times T_k$ of finitely many simple groups T_i, and since K is minimal normal in G, K is characteristically simple and the T_i are pairwise isomorphic, by Corollary 4.17. If the T_i are cyclic of prime order, then K, and hence also Γ and Π are finite. In this case, K is abelian and regular on Γ, and hence G is primitive on Γ by Corollary 3.13. Hence W is primitive and quasiprimitive on Π by Claims 5–6. It follows also that $M := \mathsf{Func}(\Delta, K)$ is an abelian normal subgroup of W which is regular on the finite set Π, and it follows from Theorem 3.12 that M is a minimal normal subgroup of W so W is innately transitive and M is an FCR-plinth (since M is finite). Therefore, we may assume that the T_i are non-abelian.

By Lemma 5.2(i), $M := \mathsf{Func}(\Delta, K)$ is a normal subgroup of W contained in B, and as K is a characteristically simple FCR-group, so is M. Thus $M = T_1 \times \cdots \times T_m$, with the T_i pairwise isomorphic non-abelian simple groups. Thus, by Lemma 5.2(iii), $M = \prod_{\delta\in\Delta}(M \cap G_\delta)$, and by Lemma 5.2(iv), each non-trivial normal subgroup N of W contained in M is a subdirect subgroup of M, that is $\{\delta f \mid f \in N\} = K$ for all $\delta \in \Delta$. Then by Theorem 4.16 (iv), N is a direct product of some of the simple direct factors of M, and by Theorem 4.16 (iii) it follows that $N = M$. Thus M is a minimal normal subgroup of W, and M is a transitive FCR-group, so W is innately transitive with FCR-plinth M. □

Now, by Claims 5–7, part (ii) implies part (i). □

6
Twisted wreath products

Several classes of innately transitive permutation groups can be constructed using the twisted wreath product construction introduced by B. H. Neumann (Neumann 1963). A good treatment of this construction can also be found in Suzuki's book (Suzuki 1982), and our exposition owes much to both Neumann and Suzuki. The relationship between twisted wreath products and primitive permutation groups was explored by Baddeley in (Baddeley 1993a).

A twisted wreath product has a natural action in which its base group is a regular normal subgroup. If this action is faithful, then a twisted wreath product can be embedded into a holomorph by Theorem 3.10. Twisted wreath products also are important examples of groups that preserve normal cartesian decompositions (defined in Section 4.7). The twisted wreath product groups constructed in this chapter are not necessarily finite. In particular we obtain a primitivity criterion (Proposition 6.10) for twisted wreath products with a non-abelian base group which is FCR.

6.1 The definition of twisted wreath products

Let T and P be groups, and assume that Q is a proper subgroup of P of finite index. In this chapter, we maintain our assumptions about the finiteness of the number of partitions in a cartesian decomposition, and this is why we assume that $|P\colon Q|$ is finite. Suppose that $\varphi\colon Q \to \mathsf{Aut}(T)$ is a homomorphism. Consider the group $\mathsf{Func}(P,T)$ of functions $f\colon P \to T$ introduced in Section 4.9 and recall that the product of functions f and g is defined pointwise; that is, for each $p \in P$, $p(fg) = (pf)(pg)$. By Lemma 4.32, this group is isomorphic to the unrestricted direct product of $|P|$ copies of T (see Sections 3.4 and 4.9). Considering an element of this direct product as a function $f\colon P \to T$ simplifies notation.

6.1.1 The base group

Let B denote the subset

$$B = \{f \in \mathsf{Func}(P,T) \mid (pq)f = (pf)(q\varphi) \text{ for all } p \in P,\ q \in Q\}.$$

(Recall that $q\varphi \in \mathsf{Aut}(T)$ and $pf \in T$ so $(pf)(q\varphi) \in T$.) Let $\mathcal{T}$ be a set of representatives of the *left* cosets of Q in P, such that $1 \in \mathcal{T}$. Note that $\mathcal{T}$ is finite and that each element of P has a unique representation of the form rq with $r \in \mathcal{T}$ and $q \in Q$. For each $c \in \mathcal{T}$ and $t \in T$, define $f_{c,t} \in \mathsf{Func}(P,T)$ by

$$f_{c,t}\colon rq \mapsto \begin{cases} 1 & \text{if } r \neq c \\ t(q\varphi) & \text{if } r = c \end{cases} \tag{6.1}$$

for $r \in \mathcal{T}$ and $q \in Q$. The definition of the elements $f_{c,t}$ implies that

$$(f_{c,t})^{-1} = f_{c,t^{-1}} \quad \text{and} \quad f_{c,s}f_{c,t} = f_{c,st}, \tag{6.2}$$

so that, for each $c \in \mathcal{T}$, $f_{c,1}$ is equal to the identity element of $\mathsf{Func}(P,T)$. Further, if $c \neq c'$ then $f_{c,s}f_{c',t} = f_{c',t}f_{c,s}$ for all t, s. Recall, before the next result, the definition of a direct decomposition of a group given after Lemma 4.2.

Lemma 6.1 (i) *An element $f \in B$ is determined by the set of images $\{rf \mid r \in \mathcal{T}\}$;*
(ii) *B is a subgroup of $\mathsf{Func}(P,T)$;*
(iii) *for each $c \in \mathcal{T}$, $T_c := \{f_{c,t} \mid t \in T\} \leqslant B$ and the map $t \mapsto f_{c,t}$ is an isomorphism between T and T_c;*
(iv) *$\{T_c \mid c \in \mathcal{T}\}$ is a direct decomposition for B, and so $B \cong T^k$ where $k = |P\colon Q|$. In particular B is FCR if and only if T is.*

Proof (i) Let $f \in B$. Then, for all $r \in \mathcal{T}$ and $q \in Q$ the membership condition for B implies that $(rq)f = (rf)(q\varphi)$ and hence f is determined by its values on $\mathcal{T}$.

(ii) Let $f_1, f_2 \in B$ and $p \in P$, $q \in Q$. Then

$$\begin{aligned} (pq)(f_1f_2^{-1}) &= (pq)f_1(pq)f_2^{-1} = (pq)f_1((pq)f_2)^{-1} \\ &= ((pf_1)(q\varphi))((pf_2)(q\varphi))^{-1} = ((pf_1)(q\varphi))((pf_2)^{-1}(q\varphi)) \\ &= ((pf_1)(pf_2^{-1}))(q\varphi) = p(f_1f_2^{-1})(q\varphi), \end{aligned}$$

and hence $f_1 f_2^{-1} \in B$. It follows that B is a subgroup of $\mathsf{Func}(P,T)$.

(iii) We prove first that each $f = f_{c,t}$ lies in B. Let $p \in P$ and $q \in Q$. Then $p = rq'$ for unique $r \in \mathcal{T}$ and $q' \in Q$, and the elements p and $pq = r(q'q)$ lie in the same left Q-coset. If $r \neq c$ then by (6.1), both $(pq)f = 1$ and $(pf)(q\varphi) = 1(q\varphi) = 1$. Suppose now that $r = c$. Then, again using (6.1),

$$(pq)f = (r(q'q))f = t((q'q)\varphi) = t(q'\varphi\, q\varphi) = (t(q'\varphi))(q\varphi) = (pf)(q\varphi).$$

Thus $f \in B$, and hence $T_c \subseteq B$. Now let $t, s \in T$. By (6.1), $(f_{c,s})^{-1} = f_{c,s^{-1}}$ and the product $f_{c,t}(f_{c,s})^{-1}$ maps each element of $P \setminus cQ$ to 1, and each element cq, for $q \in Q$, to $(t(q\varphi))(s^{-1}(q\varphi)) = (ts^{-1})(q\varphi) = (cq)f_{c,ts^{-1}}$. Hence $f_{c,t}(f_{c,s})^{-1} = f_{c,ts^{-1}}$ and it follows that T_c is a subgroup and the map $f_{c,t} \mapsto t$ is an isomorphism $T_c \to T$.

(iv) Let $\mathcal{T} = \{c_1, \ldots, c_k\}$. We prove that the map τ defined by $\tau\colon (f_{c_1,t_1}, \ldots, f_{c_k,t_k}) \mapsto f_{c_1,t_1} \cdots f_{c_k,t_k}$ (for all $f_{c_i,t_i} \in T_i$) is an isomorphism $T_{c_1} \times \cdots \times T_{c_k} \to B$. It follows from (6.1) that, for $i \neq j$ and $s, t \in T$, $f_{c_i,t} f_{c_j,s} = f_{c_j,s} f_{c_i,t}$. This combined with the product formula $f_{c,t} f_{c,s} = f_{c,ts}$ proved above implies that τ is a group homomorphism. Moreover it follows from (6.1) that $(f_{c_1,t_1}, \ldots, f_{c_k,t_k}) \in \ker \tau$ if and only if each $t_i = 1$, or equivalently $(f_{c_1,t_1}, \ldots, f_{c_k,t_k})$ is the identity. Thus τ is one-to-one. To prove that τ is onto let $f \in B$ and, for each i set $t_i := c_i f$. Then by (6.1), for each $q \in Q$, $(c_i q) f_{c_i,t_i} = t_i(q\varphi) = (c_i f)(q\varphi)$ which equals $(c_i q)f$ since $f \in B$. Let $x = (f_{c_1,t_1}, \ldots, f_{c_k,t_k})$. Then, for each i and each $q \in Q$, $x\tau$ maps $c_i q$ to $(c_i q) f_{c_i,t_i} = (c_i q) f$. Hence $x\tau = f$. Thus τ is onto and hence is an isomorphism. By definition then, $\{T_c \,|\, c \in \mathcal{T}\}$ is a direct decomposition for B. □

6.1.2 *An action of P on the base group*

Recall that ϱ_1^B denotes the right regular action of B as defined at the beginning of Section 2.3. So $B\varrho_1^B$ is a regular subgroup of $\mathsf{Sym}\, B$. In the next section we define another subgroup of $\mathsf{Sym}\, B$.

For $p \in P$, define the map $\hat{p}\colon B \to B$ by

$$x(f\hat{p}) = pxf \quad \text{for all} \quad x \in P \text{ and } f \in B. \tag{6.3}$$

We show that the image $f\hat{p}$ does indeed lie in B, and we prove that $p \mapsto \hat{p}$ defines an action of P on B.

Lemma 6.2 (i) *For $p \in P$, $\hat{p}$ is a well defined element of* $\mathsf{Aut}(B)$.

(ii) *If* $t \in T$, $s \in \mathcal{T}$ *and* $p \in P$ *then* $f_{s,t}\hat{p} = f_{r,t(q\varphi)}$ *where* $r \in \mathcal{T}$ *and* $q \in Q$ *are such that* $p^{-1}s = rq^{-1}$.

(iii) $\hat{p} = 1$ *if and only if* $p \in \mathsf{Core}_P(\ker\varphi)$.

(iv) *The map* $\vartheta \colon p \mapsto \hat{p}$ *defines an action* $\vartheta \colon P \to \mathsf{Sym}\, B$ *such that* $P\vartheta \leqslant \mathsf{Aut}(B)$ *and* $\ker\vartheta = \mathsf{Core}_P(\ker\varphi)$.

Proof (i) Let $p_0 \in P$ and $q \in Q$. Then $(p_0q)(f\hat{p}) = (p(p_0q))f$ by the definition of $\hat{p}$. This equals $((pp_0)f)(q\varphi)$ since $f \in B$, which in turn is equal to $(p_0(f\hat{p}))(q\varphi)$, again by the definition of $\hat{p}$. Thus $f\hat{p}$ lies in B.

Clearly the map $\hat{p} \colon B \to B$ is one-to-one. Let $f \in B$ and $g \in \mathsf{Func}(P, T)$ defined as $xg = (p^{-1}x)f$. Then clearly, $g \in B$, and $g\hat{p} = f$, and so $\hat{p}$ is onto. Thus $\hat{p} \in \mathsf{Sym}\, B$. Let $f_1, f_2 \in B$ and $x \in P$. Then, using the definitions of $\hat{p}$ and multiplication in B,

$$x((f_1f_2)\hat{p}) = (px)(f_1f_2) = (pxf_1)(pxf_2) = (x(f_1\hat{p}))(x(f_2\hat{p}))$$

which, finally, is equal to $x(f_1\hat{p}\, f_2\hat{p})$. Thus $\hat{p}$ is a homomorphism and hence lies in $\mathsf{Aut}(B)$.

(ii) Let $t \in T$, $s \in \mathcal{T}$ and $p \in P$. Then there are unique $q \in Q$ and $r \in \mathcal{T}$ such that $p^{-1}s = rq^{-1}$; that is, $pr = sq$. Note that $x \in rQ$ if and only if $px = pr(r^{-1}x) = sq(r^{-1}x) \in sQ$. Thus

$$xf_{s,t}\hat{p} = (px)f_{s,t} = \begin{cases} 1 & \text{if } px \notin sQ, \text{ that is, } x \notin rQ \\ t((qr^{-1}x)\varphi) & \text{if } px \in sQ, \text{ that is, } x \in rQ. \end{cases}$$

Since $(qr^{-1}x)\varphi = (q\varphi)((r^{-1}x)\varphi)$, it follows from (6.1) that $f_{s,t}\hat{p} = f_{r,t(q\varphi)}$.

(iii) Suppose that $p \in \mathsf{Core}_P(\ker\varphi)$ and let us compute $f_{s,t}\hat{p}$. We have that $p^{-1}s = s(p^{-1})^s$. Since $(p^{-1})^s \in Q$, part (ii) implies that $f_{s,t}\hat{p} = f_{s,t(p^s\varphi)}$. Since $p \in \mathsf{Core}_P(\ker\varphi)$, we find that $p^s \in \ker\varphi$, and so $f_{s,t}\hat{p} = f_{s,t}$. By Lemma 6.1(iv), the functions $f_{s,t}$ generate B, and hence, since $\hat{p} \in \mathsf{Aut}(B)$, we conclude that $f\hat{p} = f$ for all $f \in B$, that is, $\hat{p} = 1$.

Conversely suppose that $\hat{p} = 1$. Then, for all $f \in B$ and $x \in P$, $(px)f = xf$. Choose $t \in T \setminus \{1\}$. If $p \notin Q$ then by (6.1), $1 = pf_{1,t} = 1f_{1,t} = t$, which is a contradiction. Thus $p \in Q$. Then, for all $r \in \mathcal{T}$, $(pr)f_{r,t} = rf_{r,t} = t \neq 1$ and hence $pr \in rQ$, so that $p^r \in Q$. Thus $t = (pr)f_{r,t} = (rp^r)f_{r,t} = t(p^r\varphi)$, and since this holds for all $t \neq 1$, it follows that $p^r \in \ker\varphi$. This implies that p lies in $r(\ker\varphi)r^{-1}$ for all $r \in \mathcal{T}$, and hence, since $\ker\varphi$ is normal in Q, that $p \in \mathsf{Core}_P(\ker\varphi)$.

(iv) It is easy to check that $\widehat{p_1}\widehat{p_2} = \widehat{p_1p_2}$ and hence $\vartheta \colon p \mapsto \hat{p}$ defines an action of P on B with image contained in $\mathsf{Aut}(B)$. By part (iii), $\hat{p} = 1$ if and only if $p \in \mathsf{Core}_P(\ker\varphi)$, so $\ker\vartheta = \mathsf{Core}_P(\ker\varphi)$. □

It follows from Lemma 3.8 that $P\vartheta$ normalises $B\varrho_1^B$. Moreover these two groups generate their semidirect product $(B\varrho_1^B) \rtimes (P\vartheta)$ with respect to the natural action of $P\vartheta$ as a subgroup of $\mathsf{Aut}(B)$. We have the following explicit information about the action.

Corollary 6.3 (i) *For $p \in P$ and $f \in B$, $\hat{p}^{-1}(f\varrho_1^B)\hat{p} = (f\hat{p})\varrho_1^B$.*

(ii) *The group $P\vartheta$ induces a transitive action on the direct decomposition $\{T_c \mid c \in \mathcal{T}\}$ of B. Further the stabiliser of T_1 under this action is $Q\vartheta$.*

Proof (i) Let $f' \in B$. We are required to prove that $f'(f\hat{p})\varrho_1^B = f'(\hat{p}^{-1}(f\varrho_1^B)\hat{p})$. The left-hand side equals $f'(f\hat{p})$ and is the map that sends $x \in P$ to $(xf')((px)f)$. We now consider the right-hand side. Set $f'' := f'\hat{p}^{-1}$, the map satisfying $xf'' = (p^{-1}x)f'$ for $x \in P$. Then the right-hand side maps $x \in P$ to

$$x(f''(f\varrho_1^B)\hat{p}) = x((f''f)\hat{p}) = (px)(f''f) = ((px)f'')((px)f)$$

which equals $(xf')((px)f) = x(f'(f\hat{p}))$. The required equality follows.

(ii) Let $p \in P$ and $s \in \mathcal{T}$. By Lemma 6.2(ii), the image of T_s under $\hat{p}$ is contained in T_r where $p^{-1}s = rq^{-1}$ with some $q \in Q$. The same argument shows the other inclusion and hence $T_s\hat{p} = T_r$. It is not difficult to check that distinct s correspond to distinct r and so $\hat{p}$ induces a permutation of the T_c. For each $r \in \mathcal{T}$, choosing $p = r^{-1} \in P$ we have that $\hat{p}$ maps T_1 to T_r and hence P acts transitively. If $s = 1$, then the image of T_1 under $\hat{p}$ is T_r with $p^{-1} = rq^{-1}$, and $r = 1$ if and only if $p \in Q$. This shows that the normaliser of T_1 in $P\vartheta$ is $Q\vartheta$. □

6.1.3 The twisted wreath product $T \operatorname{twr}_\varphi P$

We regard the map ϑ defined in Lemma 6.2 as a homomorphism $\vartheta\colon P \to \mathsf{Aut}(B)$, and we define the *twisted wreath product* of T and P as the semidirect product:

$$T \operatorname{twr}_\varphi P = B \rtimes_\vartheta P.$$

The groups B and P are usually called the *base group* and the *top group* of $T \operatorname{twr}_\varphi P$, respectively. The subgroup Q of P is referred to as the *twisting subgroup*. While the subgroup Q does not appear explicitly in the notation $T \operatorname{twr}_\varphi P$, it is part of the input data since Q is the domain of φ. The group T is often an FCR-group, and in this case B is also an FCR-group, by Lemma 6.1(iv).

We lift the right regular action ϱ_1^B of B and the action ϑ of P to an action $\psi\colon T \operatorname{twr}_\varphi P \to \operatorname{Sym} B$ called the *base group action*. Define, for $f \in B$ and $p \in P$,

$$\psi\colon fp \mapsto f\varrho_1^B(p\vartheta)$$

and recall that $p\vartheta = \hat{p}$. Normal cartesian decompositions were discussed in Section 4.7.

Lemma 6.4 *The map ψ defines an action of $T \operatorname{twr}_\varphi P$ on B with image $(B\varrho_1^B) \rtimes (P\vartheta)$ and kernel $\mathsf{Core}_P(\ker\varphi)$ such that $\psi|_B = \varrho_1^B$ and $\psi|_P = \vartheta$. Further, the image of $T \operatorname{twr}_\varphi P$ in $\operatorname{Sym} B$ preserves a normal cartesian decomposition of B, which corresponds to the direct decomposition of B given in Lemma 6.1(iv).*

Proof The heart of the proof is to show that ψ is a homomorphism. Let $f,\ g \in B$ and $p,\ r \in P$. Note that, in the semidirect product $T \operatorname{twr}_\varphi P$, the product $(fp)(gr) = (f(g\hat{p}^{-1}))pr$ and hence, by the definition of ψ, $((fp)(gr))\psi = (f(g\hat{p}^{-1}))\varrho_1^B\hat{p}\hat{r}$. On the other hand, by Corollary 6.3(i), and since ϑ is an action (Lemma 6.2),

$$\begin{aligned}(fp)\psi(gr)\psi &= f\varrho_1^B\hat{p}g\varrho_1^B\hat{r} = f\varrho_1^B\hat{p}(g\varrho_1^B)\hat{p}^{-1}(\hat{p}\hat{r})\\ &= f\varrho_1^B(g\hat{p}^{-1})\varrho_1^B\hat{p}\hat{r} = (f(g\hat{p}^{-1}))\varrho_1^B\hat{p}\hat{r}.\end{aligned}$$

Thus $(fp)\psi(gr)\psi = ((fp)(gr))\psi$, so ψ is a homomorphism. It follows from the definition of ψ that $\psi|_B = \varrho_1^B$ and $\psi|_P = \vartheta$ and that the image of ψ is $(B\varrho_1^B) \rtimes (P\vartheta)$. Finally an element $f\varrho_1^B(p\vartheta)$ is the identity if and only if both $f\varrho_1^B = 1$ and $p\vartheta = 1$, or equivalently, if and only if $f = 1$ and $p \in \ker\vartheta$. By Lemma 6.2, $\ker\vartheta = \mathsf{Core}_P(\ker\varphi)$. Thus ψ has kernel $\mathsf{Core}_P(\ker\varphi)$.

Suppose that $W = T \operatorname{twr}_\varphi P$. Then $W\psi \leqslant \operatorname{Sym} B$. Since the kernel $\mathsf{Core}_P(\ker\varphi)$ of the action on B has trivial intersection with B, the subgroup $B\psi$ is isomorphic to B and it admits a direct decomposition as given in Lemma 6.1(iv). Since a point stabiliser in $W\psi$ is $P\psi$, which intersects trivially with $B\psi$, we have that the conditions of Example 4.23 are valid, and we can construct a $W\psi$-invariant, $B\psi$-normal cartesian decomposition of B. □

Lemma 6.5 *Let $\tau_1\colon T \to T_1$ given by $t \mapsto f_{1,t}$ (see (6.1)). Considering Q and T_1 subgroups of the twisted wreath product $T \operatorname{twr}_\varphi P$,*

(τ_1, id_Q) is a permutational isomorphism between the Q-action φ on T the conjugation action of Q on T_1.

Proof Let $t \in T$, $q \in Q$ and recall that ϑ is the action of P on the base group B as defined in Lemma 6.2. We need to show that $(t(q\varphi))\tau_1 = (t\tau_1)^q$. Now $(t\tau_1)^q = (f_{1,t})^q$, which is equal to $f_{1,t}(q\vartheta)$, by the definition of the semidirect product. Part (ii) of Corollary 6.3 implies that $f_{1,t}(q\vartheta) \in T_1$ and, by Lemma 6.2(ii), this is equal to $f_{1,t(q\varphi)}$. Thus $(t\tau_1)^q = f_{1,t(q\varphi)} = (t(q\varphi))\tau_1$. □

6.2 Internal characterisation of twisted wreath products

By the definition given in Section 6.1.3, a twisted wreath product is a semidirect product $B \rtimes_\vartheta P$ and, by Lemma 6.1 and Corollary 6.3, B has a finite P-invariant direct decomposition with $|P\colon Q|$ factors on which P acts transitively. We prove that these conditions are sufficient for the group to be isomorphic to a twisted wreath product.

Abstract group criteria: Suppose that G is a group, P is a subgroup and M is a normal subgroup of G such that $G = MP$ and $M \cap P = 1$. Suppose further that T is a subgroup of M such that the set $\mathcal{D} = \{T^g \mid g \in G\}$ of G-conjugates of T is finite and is a direct decomposition of M.

The following result can be found in (Bercov 1967, Section 2) and (Lafuente 1984).

Lemma 6.6 *Suppose that the abstract group criteria above hold, let $Q = \mathsf{N}_P(T)$, and let $\varphi\colon Q \to \mathsf{Aut}(T)$ be the homomorphism induced by the conjugation action of Q on T. Then there is an isomorphism $\alpha\colon G \to T \operatorname{twr}_\varphi P$ that maps M onto the base group, P onto the top group of $T \operatorname{twr}_\varphi P$, and $\mathcal{D}$ onto the direct decomposition of B given in Lemma 6.1(iv).*

Proof By definition, $T \operatorname{twr}_\varphi P = B \rtimes_\vartheta P$ with B as in Section 6.1.1. Also, as in Section 6.1.1, we choose a set $\mathcal{T}$ of left coset representatives for Q in P with $1 \in \mathcal{T}$. Since M normalises each element of $\mathcal{D}$ and $G = MP$, it follows that $\mathcal{D} = \{T^{p^{-1}} \mid p \in P\}$, and hence, since Q normalises T, that $\mathcal{D} = \{T^{c^{-1}} \mid c \in \mathcal{T}\}$ is of size $|\mathcal{T}|$. As $|\mathcal{D}|$ is assumed to be finite, Q has finite index in P and $|\mathcal{T}|$ is finite. We first define an isomorphism $\alpha_1\colon M \to B$ that, for each $c \in \mathcal{T}$, maps $T^{c^{-1}}$

to the subgroup T_c in the direct decomposition $\{T_c \mid c \in \mathcal{T}\}$ of B in Lemma 6.1(iv). For $t \in T$ and $c \in \mathcal{T}$, let

$$\alpha_1 \colon t^{c^{-1}} \mapsto f_{c,t}, \text{ with } f_{c,t} \text{ as in (6.1).} \tag{6.4}$$

Since $\{T^{c^{-1}} \mid c \in \mathcal{T}\}$ is a direct decomposition of M, this extends to a well defined map $\alpha_1 \colon M \to B$. It is clearly an isomorphism with the required property.

Now we define $\alpha \colon G \to T \operatorname{twr}_\varphi P$ by

$$(xp)\alpha := (x\alpha_1)p \quad \text{for} \quad x \in M,\ p \in P. \tag{6.5}$$

(On the right-hand side, p denotes an element in the top group of $T \operatorname{twr}_\varphi P$.) Since $G = MP$ and $M \cap P = 1$, the map α is well defined, and clearly α maps M to B, P to the top group, and $\mathcal{D}$ to $\{T_c \mid c \in \mathcal{T}\}$. Proving that α is a group homomorphism is equivalent to proving that $(x^p)\alpha_1 = (x\alpha_1)^p$ for all $x \in M$ and $p \in P$. Given the property of α_1 established in the previous paragraph, it is sufficient to prove this for all $x \in \bigcup_{c \in \mathcal{T}} T^{c^{-1}}$ and $p \in P$. So let $x = t^{c^{-1}}$ with $t \in T$ and $c \in \mathcal{T}$. There exist unique $r \in \mathcal{T}$ and $q \in Q$ such that $p^{-1}c = rq^{-1}$, or equivalently, $c^{-1}p = qr^{-1}$. Note that, by the definition of φ, $t^q = t(q\varphi) \in T$, and so

$$(x^p)\alpha_1 = (t^{c^{-1}p})\alpha_1 = (t^{qr^{-1}})\alpha_1 = f_{r,t(q\varphi)}.$$

Also, recalling that the action of P on B in $T \operatorname{twr}_\varphi P$ is via the map $\vartheta \colon p \mapsto \hat{p}$, we find that

$$(x\alpha_1)^p = (t^{c^{-1}}\alpha_1)^p = (f_{c,t})^p = f_{c,t}\,\hat{p}.$$

By Lemma 6.2(ii), these elements are equal, and it follows that α is a homomorphism. Since α_1 is onto, clearly the map α is also onto. Also $xp \in \ker \alpha$ if and only if $x\alpha_1 p = 1$, which implies $x = 1$ and $p = 1$. Thus α is an isomorphism. $\square$

Now we characterise as permutation groups those groups that arise as images of base group actions of twisted wreath products (as defined in Section 6.1.3). Such permutation groups have a regular normal subgroup admitting a direct decomposition on which the permutation group acts transitively. We show that these properties are sufficient as well as necessary for a permutation group to be the image of a twisted wreath product in its base group action.

Permutation group criteria: Suppose that $G \leqslant \operatorname{Sym} \Omega$ is a permutation group on Ω with a regular normal subgroup M. Suppose further that M has a subgroup T such that the set $\mathcal{D} = \{T^g \mid g \in G\}$ of G-conjugates is a finite direct decomposition of M.

Lemma 6.7 *Suppose that the permutation group criteria above hold, and let $\omega \in \Omega$, $P = G_\omega$, $Q = \mathsf{N}_P(T)$, and $\varphi\colon Q \to \mathsf{Aut}(T)$ be the homomorphism induced by the conjugation action of Q on T. Then*

(i) $\mathsf{Core}_P(\ker\varphi) = 1$ *so that $T \operatorname{twr}_\varphi P = B \rtimes_\vartheta P$ acts faithfully on B in its base group action.*

(ii) *Moreover, there exists a permutational isomorphism from G (considered as a subgroup of $\mathsf{Sym}\,\Omega$) to $T \operatorname{twr}_\varphi P$ (considered as a subgroup of $\mathsf{Sym}\,B$) that maps M onto the base group B, P onto the top group, and $\mathcal{D}$ onto the direct decomposition of B given in Lemma 6.1(iv).*

Proof Since M is regular on Ω, $M \cap P = 1$ and $G = MP$. Moreover, by Theorem 3.10, we may assume that $\Omega = M$, that M acts by right multiplication and P acts by conjugation, that $\omega = 1_M$, and hence that the conjugation action of P allows us to assume $P \leqslant \mathsf{Aut}(M)$ in its natural action. In particular Lemma 6.6 applies.

Let $\mathcal{T}$ be a set of left coset representatives for Q in P. As in the proof of Lemma 6.6, $\mathcal{D} = \{T^{c^{-1}} \mid c \in \mathcal{T}\}$. Let $p \in \mathsf{Core}_P(\ker\varphi)$. Then $p \in (\ker\varphi)^{c^{-1}}$ for each $c \in \mathcal{T}$, and hence p centralises $T^{c^{-1}}$ for each c. It follows that p centralises M and since $P \leqslant \mathsf{Aut}(M)$, this implies that $p = 1$. Thus $\mathsf{Core}_P(\ker\varphi) = 1$. By Lemma 6.4, we may therefore identify $T \operatorname{twr}_\varphi P$ with its image under the base group action in $\mathsf{Sym}\,B$.

We claim that (α_1, α) is a permutational isomorphism with the required properties, where $\alpha_1\colon M \to B$ and $\alpha\colon G \to T \operatorname{twr}_\varphi P$ are defined in (6.4)–(6.5). Thus α_1 is a bijection, α is a group isomorphism, the properties $M\alpha = B$, $P\alpha = P$ (the top group of $T \operatorname{twr}_\varphi P$) hold and $\mathcal{D}$ is mapped to the direct decomposition of B given in Lemma 6.1(iv).

It remains to prove the permutational isomorphism property: for all $g \in G$ and $\omega \in M$, $(\omega g)\alpha_1 = (\omega\alpha_1)g\alpha$. Since α is an isomorphism it suffices to prove this for (a) $g \in P$ and (b) $g \in \dot{\bigcup}_{c\in\mathcal{T}} T^{c^{-1}}$. First consider $g = p \in P$. In this case we first take $\omega = t^{c^{-1}}$ with $c \in \mathcal{T}$ and $t \in T$. There exist unique $r \in \mathcal{T}$ and $q \in Q$ such that $p^{-1}c = rq^{-1}$ and hence $c^{-1}p = qr^{-1}$. Thus

$$(\omega g)\alpha_1 = (t^{c^{-1}p})\alpha_1 = (t^{qr^{-1}})\alpha_1 = ((t(q\varphi))^{r^{-1}})\alpha_1 = f_{r,t(q\varphi)}$$

and $(\omega\alpha_1)g\alpha = f_{c,t}\hat{p}$. These two elements are equal by Lemma 6.2(ii). The proof of the equality for general $\omega \in M$ follows using the properties of the direct decompositions $\mathcal{D}$ of M and $\{T_1 \mid c \in \mathcal{T}\}$ of B. Finally consider case (b): here $g = t^{c^{-1}}$ for some $c \in \mathcal{T}$ and $t \in T$, g acts as right multiplication on M, and, by (6.5), $g\alpha$ acts by right multiplication by $f_{c,t}$ on B. Again, it is sufficient to consider the case of $\omega = s^{r^{-1}} \in T^{r^{-1}}$, for some $r \in \mathcal{T}$ and $s \in T$. We have

$$(\omega g)\alpha_1 = (s^{r^{-1}} t^{c^{-1}})\alpha_1 = (s^{r^{-1}})\alpha_1 (t^{c^{-1}})\alpha_1 = f_{r,s} f_{c,t}$$

and also $(\omega\alpha_1)g\alpha = f_{r,s} f_{c,t}$. □

The 'permutation group criteria' hold in particular for permutation groups G containing a non-abelian minimal normal FCR-subgroup M that acts regularly. Recall that an FCR-group is a direct product of finitely many simple groups; see Definition 1.2.

Corollary 6.8 *Suppose that G is a permutation group on a set Ω and M is a minimal normal* FCR*-subgroup of G such that M is non-abelian and regular. Let T be a minimal normal subgroup of M. Then the permutation group criteria hold relative to $\{T^g \mid g \in G\}$ and, for $\omega \in \Omega$, G is permutationally isomorphic to the twisted wreath product $T \operatorname{twr}_\varphi G_\omega$ in its faithful base group action.*

Proof By Lemma 3.14(ii), M is characteristically simple, and by Corollary 4.17, there is a direct decomposition $\mathcal{D} := \{S^g \mid g \in G\}$ of M for some simple normal subgroup S of M. Moreover by Theorem 4.16, T is one of the S^g and we may assume that $T = S$. Thus the permutation group criteria hold relative to $\{T^g \mid g \in G\}$. The final assertion now follows from Lemma 6.7. □

Thus any permutation group with a regular and non-abelian minimal normal FCR-subgroup is isomorphic to a twisted wreath product in its faithful base group action. This indicates the importance of twisted wreath products in the theory of permutation groups.

6.3 Primitive, quasiprimitive and innately transitive twisted wreath products

In this section we consider the base group action of a twisted wreath product $T \operatorname{twr}_\varphi P$ where T, P, and $\varphi\colon Q \to \mathsf{Aut}(T)$ are as in the beginning of Section 6.1. Since the base group of a twisted wreath product is regular, we easily obtain from Theorem 3.12 criteria for such groups to be innately transitive, quasiprimitive, or primitive. Conditions of primitivity for finite groups were also studied (Baddeley 1993a, Section 3). Baddeley's conclusions are similar to ours.

Lemma 6.9 *Assume that* $\mathsf{Core}_P(\ker\varphi) = 1$ *and let* W *denote the twisted wreath product* $T\,\mathrm{twr}_\varphi\,P$. *Let* B *be the base group of* W *as defined in Section 6.1.1.*

(i) *W is innately transitive if and only if B is a minimal normal subgroup of W.*

(ii) *Assuming that B is a minimal normal subgroup of W, the group W is quasiprimitive if and only if either $P\vartheta \cap \mathsf{Inn}(B) = 1$ or $\mathsf{Inn}(B) \leqslant P\vartheta$, where ϑ is the action of P on B defined in Lemma 6.2.*

(iii) *The group W is primitive if and only if no proper, non-trivial subgroup of B is normalised by P.*

Proof By Lemma 6.4, the condition $\mathsf{Core}_P(\ker\varphi) = 1$ implies that W is faithful in its base group action. Thus W can be considered as a permutation group. As the base group is a regular normal subgroup of W, statements (i)–(ii) follow from Theorem 3.12(i)–(ii). To prove (iii), observe that P is a point stabiliser in W under the base group action and use Theorem 3.12(iii). □

The primitivity criteria for twisted wreath products are variations of the primitivity criteria of Aschbacher and Scott (Aschbacher & Scott 1985) for finite groups. First we state a result that holds also for infinite groups.

Proposition 6.10 *Suppose that T, P, Q, and φ are as above and let $W = T\,\mathrm{twr}_\varphi\,P$.*

(i) *If W is primitive in its base group action then*

   (a) *no proper and non-trivial subgroup of T is invariant under $Q\varphi$; and*

   (b) *φ cannot be extended to a strictly larger subgroup of P.*

(ii) *If T is a non-abelian simple group (not necessarily finite), then (a) and (b) in part (i) imply that W is primitive.*

Proof Let us suppose that B is the base group of W and let $T_1,\dots,T_k$ be the factors in the direct decomposition of B given in Lemma 6.1(iv). By Lemma 6.9(iii), W is primitive if and only if 1 and B are the only subgroups of B normalised by P. By Lemma 6.1(iii), the map $\tau_1\colon t\mapsto f_{1,t}$ (see (6.1) for the definition of $f_{1,t}$) with $t\in T$ is an isomorphism between T and T_1 and, by Lemma 6.5, (τ_1,id_Q) is a permutational isomorphism between the Q-action φ on T and the conjugation action of Q on T_1.

(i) Suppose first that W is primitive in the base group action. Let L be a subgroup of T normalised by $Q\varphi$. Then $L_1 = L\tau_1$ is a subgroup of T_1 normalised by Q. For $i\in\underline{k}$ choose an element $g_i\in P$ such that $T_1^{g_i}=T_i$. Such elements exist by Corollary 6.3(ii). We may assume without loss of generality that $g_1=1$. For $i\in\underline{k}$, set $L_i=(L_1)^{g_i}$. We claim that L_i is independent on the choice of g_i. Indeed, let g_i, $h_i\in P$ such that $(T_1)^{g_i}=(T_1)^{h_i}=T_i$. Then $g_ih_i^{-1}\in \mathsf{N}_P(T_1)=Q$, and so L_1 is normalised by $g_ih_i^{-1}$. Hence $(L_1)^{g_i}=(L_1)^{h_i}$. Now set $\overline{L}=L_1\times\cdots\times L_k$ and note that $\overline{L}\cap P=1$ and that $\overline{L}$ is normalised by P. Theorem 3.12(iii) implies that either $\overline{L}=1$, in which case $L_1=1$, or $\overline{L}=B$, which gives that $L_1=T_1$. Thus either $L=1$ or $L=T$. Therefore 1 and T are the only $Q\varphi$-invariant subgroups of T.

If $\overline{Q}$ is a subgroup of P properly containing Q such that φ can be extended to a homomorphism $\overline{\varphi}\colon\overline{Q}\to\mathsf{Aut}(T)$, then the definition of the twisted wreath product implies that $\overline{W}=T\,\mathsf{twr}_{\overline{\varphi}}P$ can be embedded into $W=T\,\mathrm{twr}_\varphi P$ such that P is a proper subgroup of $\overline{W}$. This shows that in this case, P cannot be a maximal subgroup of W, and so W cannot be primitive (Corollary 2.15). Thus such a $\overline{Q}$ cannot exist. Thus if W is primitive then (a) and (b) are valid.

(ii) Assume now that (a) and (b) are valid for P and that T is a non-abelian simple group. Suppose that $\overline{L}$ is a non-trivial P-invariant subgroup of B. We aim to prove that $\overline{L}=B$, and then Lemma 6.9(iii) will imply that W is primitive. We assume as above that $B=T_1\times\cdots\times T_k$, and in this case the T_i are non-abelian simple factors of B. Let us note that an argument in the first part of the proof implies that Q normalises T_1. Let $\sigma_i\colon B\to T_i$ be the i-th coordinate projection and set $L_i=\overline{L}\sigma_i$. Then $\overline{L}\leqslant L_1\times\cdots\times L_k$. As $\overline{L}$ is normalised by P, and as P is transitive on the simple direct factors T_i of B, we obtain that the L_i are conjugate under P. Since $\overline{L}$ is non-trivial, at least one L_i is non-trivial, which implies that each of the L_i is non-trivial and L_1 is normalised by Q in T_1. Hence $L=L_1\tau_1^{-1}$ is normalised by $Q\varphi$ in T. Condition (a) implies that $L=T$, and hence $L_i=T_i$ for all i. By Theorem 4.16, $\overline{L}$ is the direct product of disjoint full strips. Suppose that S_1 is the strip in $\overline{L}$ that covers T_1 and set $\overline{Q} = \mathsf{N}_P(S_1)$. The group P permutes the strips of $\overline{L}$, and so P permutes the supports of the strips. Further, for $p \in P$, we have that either $(S_1)^p = S_1$ or $(S_1)^p \cap S_1 = 1$. Since Q stabilises T_1, we obtain, for all $q \in Q = \mathsf{N}_P(T_1)$, that $T_1 \in \operatorname{Supp} S_1 \cap \operatorname{Supp}(S_1)^q$, and hence $(S_1)^q = S_1$. Thus $q \in \mathsf{N}_P(S_1) = \overline{Q}$, and so $Q \leqslant \overline{Q}$. Let $\overline{\varphi}\colon \overline{Q} \to \mathsf{Aut}(S_1)$ be the homomorphism induced by the $\overline{Q}$-action on S_1. Then the restriction $\sigma_1|_{S_1}$ is an isomorphism between S_1 and T_1. Let $\widehat{\varphi}\colon \overline{Q} \to \mathsf{Aut}(T_1)$ be the homomorphism given by

$$t(q\widehat{\varphi}) = t\sigma_1^{-1}(q\overline{\varphi})\sigma_1.$$

As Q normalises T_1, for $q \in Q$ and $t \in T_1$ the transformation σ_1 commutes with conjugation by q, and so

$$t(q\widehat{\varphi}) = t\sigma_1^{-1}(q\overline{\varphi})\sigma_1 = (t\sigma_1^{-1})^q\sigma_1 = t^q.$$

Thus $\widehat{\varphi}$ is an extension of the conjugation action of Q on T_1 to a subgroup $\overline{Q}$ containing Q. Since this conjugation action is equivalent to the Q-action φ on T, this shows that φ can also be extended to $\overline{Q}$. Therefore condition (b) implies that $\overline{Q} = Q$. If $S_1 \neq T_1$, then S_1 covers another minimal normal subgroup, T_2 say, of B. Since P is transitive on the T_i, there is an element $p \in P$ that maps T_1 to T_2, and p normalises S_1; that is $p \in \overline{Q}$. Then $p \in Q = \mathsf{N}_P(T_1)$, which is a contradiction. Therefore $S_1 = T_1$, that is, $\overline{L} = B$. Thus, as explained above, W must be primitive. □

6.4 Finite primitive twisted wreath products

Theorem 3.22 applies for finite twisted wreath products that are primitive in the base group action. A stronger result will be proved in Theorem 6.15.

Theorem 6.11 *Let T be a finite non-abelian simple group, let P be a finite group, and let $\varphi\colon Q \to \mathsf{Aut}(T)$ be a homomorphism for some subgroup $Q < P$ such that $\mathsf{Core}_P(\ker\varphi) = 1$. Suppose that $W = T \operatorname{twr}_\varphi P$ is primitive and let B be the base group of W. Then the following are valid.*

(i) *Each non-trivial normal subgroup of P is insoluble.*
(ii) *If B is the unique minimal normal subgroup of W, then action of P on the set of simple direct factors of B is faithful.*

Proof (i) Since B is a regular non-abelian minimal normal subgroup of W, this follows from Theorem 3.22(ii).

(ii) Let K be the kernel of the P-action on the set of simple direct factors of B. Then $K \trianglelefteq P$ and we claim that K is soluble. Suppose that $\vartheta \colon P \to \mathsf{Aut}(B)$ is the P-action on B as defined in Lemma 6.2(iv). Then Lemma 6.9 implies that either $\mathsf{Inn}(B) \leqslant P\vartheta$ or $P\vartheta \cap \mathsf{Inn}(B) = 1$. However, if $\mathsf{Inn}(B) \leqslant P\vartheta$, $\mathsf{Hol}\, B \leqslant W$, and W contains a second minimal normal subgroup that is formed by the permutations that are induced by left multiplications by elements of B. Thus $P\vartheta \cap \mathsf{Inn}(B) = 1$ must hold. In particular, K induces a group of outer automorphisms of $B \cong T^{|P:Q|}$ normalising each of the simple direct factors of B. Therefore B is isomorphic to a subgroup of $\mathsf{Out}(T)^{|P:Q|}$ which is a soluble group by the Schreier Conjecture (see (Dixon & Mortimer 1996, p. 133)). Now part (i) implies that $K = 1$, and hence the P-action on the simple direct factors of B is faithful. □

Next we rephrase the condition in Proposition 6.10(i)(a) having in mind the case of finite simple groups. The statement in part (ii) should be read as 'A or (B and C)'.

Lemma 6.12 *Let T be a non-abelian simple group and X a subgroup of $\mathsf{Aut}(T)$. Then the following are equivalent.*

(i) *No non-trivial proper subgroup of T is invariant under X.*
(ii) $\mathsf{Inn}(T) \leqslant X$, *or* $\mathsf{Inn}(T) \cap X = 1$ *and X is maximal in* $\mathsf{Inn}(T)X$.

In particular, if T is finite, then (i) holds if and only if $\mathsf{Inn}(T) \leqslant X$.

Proof The fact that (ii) implies (i) is easy. Indeed, if $\mathsf{Inn}(T) \leqslant X$, then any X-invariant subgroup of T is normal in T, and, since T is simple, such a subgroup is either 1 or T. To prove that the second alternative of (ii) implies (i), suppose that $\mathsf{Inn}(T) \cap X = 1$ and that X is maximal in $\mathsf{Inn}(T)X$. Let L be an X-invariant subgroup of T and let H denote the subgroup of $\mathsf{Inn}(T)$ that contains the automorphisms that are induced by conjugation by elements of L. As L is X-invariant, H is normalised by X. Hence, HX is a subgroup of $\mathsf{Aut}(T)$ containing X and is contained in $\mathsf{Inn}(T)X$. Hence either $HX = X$ or $HX = \mathsf{Inn}(T)X$. In the former case, $H \leqslant X$, which implies, as $X \cap \mathsf{Inn}(T) = 1$, that $H = 1$, and, in turn, that $L = 1$. In the latter $L = T$.

Let us now prove that (i) implies (ii). Suppose that $\mathsf{Inn}(T) \not\leqslant X$. Let $R = \mathsf{Inn}(T)X$. Then $\mathsf{Inn}(T) \cap X$ is normalised by $\mathsf{Inn}(T)$ and the corresponding subgroup in T is X-invariant. As we assume (i) and that $\mathsf{Inn}(T) \nleqslant X$, we obtain that $\mathsf{Inn}(T) \cap X = 1$, and so R is a semidirect product $\mathsf{Inn}(T) \rtimes X$. Suppose that H is a subgroup of $\mathsf{Aut}(T)$ such that $X < H \leqslant R$ and let $h \in H \setminus X$. Then $\mathsf{Inn}(T)h = \mathsf{Inn}(T)x$ for some $x \in X$, since $\mathsf{Inn}(T)H = \mathsf{Inn}(T)X$. Hence $hx^{-1} \in \mathsf{Inn}(T) \cap H$ and $hx^{-1} \neq 1$ (since it lies in $H \setminus X$). Thus $\mathsf{Inn}(T) \cap H$ is a non-trivial subgroup of $\mathsf{Inn}(T)$ normalised by H, and, as $X < H$, it is also normalised by X. Hence the corresponding subgroup in T is X-invariant, and hence it must be T. Thus $\mathsf{Inn}(T) \cap H = \mathsf{Inn}(T)$; that is, $\mathsf{Inn}(T) \leqslant H$. This shows that $H = R$, and so X is maximal in R.

Let us prove the final statement of the lemma for finite T; we are left to show that (i) implies that $\mathsf{Inn}(T) \leqslant X$. Hence assume (i) and that T is finite; set $R = \mathsf{Inn}(T)X$ as above. If X were a maximal subgroup of R, then R under its action on the coset space $[R\colon X]$ would be a finite primitive group with a unique non-abelian simple minimal normal subgroup which is regular. By Corollary 3.23, no such group exists. This shows that X is not maximal in $\mathsf{Inn}(T)X$ and in this case $\mathsf{Inn}(T) \leqslant X$ must be true. □

Combining Proposition 6.10 and Lemma 6.12 we obtain the following corollary. The first condition of the corollary was found independently by (Aschbacher & Scott 1985, Theorem 1(C)(1)) and (Kovács 1986, Corollary 5.2). The second condition, which may appear somewhat mysterious, shows up explicitly in (Kovács 1986, Theorem 4.3(c)) and it is also a consequence of (Aschbacher & Scott 1985, Theorem 1(C)(1)); see also the discussion in (Praeger & Schneider 2017a, Section 3).

Corollary 6.13 *Suppose that T is a non-abelian finite simple group. Then $W = T \operatorname{twr}_\varphi P$ is primitive in its base group action if and only if both*

(i) $\mathsf{Inn}(T) \leqslant Q\varphi$; *and*
(ii) φ *cannot be extended to a strictly larger subgroup of* P.

Using Example 3.24, it is easy to construct examples to show that the conclusions of Corollary 6.13 may not hold if T is an infinite simple group.

Example 6.14 Let $G = T \rtimes \langle \beta \rangle$ be the primitive group on Ω constructed in Example 3.24 with socle T where T is an infinite simple group acting regularly. Then, for $k \geqslant 2$, the wreath product $W = G \wr \mathsf{S}_k$

acting on Ω^k is primitive (Theorem 5.18) and has a unique minimal normal subgroup $M = T^k$ which is non-abelian and regular. Let $\omega \in \Omega$ be fixed. For $\alpha = (\omega, \ldots, \omega)$, $G = T^k \rtimes G_\alpha \cong T \operatorname{twr} (\langle\beta\rangle^k \rtimes \mathsf{S}_k)$. However, the top group $\langle\beta\rangle^k \rtimes \mathsf{S}_k$ only induces the cyclic group $\langle\beta\rangle$ in $\mathsf{Aut}(T)$, and this induced subgroup clearly does not contain $\mathsf{Inn}(T)$.

Finally we prove the following result that first appeared in the unpublished note (Förster & Kovács 1989). Recall *Dedekind's Modular Law* that states, for subgroups X, Y, Z of a group with $Z \leqslant X$, the equality $(X \cap Y)Z = X \cap YZ$.

Theorem 6.15 *Let T be a finite non-abelian simple group, let P be a finite group, and let $\varphi\colon Q \to \mathsf{Aut}(T)$ be a homomorphism for some subgroup $Q < P$ such that $\mathsf{Core}_P(\ker\varphi) = 1$. If P is a maximal subgroup of $T \operatorname{twr} P$, then P contains a unique minimal normal subgroup that is non-abelian.*

Proof The conditions imply that $T \operatorname{twr} P$ is faithful and primitive in the base group action, and so the conclusions (i) and (ii) of Corollary 6.13 are valid. Set $C = \ker\varphi$. Note that Q/C can be considered as a subgroup of $\mathsf{Aut}(T)$ that, by Corollary 6.13(i), contains $\mathsf{Inn}(T)$. Thus Q/C is an almost simple group with socle isomorphic to T. Since Q/C has a unique minimal normal subgroup, which is isomorphic to T, there is a unique normal subgroup D of Q such that $C \leqslant D$ and $D/C = \mathsf{Soc}(Q/C)$. Then D/C is the unique minimal normal subgroup of Q/C and $D/C \cong T$.

Set $N = \mathsf{N}_P(C) \cap \mathsf{N}_P(D)$. We claim that $N = Q$. The group Q is clearly contained in N, and so we only need to check the reverse inclusion. Suppose that $n \in N$. Then conjugation by n induces an automorphism of D/C, and hence it induces an automorphism on T. We make this explicit as follows. Let $\iota\colon T \to \mathsf{Inn}(T)$ be the natural isomorphism. Using this notation, we have

$$t\alpha\iota = (t\iota)^\alpha \quad \text{for all } t \in T \text{ and } \alpha \in \mathsf{Aut}(T). \tag{6.6}$$

The restriction to D of $\varphi\iota^{-1}$ is a surjective homomorphism from D to T with kernel C. If $n \in N$, then the automorphism ψ_n induced by n on T can be given as follows. If $t = d\varphi\iota^{-1} \in T$ with $d \in D$, then $t\psi_n = d\varphi\iota^{-1}\psi_n = (d^n)\varphi\iota^{-1}$. Since D is normalised by N and $C = \ker\varphi$, the map $\psi\colon N \to \mathsf{Aut}(T)$ given by $\psi\colon n \mapsto \psi_n$ is well defined and is a homomorphism. We claim that $\psi|_Q = \varphi$. Let $q \in Q$

and $t \in T$ such that $t = d\varphi\iota^{-1}$ for some $d \in D$ as above. Then

$$t\psi_q\iota = d\varphi\iota^{-1}\psi_q\iota = (d^q)\varphi\iota^{-1}\iota = (d\varphi)^{q\varphi}.$$

On the other hand, using (6.6),

$$t(q\varphi)\iota = (t\iota)^{q\varphi} = (d\varphi\iota^{-1}\iota)^{q\varphi} = (d\varphi)^{q\varphi}.$$

Since ι is a bijection, $t\psi_q = t(q\varphi)$ for all $t \in T$. Thus $\psi_q = q\varphi$; that is, ψ is an extension of φ to the subgroup N containing Q. By Corollary 6.13(ii), $N = Q$ as claimed.

Suppose that X is a subgroup of P properly containing C such that X is normalised by Q. We claim that $D \leqslant X$. Suppose that this is not the case and $D \nleqslant X$. Then $(Q \cap X)/C$ is a normal subgroup of Q/C which does not contain $D/C = \mathsf{Soc}\, Q/C$. Hence $(Q \cap X)/C = C/C$; that is $Q \cap X = C$. Thus we obtain a well defined homomorphism $\psi\colon QX \to \mathsf{Aut}(T)$, $qx \mapsto q\varphi$ for all $q \in Q$ and $x \in X$. Clearly, ψ extends φ, and hence, by Corollary 6.13, $QX = Q$. That is $X \leqslant Q$, which is a contradiction to $Q \cap X = C$.

Let Y_1 and Y_2 be minimal normal subgroups of P. Since C is core-free, we have that $C \cap Y_i = 1$ and Y_iC is a subgroup that satisfies the conditions for X in the previous paragraph. Hence $D \leqslant Y_iC$. By Dedekind's Modular Law, $(D \cap Y_i)C = D \cap Y_iC = D$, and so

$$T \cong D/C = (D \cap Y_i)C/C \cong (D \cap Y_i)/(D \cap Y_i \cap C) \cong D \cap Y_i.$$

In particular, Y_i is non-abelian. Suppose that ab, $cd \in D$ with $a \in D \cap Y_1$, $c \in D \cap Y_2$, and b, $d \in C$. Then, using the well-known commutator identities (see (Robinson 1996, 5.1.5)),

$$[ab, cd] = [a, d]^b [a, c]^{bd} [b, d] [b, c]^d.$$

Since a, $c \in D$ and b, $d \in C$, each of the elements $[a, d]^b$, $[b, d]$, $[b, c]^d$ belongs to C. Further, if $Y_1 \neq Y_2$, then $[a, c]^{bd} = 1$. Hence if $Y_1 \neq Y_2$, then the commutator subgroup D' is contained in C, which is a contradiction, as $D'C/C = D/C$. Hence $Y_1 = Y_2$, and P has a unique non-abelian minimal normal subgroup. □

Corollary 6.16 *Suppose that G is a finite primitive permutation group such that G contains a unique minimal normal subgroup that is non-abelian and regular. Then a point stabiliser G_ω contains a unique minimal normal subgroup that is non-abelian.*

Proof By Lemma 6.7, G is permutationally isomorphic to $T \operatorname{twr}_{\varphi} P$ for some finite simple group T, subgroup $Q < P$, and homomorphism $\varphi\colon Q \to \mathsf{Aut}(T)$, such that $\mathsf{Core}_P(\ker\varphi) = 1$. Now the result follows from Theorem 6.15. □

6.5 Twisted wreath products with minimal normal FCR-subgroups

In applications of twisted wreath products for the theory of permutation groups, the base group B is often a characteristically simple FCR-group. This is the case, for instance, when B is minimal normal and finite. This situation is investigated in the next result.

Proposition 6.17 *Assume that T is non-abelian and suppose that $\mathsf{Core}_P(\ker\varphi) = 1$. Let $W = T \operatorname{twr}_{\varphi} P$ and let B be the base group of W as defined in Section 6.1.1.*

(i) *Then W is innately transitive with* FCR*-plinth in its base group action if and only if T is a characteristically simple* FCR*-group and $Q\varphi$ is transitive on the simple direct factors of T.*

(ii) *Further, in this case, if S is a simple direct factor of T and $\psi\colon \mathsf{N}_P(S) \to \mathsf{Aut}(S)$ is the homomorphism induced by the restriction $\varphi|_{\mathsf{N}_P(S)}$, then W is permutationally isomorphic to $S \operatorname{twr}_{\psi} P$.*

Proof The condition $\mathsf{Core}_P(\ker\varphi) = 1$ implies that the base group action of W is faithful, and so we may consider W as a permutation group acting on B. By Lemma 6.4, B is a regular normal subgroup of W and P is a point stabiliser in W.

Assume first that W is innately transitive with FCR-plinth. Then, by Lemma 6.9, B is a minimal normal subgroup of W. By Theorem 3.6 either B is the unique transitive minimal normal subgroup of W or W has two transitive minimal normal subgroups, one of them being B, and they are isomorphic by Corollary 3.11. As W is innately transitive with FCR-plinth, each transitive minimal normal subgroup, and in particular B, must be FCR. Therefore, since T is non-abelian, B must be a non-abelian minimal normal subgroup of W and so B must be a direct product of finitely many pairwise isomorphic non-abelian simple groups. As $B \cong T^k$ where $k = |P\colon Q|$, Theorem 4.16 implies that T in this case also must be a direct product of pairwise isomorphic simple groups. Suppose that $T = S^r$ where S is simple.

Recall that $\mathcal{T}$ is a left transversal for Q in P such that $1 \in \mathcal{T}$ and consider the factor T_1 defined in Lemma 6.1 (that is, T_c with $c = 1$). Then $T_1 \cong T \cong S^r$. Let S_1 and S_2 be two minimal normal subgroups of T_1. As B is a minimal normal subgroup of W, there is an element of $g \in W$ such that $(S_1)^g = S_2$. Since B acts trivially on the simple direct factors of B and $W = BP$, we may assume that $g \in P$. Since, by Corollary 6.3, P preserves the direct decomposition $\{T_c \mid c \in \mathcal{T}\}$, this implies that p must normalise T_1. As Lemma 6.2(ii) implies that $\mathsf{N}_P(T_1) = Q$, we find that $p \in Q$. Hence Q must be transitive on the simple direct factors of T_1. By Lemma 6.5, the Q-action by conjugation on T_1 is equivalent to the twisting action φ on T, and hence Q must act transitively on the simple direct factors of T.

Conversely, assume that T is a characteristically simple FCR-group and Q acts transitively by conjugation on the simple direct factors of T. As argued above, in this case Q acts transitively on the simple direct factors of T_1. Since P is transitive on the direct decomposition $\{T_c \mid c \in \mathcal{T}\}$, this implies that P is transitive on the simple direct factors of the base group B. Thus B is a transitive minimal normal FCR-subgroup of W, and hence W is innately transitive with FCR-plinth B. This proves part (i).

(ii) Consider the isomorphism $\tau_1 \colon t \mapsto f_{1,t}$ from T to T_1. By Lemma 6.5, (τ_1, id_Q) is a permutational isomorphism between the Q-action φ on T and the conjugation action of Q on T_1. Set $S_1 = S\tau_1$, and notice that the permutation group criteria defined before Lemma 6.7 hold with respect to the direct decomposition $\{(S_1)^g \mid g \in W\}$ of B. Hence Lemma 6.7 implies that W is permutationally isomorphic to $S_1 \operatorname{twr}_{\bar{\psi}} P$ where $\bar{\psi}$ is the homomorphism $\mathsf{N}_P(S_1) \to \mathsf{Aut}(S_1)$ induced by conjugation. Now the assertion follows by invoking the permutational isomorphism induced by τ_1^{-1} between S_1 and S. □

In the case when the group W in Proposition 6.17 is innately transitive, we may decide whether it is in fact primitive or quasiprimitive using the results of Sections 6.3–6.4.

7
O'Nan–Scott Theory and the maximal subgroups of finite alternating and symmetric groups

Since the completion of the classification of finite simple groups, the O'Nan–Scott Theorem has proved to be the most effective tool in the investigation of finite primitive permutation groups. The theorem was later extended to finite quasiprimitive groups by Praeger (Praeger 1993), to finite innately transitive groups by Bamberg and Praeger (Bamberg 2003, Bamberg & Praeger 2004), and to certain families of infinite permutation groups (Macpherson & Praeger 1994, Smith 2015, Neumann *et al.* 2017). In this chapter we extend the O'Nan–Scott Theory to cover groups with a non-abelian minimal normal FCR-subgroup.

Results about the structure of finite primitive groups go back, according to Peter Neumann (Neumann 2004), to Camille Jordan's analysis of the normal subgroups of multiply transitive permutation groups in his 1870 Traité (Jordan 1870, Section 84). William Burnside, in §131 of his 1897 book (Burnside 1897),

> *...continued well beyond Jordan's theorem, ... [to give] an analysis of normal subgroups of primitive groups which can nowadays be viewed as the beginning of the O'Nan–Scott Theorem, and exploited it to prove that a 2-transitive group G has a unique minimal normal subgroup H and H is either non-abelian simple with trivial centraliser in G or elementary abelian.* (Neumann 2004, page 25)

The result quoted by Neumann on finite 2-transitive groups is stated in our Theorem 3.21 in Chapter 3. The modern form of the O'Nan–Scott Theorem dates back to 1978. By then it was clear that the announcement of the classification of the finite simple groups was imminent, and many mathematicians began to consider how this classification might influence work in other parts of mathematics, especially in finite group theory. With different applications in mind, Leonard Scott and Michael O'Nan independently proved versions of this theorem, pin-pointing various roles for the simple groups in describing the possible kinds of finite primitive permutation groups. Both Scott and O'Nan brought papers containing their result to the Santa Cruz Conference of Finite Groups in 1979 (Cooperstein & Mason 1980). According to Scott (Scott ∼1997), O'Nan commented, after seeing Scott's manuscript: 'Same damn theorem!'. A preliminary version of the conference proceedings included the papers by both Scott and O'Nan, while the final version only contained Scott's paper, which stated the theorem in an appendix (Scott 1980) (acknowledging that O'Nan had obtained the theorem independently). Several colleagues brought copies of O'Nan's manuscript to Australia to give to the first author, and she was disappointed that it was not published in the proceedings.

Influenced by this result of O'Nan and Scott, Peter Cameron gave a course at the University of Sydney, during a sabbatical visit,

> *...on its implications for the study of permutation groups. The theorem essentially provided a method for applying the [Classification of the Finite Simple Groups] to the study of primitive permutation groups.* (Cameron 2011)

This led to his highly cited survey article (Cameron 1981).

These early versions of the O'Nan–Scott Theorem were presented as characterisations of the maximal subgroups of the finite alternating and symmetric groups, as well as structural descriptions for finite primitive permutation groups (see the theorems stated in (Scott 1980, pages 329 and 328)). They were unfortunately not adequate for the latter purpose since the statement of the Theorem in (Scott 1980, page 328) failed to identify primitive groups of twisted wreath type (that is, those with a unique minimal normal subgroup which is non-abelian and regular). In addition, the theorems in (Scott 1980, pages 328–329) erroneously claimed that the number of simple factors of the unique minimal normal subgroup of a primitive group of simple diagonal type had to be a prime number.

These oversights were quickly recognised. They were first rectified by Aschbacher and Scott (Aschbacher & Scott 1985, Appendix) and, independently of their work, by Kovács (Kovács 1986). Gross and Kovács observed after Corollary 3.3 of (Gross & Kovács 1984) that a primitive group with a regular non-abelian minimal normal subgroup has the structure of a twisted wreath product. This observation has lent its name to the groups of this type. A short, self-contained proof of the theorem was then published by Liebeck, Praeger and Saxl, in particular proving additional properties of the twisted wreath types (Liebeck *et al.* 1988).

In this chapter we will give an overview of the O'Nan–Scott Theory of finite innately transitive, quasiprimitive and primitive groups. Moreover, we extend this theory to (possibly infinite) groups with a minimal normal FCR-subgroup. We use the approach of (Bamberg & Praeger 2004); that is, we describe the classes of innately transitive groups and study conditions under which the groups in the various classes are primitive and quasiprimitive. Our system of notation for these classes was suggested by Kovács to the first author.

We now introduce notation that we will use throughout this chapter. We assume that G is an innately transitive group acting on Ω with FCR-plinth M and set $C = \mathsf{C}_G(M)$. By Theorem 3.2, C is semiregular. In particular, if N is a transitive subgroup of C, then N is regular and $N = C$, by Lemma 2.11. By Lemma 3.4, all normal subgroups of G not containing M are contained in C.

We also fix a point ω in Ω. Since M is transitive, $G = MG_\omega$ by Lemma 2.11. Since M is a characteristically simple FCR-group, $M = T_1 \times \cdots \times T_k$ where the T_i are simple groups (see Lemma 3.14). Since every subgroup of an abelian group is normal, an abelian simple group must be cyclic and of prime order. Therefore, if M is abelian, then $T_i \cong C_p$, where C_p is a cyclic group of prime order p, and in this case M is finite. If M is non-abelian, then the T_i are all the minimal normal subgroups of M (Theorem 4.16(iv)) and hence G_ω permutes the set $\{T_1, \ldots, T_k\}$ transitively by conjugation.

When M is non-abelian, we let $\sigma_i \colon M \to T_i$ be the i-th coordinate projection. As G_ω is transitive on the T_i and $M_\omega \lhd G_\omega$, we find that, for all i, j, and for each g such that $T_i^g = T_j$, the map $x\sigma_i \mapsto x^g\sigma_j$ defines an isomorphism $M_\omega\sigma_i \to M_\omega\sigma_j$.

Lemma 7.1 *If G is quasiprimitive then either $C = 1$ or C is transitive. Conversely, if $C = 1$ then G is quasiprimitive; while if C is transitive, then both M and C are regular, possibly equal, FCR-groups and G is primitive (and hence quasiprimitive).*

Proof If G is quasiprimitive, then all non-trivial normal subgroups of G are transitive. As C is a normal subgroup of G, this implies that either $C = 1$ or C is transitive.

Conversely, if $C = 1$ and N is a non-trivial normal subgroup of G, then $M \leqslant N$ (Lemma 3.4), and hence N is transitive, and so G is quasiprimitive.

Suppose now that C is transitive. Then C is regular and so M is regular by Theorem 3.2. If M is abelian, then $M = C$ and G is primitive by Corollary 3.13. Thus we may assume that M is non-abelian, and so $M \cap C = 1$. We use Theorem 3.10 to construct a permutational embedding (ϑ, χ) where $\vartheta\colon \Omega \to M$ and $\chi\colon G \to \mathsf{Hol}\, M$ such that $M\chi = M\varrho$ and $G_\omega\chi \leqslant \mathsf{Aut}(M)$. Then $C\chi \leqslant \mathsf{C}_{\mathsf{Sym}\, M}(M\varrho) = M\lambda$ (Corollary 3.3). As C is transitive, $C\chi = M\lambda$, by Lemma 2.11. Let $\iota \in \mathsf{Sym}\, M$ be the element as in Lemma 3.8 that maps $m \mapsto m^{-1}$. Then $\iota \in \mathsf{N}_{\mathsf{Sym}\, M}(G\chi)$ and ι swaps $C\chi$ and $M\chi$. As $M\chi$ is a minimal normal subgroup of $G\chi$, so is $C\chi$, implying that C is a minimal normal FCR-subgroup of G. As M and C are distinct transitive minimal normal subgroups of G, Corollary 3.13, implies that G is primitive. □

7.1 Abelian plinth

Suppose that M is abelian. By Lemma 2.4, M is a regular abelian subgroup of G. Since M is a characteristically simple FCR-group, $M \cong (C_p)^k$ where C_p is a cyclic group of prime order p and k is finite (Lemma 3.14). In particular M is finite. It is customary to identify M with the k-dimensional vector space $(\mathbb{F}_p)^k$ over the field $\mathbb{F}_p$ and to use additive notation in M. We therefore denote the identity of M by '0'. Since M is regular, $G = M \rtimes G_0$ and G_0 can be viewed, by Theorem 3.10, as a subgroup of $\mathsf{Aut}(M) \cong \mathsf{GL}_n(p)$. As M is minimal normal, G_0 must be an irreducible subgroup of $\mathsf{GL}_n(p)$. Lemma 7.1 readily implies that G is primitive. The groups $(\mathbb{F}_p)^d \rtimes \mathsf{GL}_d(p)$ and $(\mathbb{F}_p)^d \rtimes \mathsf{SL}_d(p)$ are usually referred to as the *affine general linear group* and *affine special linear group* and are denoted as $\mathsf{AGL}_d(p)$ and $\mathsf{ASL}_d(p)$, respectively. Note that $\mathsf{AGL}_d(p)$ is the holomorph of $(\mathbb{F}_p)^d$ considered as an abelian group.

A primitive group with abelian FCR-plinth is said to have type HA, indicating that it is permutationally isomorphic to a subgroup of the *h*olomorph of the *a*belian group M.

7.2 Simple plinth

Suppose that $M = T_1$ is non-abelian simple and non-regular. Then, by Lemma 7.1, G is quasiprimitive if and only if $C = 1$ and G is primitive if and only if G_ω is maximal in G. The O'Nan–Scott type of G is denoted as follows.

As If G is quasiprimitive.
AS If G is primitive.

In these cases each primitive group of type AS is also a quasiprimitive group of type As. That is to say, the quasiprimitive type contains the corresponding primitive type as a proper subset. Similar inclusions occur for the other types introduced in Sections 7.3, 7.4, 7.5. The notation reflects the fact that, in these cases, the group G is an *a*lmost *s*imple group.

7.3 Regular plinth

Suppose now that M is non-abelian and regular, so $M \cap C = 1$. By Lemma 7.1, G is quasiprimitive if and only if $C = 1$ or C is a transitive FCR-group. In the latter case, G has precisely two minimal normal subgroups, M and C, and G is primitive, by Lemma 7.1. Note that G can be embedded into $\mathsf{Hol}\,M$ as in Theorem 3.10, and so the results of Section 3.3 apply giving further information on the structure of G.

Another useful way to view G is to notice that the permutation group criteria in Section 6.2 hold for G with respect to the regular minimal normal subgroup M and one of its simple factors, T_1, say (see Proposition 6.17). Setting $Q = \mathsf{N}_{G_\omega}(T_1)$ and $\varphi\colon Q \to \mathsf{Aut}(T_1)$ to be the homomorphism induced by conjugation, we obtain from Lemma 6.7 that G is permutationally isomorphic to $T_1 \operatorname{twr}_\varphi G_\omega$ considering $T_1 \operatorname{twr}_\varphi G_\omega$ as a permutation group in its base group action. We have by Lemma 6.7 that $\mathsf{Core}_{G_\omega}(\ker\varphi) = 1$. Proposition 6.10 and Corollary 6.13 imply that primitivity of G can be analysed in terms of the following conditions:

(a) Q does not leave invariant a proper, non-trivial subgroup of T_1;
(a') $\mathsf{Inn}(T_1) \leqslant Q\varphi$;
(b) φ cannot be extended to a strictly larger subgroup of G_ω.

More precisely, we obtain the following theorem. Theorem 7.2(iii) follows from Lemma 7.1.

Theorem 7.2 (i) *The group G is primitive if and only if (a) and (b) hold.*
(ii) *If T is finite, then G is primitive if and only if (a') and (b) hold.*
(iii) *If C is transitive, then G is primitive.*

In the case when G is quasiprimitive, we say that the O'Nan–Scott type of G is

As$_{\text{reg}}$ when M is simple, $C = 1$ and G is quasiprimitive;
AS$_{\text{reg}}$ when M is simple, $C = 1$ and G is primitive;
HS when M is simple and C is transitive;
HC when M is not simple and C is transitive;
Tw when M is not simple, $C = 1$ and G is quasiprimitive;
TW when M is not simple, $C = 1$ and G is primitive.

In the first two cases, G is an *a*lmost *s*imple group. Note that the case AS$_{\text{reg}}$ does occur (see for instance Example 3.24), but such a group is always infinite by Corollary 3.23. The names HS and HC refer to the fact that G, in these cases, is permutationally isomorphic to a subgroup of the *h*olomorph of a *s*imple or a *c*ompound group, respectively (see the Holomorph Embedding Theorem 3.10). The last type is called Tw or TW as groups of this type are permutationally isomorphic to a *t*wisted *w*reath product in the base group action. Recognition of the structure of these groups as twisted wreath products was noted in the discussion after Corollary 3.3 in (Gross & Kovács 1984).

7.4 Diagonal plinth

If M is non-abelian, $M = T_1 \times \cdots \times T_k$ with $k \geqslant 2$ and $M_\omega \sigma_i = T_i$ for all i, then we say that M is a *diagonal plinth.* Then M_ω is a subdirect subgroup of M. Hence, by Theorem 4.16(iii), M_ω is a direct product of pairwise disjoint full strips $X_1, \ldots, X_\ell$. Combining Lemmas 4.13 and 4.14, we obtain that $\mathsf{N}_M(M_\omega) = M_\omega$ and so Theorem 3.2 gives $C = 1$, and hence M is the unique minimal normal subgroup of G. Therefore the following result follows from Lemma 7.1.

Theorem 7.3 *An innately transitive group G with diagonal plinth M is quasiprimitive and $\mathsf{C}_G(M) = 1$.*

If M_ω is not simple, that is, $\ell \geqslant 2$, then, by Theorem 4.24, G preserves a non-trivial normal cartesian decomposition $\mathcal{E} = \{\Gamma_1, \ldots, \Gamma_\ell\}$ (see Section 4.7). Hence G can be embedded into the wreath product $W = G_0 \wr H$ where G_0 is the first component G^{Γ_1} and H is the permutation group induced by G on $\mathcal{E}$ (Theorem 5.14).

In the various cases of this section, the type of G is

Sd if M_ω is simple and G is quasiprimitive;

SD if M_ω is simple and G is primitive;
CD if M_ω is not simple and G is quasiprimitive;
CD if M_ω is not simple and G is primitive.

We say that G is of *s*imple *d*iagonal or *c*ompound *d*iagonal type, according to whether M_ω is simple or not.

Theorem 7.4 *Suppose that G is innately transitive with diagonal plinth M (and hence G is quasiprimitive of type* SD *or* CD*).*

(i) *Assume that M_ω is simple. Then G is primitive if and only if the conjugation action of G on the set $\{T_1, \ldots, T_k\}$ is primitive.*
(ii) *Suppose that M_ω is not simple. Then the component G_0 is quasiprimitive with a diagonal plinth M_0 such that a point stabiliser in M_0 is a simple group. Further, if G is primitive then G_0 is primitive.*

Proof By Theorem 7.3, G is quasiprimitive with $C = 1$.

(i) Suppose that G is primitive on Ω and let $\Delta_1, \ldots, \Delta_\ell$ be a G-invariant block system on the set $\mathcal{T} = \{T_1, \ldots, T_k\}$. For $i \in \{1, \ldots, \ell\}$, let σ_{Δ_i} denote the coordinate projection $\sigma_{\Delta_i} : M \to \prod_{T_j \in \Delta_i} T_j$ and set $X_i = M_\omega \sigma_{\Delta_i}$. Let $H = X_1 \times \cdots \times X_\ell$. Then H is a subgroup of M containing M_ω. Further, since the Δ_i are permuted by G_ω and M_ω is normalised by G_ω, the corresponding projections X_i are also permuted by G_ω. Hence H is normalised by G_ω, and so HG_ω is a subgroup of G containing G_ω. As G is primitive, G_ω is maximal in G, and so either $HG_\omega = G_\omega$ or $HG_\omega = G$. In the first case $H \leqslant M \cap G_\omega = M_\omega$ and hence $H = M_\omega$; as M_ω is simple, this implies that $\ell = 1$. In the latter case, H is a transitive subgroup of M containing a point stabiliser M_ω, which implies that $H = M$, and so $\ell = k$ and the blocks Δ_i are singletons. Thus in both cases the block system $\{\Delta_i\}$ is trivial, and so G must be primitive on $\mathcal{T}$.

Conversely, suppose now that G is primitive on $\mathcal{T}$ and let H be a subgroup of G properly containing G_ω. Then $H \cap M$ contains M_ω, and so $H \cap M$ is a subdirect subgroup of M, and hence, by Theorem 4.16(iii), is a direct product of disjoint full strips $X_1, \ldots, X_\ell$. Now G_ω normalises H (since $G_\omega \leqslant H$) and, since $M \trianglelefteq G$, G_ω normalises $H \cap M$. Thus the strips X_i are permuted by G_ω and so also their supports $\operatorname{Supp} X_i = \Delta_i$ are permuted by G_ω. As M acts trivially on $\mathcal{T}$ and $G = MG_\omega$, the Δ_i form a G-invariant partition of $\mathcal{T}$. Since G is primitive on $\mathcal{T}$, either $\Delta_1 = \mathcal{T}$ or each Δ_i is a singleton. In the former case $H \cap M = X_1$, and since M_ω is a full strip contained in $H \cap M$, it follows that $H \cap M = M_\omega$. Now, by Dedekind's Modular Law,

$$H = G \cap H = MG_\omega \cap H = (M \cap H)G_\omega = M_\omega G_\omega = G_\omega.$$

Thus, in this case, $H = G_\omega$. In the latter case $M \leqslant H$. Since $G_\omega \leqslant H$, this implies that $G = H$. Thus G is primitive.

(ii) Suppose that M_ω is not simple, and so (by Theorem 4.16(iii)), M_ω is a direct product of pairwise disjoint full strips $X_1, \ldots, X_\ell$ with $\ell \geqslant 2$. Suppose that $\{\Gamma_1, \ldots, \Gamma_\ell\}$ is the corresponding normal cartesian decomposition of Ω by Theorem 4.24. Let G_0 be the component G^{Γ_1}. Then Theorem 5.14 implies that G is permutationally isomorphic to a subgroup of $W = G_0 \wr \mathsf{S}_\ell$ acting in product action on $(\Gamma_1)^\ell$. Theorem 5.14 further implies that the subgroup $M_0 = \prod_{T_i \in \mathrm{Supp}\, X_1} T_i$ is a minimal normal subgroup of G_0 and the full strip X_1 is a point stabiliser of M_0. Hence M_0 is a diagonal plinth in G_0, and Theorem 7.3 implies that G_0 is quasiprimitive on Γ_1. If G is primitive, then W is primitive which implies, by Theorem 5.18, that G_0 is primitive. □

The (quasi)primitive groups in Theorem 7.4(i) are of type SD or SD and those in Theorem 7.4(ii) are of type CD or CD. We will give a full characterisation of quasiprimitive groups of types CD and CD using the blow-up construction in Theorem 11.13. In the following example we construct the largest primitive group of type SD that contains a given FCR-group T^k as its socle.

Example 7.5 Let A be a group with a minimal normal subgroup T that is non-abelian and simple. Also suppose that $\mathsf{C}_A(T) = 1$ (and hence T is the unique minimal normal subgroup of A) and that A induces by conjugation the full automorphism group of T. Thus, if convenient, A and T can naturally be identified with $\mathsf{Aut}(T)$ and with $\mathsf{Inn}(T)$, respectively. Let $k \geqslant 2$ and consider the following subgroup of A^k:

$$Z = \{(a_1, \ldots, a_k) \mid a_i \in A \text{ and } a_i a_j^{-1} \in T \text{ for all } i,\ j \in \underline{k}\};$$

that is, Z consists of all the k-tuples $(a_1, \ldots, a_k)$ such that the cosets $Ta_1, \ldots, Ta_k$ are all equal. Note that Z can also be considered as a subgroup of the base group of the wreath product $W = A \wr \mathsf{S}_k$. Then Z is normalised by the top group S_k of W, and we may define $Y(T, k) = Z \rtimes \mathsf{S}_k$, considering $Y(T, k)$ as a subgroup of W. Our notation reflects that the group Y depends on the simple group T and on the integer k;

however, in the rest of the discussion we denote $Y(T,k)$ simply by Y. Set $M = T^k$ and consider M as a subgroup of Y. Since $T \trianglelefteq A$, we have $M \trianglelefteq Y$. Since the conjugation action of the top group S_k induces the full symmetric group on the simple factors of M, M is a minimal normal subgroup of Y. Further, as, by Lemma 5.2(i), $\mathsf{C}_W(M) = 1$, M is the unique minimal normal subgroup of Y.

Let us define the subgroup $H \leqslant A^k$ as

$$H = \{(a, \ldots, a) \mid a \in A\}.$$

Then H is normalised by S_k and $H \rtimes \mathsf{S}_k$ can be viewed as a subgroup of Y. Since the unique minimal normal subgroup M of Y is not contained in $H \rtimes \mathsf{S}_k$, the subgroup $H \rtimes \mathsf{S}_k$ is core-free in Y, and we may consider Y as a permutation group on the right coset space $\Sigma = [Y \colon H \rtimes \mathsf{S}_k]$. Setting σ to be the trivial coset $H \rtimes \mathsf{S}_k$, we have $Y_\sigma = H \rtimes \mathsf{S}_k$, and

$$M_\sigma = (H \rtimes \mathsf{S}_k) \cap M = \{(t, \ldots, t) \mid t \in T\}.$$

Therefore M is a diagonal plinth of Y and M_σ is isomorphic to the simple group T. The group Y induces the full symmetric group in its conjugation action on the simple components of M, and hence Theorem 7.4(i) implies that Y is a primitive permutation group of SD type.

Claim. $Y = \mathsf{N}_{\mathsf{Sym}\,\Sigma}(M)$.

Proof of the claim. Set $N = \mathsf{N}_{\mathsf{Sym}\,\Sigma}(M)$ and set $D = M_\sigma = \{(t, \ldots, t) \mid t \in T\}$. As above, let σ be the trivial coset of Σ. By Theorem 3.2(i), $\mathsf{C}_{\mathsf{Sym}\,\Sigma}(M) \cong \mathsf{N}_M(M_\sigma)/M_\sigma = \mathsf{N}_M(D)/D$. By Corollary 4.15, $\mathsf{N}_M(D) = D$, and hence $\mathsf{C}_{\mathsf{Sym}\,\Sigma}(M) = 1$. Therefore N can be viewed as a subgroup of $\mathsf{Aut}(M) = \mathsf{Aut}(T^k)$ and we note that $\mathsf{Aut}(M)$ is isomorphic to $W = A \wr \mathsf{S}_k$. Then, $N_\sigma < N$, $D = M_\sigma \trianglelefteq N_\sigma$, and, by the transitivity of M, $N_\sigma M = N$. Since $\mathsf{S}_k \leqslant N_\sigma$, we obtain that $N_\sigma = (N_\sigma \cap A^k)\mathsf{S}_k$. Now $D \trianglelefteq N_\sigma \cap A^k$, and so $N_\sigma \cap A^k \leqslant \mathsf{N}_{A^k}(D)$. We claim that $\mathsf{N}_{A^k}(D) = H$. The inclusion $H \leqslant \mathsf{N}_{A^k}(D)$ is clear. If $(a_1, \ldots, a_k) \in \mathsf{N}_{A^k}(D)$, then

$$(t, \ldots, t)^{(a_1, \ldots, a_k)} = (t^{a_1}, \ldots, t^{a_k}) \in D$$

holds for all $t \in T$; that is, $t^{a_i} = t^{a_j}$ for all $t \in T$ and for all $i, j \in \underline{k}$. Since $\mathsf{C}_A(T) = 1$, this implies that $a_i = a_j$ for all $i, j \in \underline{k}$, and hence $(a_1, \ldots, a_k) \in H$. This shows that $N_\sigma \cap A^k = H$ and

$$N_\sigma = (N_\sigma \cap A^k)\mathsf{S}_k \leqslant Y.$$

Thus, $N = N_\sigma M \leqslant Y$, and so $N = Y$, as claimed. □

Finally, in this section we show that every quasiprimitive group of type SD can be permutationally embedded into one of the groups $Y(T,k)$ constructed in Example 7.5.

Proposition 7.6 *Assume that G is a quasiprimitive permutation group of* SD *type acting on Ω with socle $M = T^k$ where T is a non-abelian simple group. Then $k \geqslant 2$ and there is a permutational embedding (ϑ, χ) in which $\vartheta\colon \Omega \to \Sigma$ and $\chi\colon G \to Y(T,k)$ where $Y(T,k)$ and Σ are the group and the set, respectively, constructed in Example 7.5.*

Proof Let $\omega \in \Omega$. Since G is quasiprimitive of type SD, we have that M_ω is a full strip in M; that is, for $i = 2, \ldots, k$, there exists an automorphism $\alpha_i \in \mathsf{Aut}(T)$ such that

$$M_\omega = \{(t, t\alpha_2, \ldots, t\alpha_k) \mid t \in T\}.$$

Let D denote the straight full strip in M:

$$D = \{(t, \ldots, t) \mid t \in T\}.$$

Let us construct the set Σ and the group $Y = Y(T,k)$ following Example 7.5. Since D is a point stabiliser of M under its action on Σ, we identify Σ with the right coset space $[M\colon D]$. Let us define a map $\vartheta\colon \Omega \to \Sigma$ as follows. Let $\alpha \in \Omega$. Then there exists $m \in M$ such that $\omega m = \alpha$. Write $m = (t_1, \ldots, t_k)$ where $t_i \in T$, and set $\alpha\vartheta = D(t_1, t_2\alpha_2^{-1}, \ldots, t_k\alpha_k^{-1})$.

We show that $\vartheta\colon \Omega \to \Sigma$ is a well defined bijection. First we verify that the image $\alpha\vartheta$ is independent of the choice of the group element $m \in M$. Suppose that $m_1 = (t_1, \ldots, t_k)$ and $m_2 = (s_1, \ldots, s_k)$ are elements of M such that $\omega m_1 = \omega m_2 = \alpha$. Then

$$m_1 m_2^{-1} = (t_1 s_1^{-1}, t_2 s_2^{-1}, \ldots, t_k s_k^{-1}) \in M_\omega$$

and so $t_i s_i^{-1} = (t_1 s_1)\alpha_i$ holds for all $i \geqslant 2$. Therefore

$$\begin{aligned}(t_1, t_2\alpha_2^{-1}, \ldots, t_k\alpha_k^{-1})&(s_1, s_2^{-1}\alpha_2^{-1}, \ldots, s_k^{-1}\alpha_k^{-1}) \\ &= (t_1 s_1^{-1}, (t_2 s_2^{-1})\alpha_2^{-1}, \ldots, (t_k s_k^{-1})\alpha_k^{-1}) \\ &= (t_1 s_1^{-1}, t_1 s_1^{-1}, \ldots, t_1 s_1^{-1}) \in D.\end{aligned}$$

The last line implies that

$$D(t_1, t_2\alpha_2^{-1}, \ldots, t_k\alpha_k^{-1}) = D(s_1, s_2\alpha_2^{-1}, \ldots, s_k\alpha_k^{-1}),$$

which shows that ϑ is well defined, as claimed.

Let us now show that ϑ is a bijection. Suppose that $\alpha\vartheta = \beta\vartheta$ for some $\alpha,\ \beta \in \Omega$. Let $(t_1, \dots, t_k)$, $(s_1, \dots, s_k) \in M$ such that $\omega(t_1, \dots, t_k) = \alpha$ and $\omega(s_1, \dots, s_k) = \beta$. Then the condition $\alpha\vartheta = \beta\vartheta$ implies that

$$D(t_1, t_2\alpha_2^{-1}, \dots, t_k\alpha_k^{-1}) = D(s_1, s_2\alpha_2^{-1}, \dots, s_k\alpha_k^{-1}).$$

Note that the automorphism $(\mathrm{id}, \alpha_2, \dots, \alpha_k)$ of M maps D to M_ω and so, applying this automorphism to the last equation, we obtain

$$M_\omega(t_1, t_2, \dots, t_k) = M_\omega(s_1, s_2, \dots, s_k).$$

Hence $(t_1, t_2, \dots, t_k)(s_1, s_2, \dots, s_k)^{-1} \in M_\omega$, and so

$$\alpha = \omega(t_1, t_2, \dots, t_k) = \omega(s_1, s_2, \dots, s_k) = \beta.$$

Thus ϑ is injective. To prove that ϑ is surjective, let $(t_1, t_2, \dots, t_k) \in M$ and let $\alpha = \omega(t_1, t_2\alpha_2, \dots, t_k\alpha_k)$. Then $\alpha\vartheta = D(t_1, t_2, \dots, t_k)$. Since the t_i were chosen arbitrarily, we obtain that the image of ϑ contains all cosets of $[M\colon D]$ which implies that ϑ is surjective.

Let us define $\chi\colon G \to \mathsf{Sym}\,\Sigma$ by the rule $g\chi = \vartheta^{-1}g\vartheta$ for all $g \in G$. Then $g\chi \in \mathsf{Sym}\,\Sigma$ and (ϑ, χ) is a permutational embedding of G into $\mathsf{Sym}\,\Sigma$. We are required to show that $G\chi \leqslant Y$. This however follows from the fact that $G\chi \leqslant \mathsf{N}_{\mathsf{Sym}\,\Sigma}(M\chi) = Y$ by the claim proved at the end of Example 7.5. □

7.5 The remaining case

In this case, $M = T_1 \times \dots \times T_k$ with M non-abelian, $k \geqslant 2$ and $1 < M_\omega\sigma_i < T_i$ for all i.

Theorem 7.7 *The following hold.*

(i) *G is quasiprimitive if and only if $C = 1$.*
(ii) *If G is primitive then $M_\omega = M_\omega\sigma_1 \times \dots \times M_\omega\sigma_k$. Consequently in this case G preserves a non-trivial normal cartesian decomposition of Ω. If Γ is a partition in this cartesian decomposition, then the component G^Γ is primitive with type* AS.

Proof (i) Since M is not regular, C is intransitive. Hence, by Lemma 7.1, G is quasiprimitive if and only if $C = 1$.

(ii) Suppose that G is primitive and set $H = M_\omega\sigma_1 \times \dots \times M_\omega\sigma_k$. Then H is a subgroup of M containing M_ω. Further, G_ω normalises M_ω and so it permutes the projections $M_\omega\sigma_i$. Hence H is normalised by G_ω and so HG_ω is a subgroup of G containing G_ω. As $H \neq M$, the fact that H is a subgroup between M_ω and M implies that H is intransitive, and hence $HG_\omega \neq G$. Since G is primitive, we must have $HG_\omega = G_\omega$; that is, $H \leqslant G_\omega \cap M = M_\omega$. Thus $M_\omega = H$ as required.

By Theorem 4.24, G preserves a normal cartesian decomposition of Ω corresponding to the direct decomposition $\{T_1, \ldots, T_k\}$ of M. Let Γ be the partition of this cartesian decomposition that corresponds to T_1 and let G_0 be the component G^Γ. The argument in the proof of Theorem 7.4(ii) shows that G_0 is primitive. Further, T_1 is a minimal normal subgroup of G_0 with point stabiliser $M_\omega\sigma_1$. Hence the type of G_0 is AS. □

For a group G that satisfies the conditions of this section, the type of G is defined as follows.

PA if G is quasiprimitive;
PA if G is primitive.

We say that in these cases G is of *p*roduct *a*ction type.

In the case when G is not primitive, the subgroup HG_ω constructed in the proof of Theorem 7.7 is a proper subgroup of G and it may properly contain G_ω. In this case G is often studied by considering the G action on the G-invariant partition $\overline{\Omega}$ that corresponds to the overgroup HG_ω of G_ω (Lemma 2.14). The argument in the proof of Theorem 7.7(ii) shows that G preserves a normal cartesian decomposition of $\overline{\Omega}$.

We will give a better characterisation of quasiprimitive groups of type PA and a full characterisation of primitive groups of type PA using the blow-up construction in Theorem 11.13.

7.6 The O'Nan–Scott Theorem

Next in this chapter we state the O'Nan–Scott Theorem for (possibly infinite) quasiprimitive permutation groups with a minimal normal FCR-subgroup M.

Theorem 7.8 (O'Nan–Scott Theorem) *Let G be a quasiprimitive group on Ω with minimal normal* FCR-*subgroup $M = T_1 \times \cdots \times T_k$ where $T_i \cong T$ is a simple group and $k \geqslant 1$. Then one of the lines of Table 7.1 holds.*

	Conditions on M	Type	Reference
(1)	abelian	HA	Section 7.1
(2)	$M = T_1$, $\mathsf{C}_G(M) = 1$	AS, AS, AS$_{\rm reg}$ or AS$_{\rm reg}$	Sections 7.2, 7.3
(3)	M is non-abelian, $M \cong \mathsf{C}_G(M)$ and $k = 1$	HS	Section 7.3
(4)	M is non-abelian, $M \cong \mathsf{C}_G(M)$ and $k \geqslant 2$	HC	Section 7.3
(5)	$M_\omega = 1$, $\mathsf{C}_G(M) = 1$ and $k \geqslant 2$	TW or TW	Section 7.3
(6)	M is a diagonal plinth with M_ω simple	SD or SD	Section 7.4
(7)	M is a diagonal plinth with M_ω non-simple	CD or CD	Section 7.4
(8)	$k \geqslant 2$, $\mathsf{C}_G(M) = 1$ $1 < M\sigma_i < T$	PA or PA	Section 7.5

Table 7.1. *The O'Nan–Scott types of quasiprimitive and primitive groups with an* FCR*-plinth*

Proof If M is abelian, then line (1) holds, by the discussion in Section 7.1. Hence we may suppose that M is non-abelian and set $C = \mathsf{C}_G(M)$. Suppose first that M is regular. Then, by Lemma 7.1, either $C = 1$ or C is transitive; in the latter case $C \cong M$. If $C = 1$ and M is simple, then line (2) holds (see Sections 7.2 and 7.3), while if $C = 1$ and M is not simple, then line (5) holds (Section 7.3). Further, if M is non-abelian and regular and $C \cong M$, then either line (3) or line (4) is valid, depending on whether M is simple (Section 7.3). Suppose now that M is not regular; then $\mathsf{C}_G(M) = 1$ holds by Lemma 7.1. If M is simple then line (2) is valid (Section 7.2). If M is not simple and M_ω is a subdirect subgroup of M, then either line (6) or line (7) holds (Section 7.4). If M is not simple and M_ω is not a subdirect subgroup of M, then line (8) holds (Section 7.5). Since we have considered all possible cases, one of the lines of Table 7.1 must indeed be valid. □

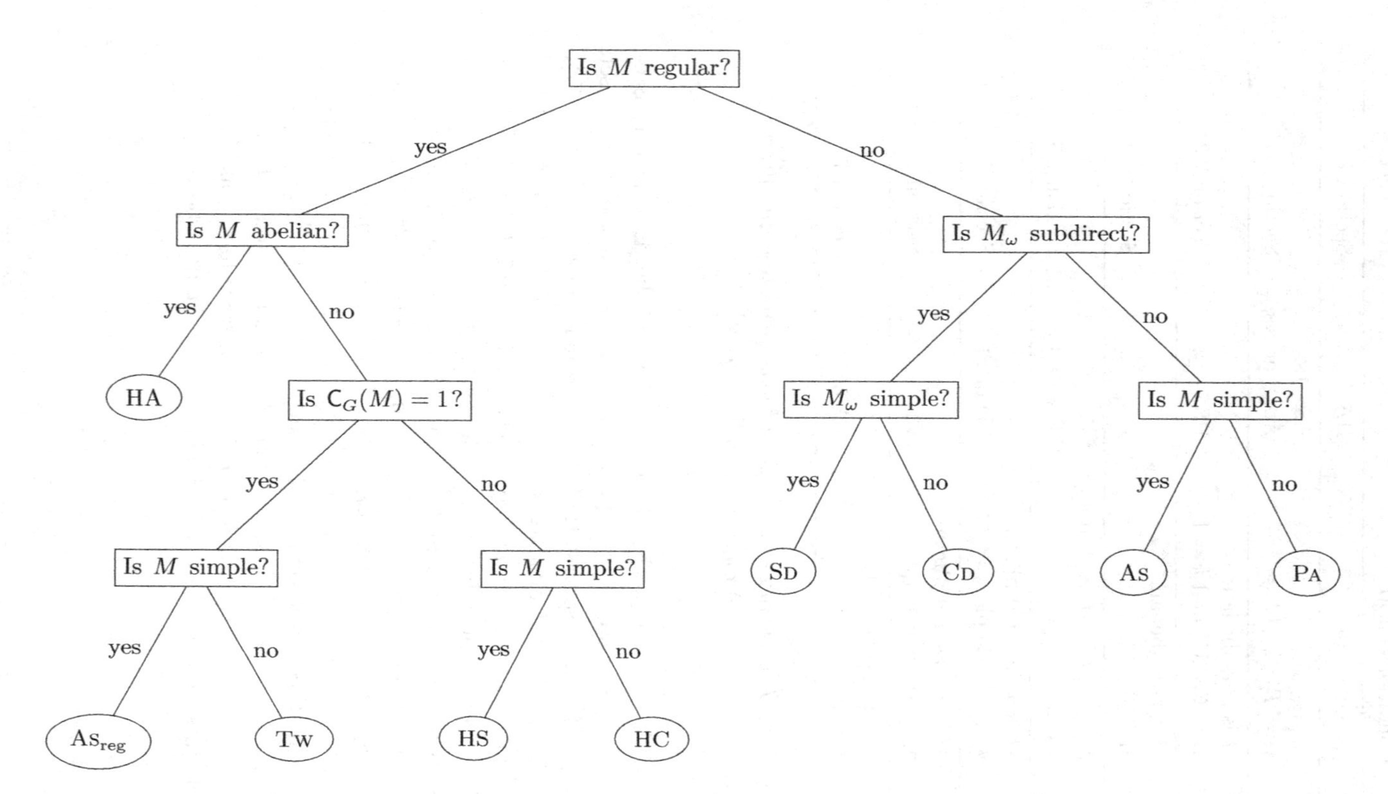

Fig. 7.1. A recognition diagram for the O'Nan–Scott type of a quasiprimitive permutation group G with FCR-socle M

The O'Nan–Scott type of a quasiprimitive permutation group G with a minimal normal subgroup M can be recognised by considering the permutational isomorphism type of M and the structure of $\mathsf{C}_G(M)$ as shown on Figure 7.1.

We obtain the following application for quasiprimitive permutation groups with finite stabilisers. These groups were already studied in Theorem 3.18. Theorem 7.9 for primitive groups is due to (Smith 2015).

Theorem 7.9 *Let G be a quasiprimitive permutation group on an infinite set Ω such that a point stabiliser G_α is finite. Then G has a unique minimal normal* FCR*-subgroup and the O'Nan–Scott type of G is one of* As, As$_{\mathrm{reg}}$, Pa, Tw. *Furthermore, each of the primitive O'Nan–Scott classes* AS, AS$_{\mathrm{reg}}$, PA, *and* TW *contains infinite permutation groups with finite stabilisers.*

Proof By Theorem 3.18, G has a unique minimal normal FCR-subgroup $M = T_1 \times \cdots \times T_k$ where the T_i are simple groups. Hence Theorem 7.8 applies and M must be as in one of the rows of Table 7.1. As noted in Section 7.1, groups of type HA are finite, and so the type of G is not HA. If the O'Nan–Scott type of G is HS, HC, Sd, Cd, then a point stabiliser M_α must be isomorphic to T^r with some $r \geqslant 1$. Since Ω is infinite and k is finite, T must be infinite, and in these cases point stabilisers are infinite. Thus these cases do not occur. Hence the statement concerning the O'Nan–Scott type of G is valid.

Let us now verify the statement on the existence of primitive permutation groups in the O'Nan–Scott classes AS, AS$_{\mathrm{reg}}$, PA, and TW. For certain large enough primes p, there exist infinite simple groups G such that a proper, non-trivial subgroup of G is cyclic of order p (see (Robinson 1996, Exercise 14.4.1)). Considering such a group G and a proper, non-trivial subgroup $H < G$, we have that H is a maximal subgroup of G and the coset action of G on the right coset space $[G\colon H]$ is primitive. In this action, G is a primitive permutation group with finite point stabiliser H, and the O'Nan–Scott type of G is AS. Primitive groups of O'Nan–Scott type AS$_{\mathrm{reg}}$ with finite stabilisers were constructed in Example 3.24. If G is a primitive group of type AS or AS$_{\mathrm{reg}}$ such that a stabiliser G_α is finite, then, by Theorem 5.18, for $k \geqslant 2$, the wreath product $G \wr \mathsf{S}_k$ in product action will be a primitive group of type PA or TW (respectively) also with finite stabilisers. □

7.6.1 A comparison of the O'Nan–Scott type subdivisions

One can find many different versions of the O'Nan–Scott Theorem for finite primitive groups in the literature, the first appearing in (Scott 1980, p. 328). To assist the reader, we provide in Table 7.2 a glossary of the different names and type-subdivisions given in the statements of the O'Nan–Scott Theorem in (Baddeley & Praeger 2003, pp 303–4), (Dixon & Mortimer 1996, Section 4.5–4.7), (Liebeck *et al.* 1988), (Cameron 1999, Sections 4.4–4.5), (Buekenhout 1988), and (Aschbacher & Scott 1985, Theorem 1). The columns that correspond to these subdivisions are labelled as BP, DM, LPS, C, B, AS, respectively. The last column of the table shows the section where the particular type is described in this book.

A discussion of maximal subgroups of finite symmetric and alternating groups can be found in (Smith 2018, Section 6.1). The O'Nan–Scott Theorem is stated in (Smith 2018, Theorem 6.1.3) using the formulation of (Liebeck *et al.* 1988), and a brief comparison with the version used in this chapter is provided in (Smith 2018, Remark 6.1.4). Smith's book gives a wide range of applications of the finite simple groups. A rough version of the O'Nan–Scott Theorem is also stated in Wilson's book (Wilson 2009, Chapter 2.6), which contains a wealth of information about the finite simple groups.

7.7 Maximal subgroups of finite symmetric and alternating groups

A description of the maximal subgroups of finite alternating and symmetric groups can be obtained using the O'Nan–Scott Theorem for finite primitive groups. Let $n \geqslant 3$ and let us denote by X one of the groups S_n or A_n. Set $\Omega = \underline{n}$. The maximal subgroups which are not primitive are easy to describe. On the other hand, it can be difficult to decide if a primitive subgroup of X is maximal. Let H be a proper subgroup of X. If H is intransitive, then H stabilises a subset Δ of Ω such that $|\Delta| \leqslant |\Omega|/2$. If $|\Delta| < |\Omega|/2$, then H is contained in the full stabiliser $(\mathsf{Sym}\,\Delta \times \mathsf{Sym}\,(\Omega \setminus \Delta)) \cap X$ in X of Δ. If $|\Delta| = |\Omega|/2$, then H is contained in the full stabiliser in X of the partition $\Omega = \Delta \cup (\Omega \setminus \Delta)$. By Lemma 5.4(iv), the stabiliser of this partition is permutationally isomorphic to $(\mathsf{Sym}\,\Delta \wr \mathsf{S}_2) \cap X$ acting in imprimitive action on the set $\Delta \times \{1, 2\}$. In each of these cases, H is a maximal subgroup of X; see (Liebeck *et al.* 1987, Theorem, page 366).

BP	DM	LPS	C	B	AS	Section
HA	Affine	I	Affine	Affine	(A)	7.1
AS	Almost simple	II	Almost simple	Simple	(C)(3)	7.2
SD	Diagonal	III(a)(i)	Diagonal	Diagonal	(C)(2)	7.4
HS	Diagonal	III(a)(ii)	Diagonal	Biregular	(B)	7.3
PA	Product	III(b)(i)	non-basic	Cartesian semi-simple	(C)(3)	7.5
HC	Product	III(b)(ii)	non-basic	Cartesian semi-simple	(B)	7.3
CD	Product	III(b)(ii)	non-basic	Cartesian semi-simple	(C)(2)	7.4
TW	Regular non-abelian	III(c)	Twisted wreath product (non-basic)	Cartesian semi-simple	(C)(1)	7.3

Table 7.2. *Various O'Nan–Scott type subdivisions for finite primitive permutation groups; see Section 7.6.1 for the meaning of the column headers*

If H is a transitive, but imprimitive, subgroup of X, then H stabilises a partition $\Omega = \Delta_1 \cup \cdots \cup \Delta_\ell$ of Ω such that $|\Delta_i| = |\Delta_j|$ for all i, j. Hence H is contained in the full stabiliser of this partition in X. As in the previous paragraph, this full stabiliser is permutationally isomorphic to $\mathsf{Sym}\,\Delta_1 \wr \mathsf{S}_\ell$ acting in imprimitive action on $\Delta_1 \times \underline{\ell}$ and is usually a maximal subgroup of X; see Theorem 7.12.

Let us consider the case when H is primitive.

Lemma 7.10 *Let $n \geqslant 3$, let X be S_n or A_n, and let H be a proper primitive subgroup of X.*

(i) *If H has type* HS, *then H is contained in a subgroup of the form $K \cap X$ where K is a primitive group of type* SD.

(ii) *Let H be of type* CD, HC, *or* TW. *Then H stabilises a normal cartesian decomposition $\mathcal{E}$ of Ω with $|\mathcal{E}| \geqslant 2$. Therefore H is properly contained in the full stabiliser of $\mathcal{E}$ in $\mathsf{Sym}\,\Omega$ which is permutationally isomorphic to $W = \mathsf{Sym}\,\Gamma \wr \mathsf{S}_\ell$ (with $\ell = |\mathcal{E}|$ and $\Gamma \in \mathcal{E}$) acting in product action on Γ^ℓ. In particular, H is contained in $W \cap X$ where W is a primitive group of type* PA.

Proof Set $\Omega = \underline{n}$ and let $\omega \in \Omega$.

(i) Suppose that the type of H is HS. Then H has two transitive minimal normal subgroups T_1 and T_2 and they are both regular. By Corollary 3.11, there is an involution π in $\mathsf{Sym}\,\Omega$ such that conjugation by π swaps T_1 and T_2 and centralises the stabiliser H_ω. Since $T_1 \times T_2$ is transitive, $H = (T_1 \times T_2)H_\omega$, and so π normalises H. Thus H is contained in the subgroup $K = \langle H, \pi\rangle$. The group K is primitive and it has a unique minimal normal subgroup $M = T_1 \times T_2$. If $t_1 \in T_1$, then, since T_2 is regular, there exists a unique element $t_2 \in T_2$ such that $\omega t_1 = \omega t_2$. Hence $\omega t_1 t_2^{-1} = \omega$. Thus $M_\omega \sigma_1 = T_1$ where σ_1 is the first coordinate projection $\sigma_1 \colon M \to T_1$. Therefore K is primitive with O'Nan–Scott type SD. Hence H is contained in $K \cap X$ where K is a primitive group of type SD.

(ii) Suppose now that H is primitive of one of the types HC, CD, or TW. We show first that H stabilises a normal cartesian decomposition $\mathcal{E}$ of Ω with $|\mathcal{E}| \geqslant 2$. If H has type HC or TW, then H contains a regular and non-abelian minimal normal subgroup, and so this claim follows from Corollary 4.25. If the type of H is CD, then the existence of this cartesian decomposition was observed in Section 7.4. In each of these cases, H can be embedded into the full stabiliser W of $\mathcal{E}$ in $\mathsf{Sym}\,\Omega$.

Since, by Lemma 5.10, W is isomorphic to $\mathsf{Sym}\,\Gamma \wr \mathsf{S}_\ell$ with $\ell \geqslant 2$ acting in product action on Γ^ℓ, we find that H is a proper subgroup of $W \cap X$ where W is a primitive group of type PA. □

As a consequence of the discussion in this section and of Lemma 7.10, we obtain the following descriptions of the maximal subgroups of S_n and A_n. We note that $\mathsf{S}_2 \wr \mathsf{S}_\ell$ is the wreath product of a cyclic group of order two and a full symmetric group and hence, for $\ell \geqslant 2$, it is imprimitive in its product action by Theorem 5.18. Further, $\mathsf{S}_k \wr \mathsf{S}_\ell$ with $k \in \{3, 4\}$ and $\ell \geqslant 2$ is a primitive permutation group in its product action with an abelian minimal normal subgroup, and hence it is contained in $\mathsf{AGL}_d(p)$ for suitable d and p.

Theorem 7.11 *Let $n \geqslant 3$, let X be S_n or A_n, and let H be a maximal subgroup of X different from A_n. Then there is a subgroup K of S_n such that $H = K \cap X$ and one of the following is valid for K.*

(i) *K is the full stabiliser in $\mathsf{Sym}\,\Omega$ of a non-empty subset Δ of Ω with $|\Delta| < |\Omega|/2$ and $K = \mathsf{Sym}\,\Delta \times \mathsf{Sym}(\Omega \setminus \Delta)$.*

(ii) *K is the full stabiliser in $\mathsf{Sym}\,\Omega$ of a partition $\Omega = \Delta_1 \cup \cdots \cup \Delta_\ell$ with $|\Delta_i| = |\Delta_j| \geqslant 2$ for all i, j, and $\ell \geqslant 2$. In this case K is permutationally isomorphic to $\mathsf{Sym}\,\Delta_1 \wr \mathsf{S}_\ell$ in imprimitive action.*

(iii) *$n = p^k$ with some prime p and $k \geqslant 1$ and K is permutationally isomorphic to $\mathsf{AGL}_k(p)$.*

(iv) *$n = k^\ell$ with some $k \geqslant 5$ and $\ell \geqslant 2$, and K is permutationally isomorphic to $\mathsf{S}_k \wr \mathsf{S}_\ell$ acting in its product action on $\mathsf{Func}(\underline{\ell}, \underline{k})$.*

(v) *$n = |T|^k$ for some non-abelian finite simple group T and for some $k \geqslant 1$ and K is permutationally isomorphic to the group $Y(T, k+1)$ constructed in Example 7.5.*

(vi) *K is a primitive permutation group of type AS and $K = \mathsf{N}_{\mathsf{S}_n}(T)$ where T is a finite simple group acting transitively on $\underline{n}$.*

Deciding which of the subgroups listed in Theorem 7.11 is maximal is, in general, a difficult problem whose solution relies on the finite simple group classification. We state the following result that explicitly lists the exceptions for all cases when H is in cases (i)–(v) of Theorem 7.11. The proof of this result along with the exceptions for case (vi) can be found in (Liebeck *et al.* 1987, Theorem, page 366). Some of the exceptions for case (vi) are presented in Theorem 8.18.

Theorem 7.12 *Let $n \geqslant 3$, let X be S_n or A_n, let K be a subgroup of*

	Conditions on M	Comment	Section
(1)	abelian	always primitive of type HA	7.1
(2)	M is non-abelian and simple	not quasiprimitive $\Leftrightarrow$ $C \neq 1$ and C is intransitive	7.2, 7.3
(3)	M is non-abelian non-simple and regular	not quasiprimitive $\Leftrightarrow$ $C \neq 1$ and C is intransitive	7.3
(4)	M is a diagonal plinth	always quasiprimitive	7.4
(5)	M is non-abelian, non-simple, non-regular and M_ω is not subdirect	not quasiprimitive $\Leftrightarrow$ $C \neq 1$	7.5

Table 7.3. *O'Nan–Scott theory for innately transitive groups G with an* FCR*-plinth M, and $C = \mathsf{C}_G(M)$*

S_n *as in one of the cases (i)–(v) in Theorem 7.11, and let $H = X \cap K$. Then the following is valid.*

(i) *If K is as in Theorem 7.11(i), then H is a maximal subgroup of X.*

(ii) *If K is as in Theorem 7.11(ii), then H is a maximal subgroup of X unless $X = \mathsf{A}_8$ and $H = (\mathsf{S}_2 \wr \mathsf{S}_4) \cap \mathsf{A}_8$. In the exceptional case, H is properly contained in the affine group $\mathsf{AGL}_3(2)$.*

(iii) *If K is as in Theorem 7.11(iii), then H is a maximal subgroup of X unless $X = \mathsf{A}_n$, $k = 1$, and $p \in \{7, 11, 17, 23\}$. In the exceptional cases, H is contained in a primitive permutation group of type AS.*

(iv) *If K is as in Theorem 7.11(iv), then H is a maximal subgroup of X unless $X = \mathsf{A}_n$, $k \equiv 2 \pmod 4$ and $\ell = 2$. In the exceptional cases, H is contained in an imprimitive maximal subgroup of A_n as in part (ii).*

(v) *If K is as in Theorem 7.11(v), then H is maximal in X.*

Remark 7.13 Similarly to primitive and quasiprimitive groups, the class of innately transitive groups with an FCR-socle can also be subdivided into O'Nan–Scott classes and such a subdivision is displayed in Table 7.3. In this table, G is an innately transitive permutation group with an FCR-plinth M and $C = \mathsf{C}_G(M)$. A finer subdivision for finite innately transitive groups was presented by (Bamberg 2003, Bamberg & Praeger 2004).

Part II

Innately transitive groups: Factorisations and cartesian decompositions

8
Cartesian factorisations

The theme of this book is to investigate cartesian decompositions that are invariant under some permutation group. In this chapter we establish (Theorem 8.2), for a not necessarily finite innately transitive group G on Ω with plinth M, a one-to-one correspondence between the set of G-invariant cartesian decompositions of Ω and certain sets of subgroups of G, namely certain factorisations of M. This correspondence is analogous to the one-to-one correspondence given in Lemma 2.14 between the set of G-invariant partitions of Ω for a transitive group G and the lattice of subgroups between G and a point stabiliser.

8.1 Cartesian factorisations of transitive permutation groups

We first recall the definition of cartesian factorisations introduced in Section 1.3.

Definition 8.1 Let M be a transitive permutation group on a set Ω and let $\omega \in \Omega$. We say that a non-empty set $\{K_1, \ldots, K_\ell\}$ of proper subgroups of M is a *cartesian factorisation* of M with respect to ω if each K_i is a proper subgroup of M,

$$\bigcap_{i=1}^{\ell} K_i = M_\omega \quad \text{and} \tag{8.1}$$

$$K_i \left(\bigcap_{j \neq i} K_j \right) = M \quad \text{for all} \quad i \in \{1, \ldots, \ell\}. \tag{8.2}$$

A cartesian factorisation is said to be *homogeneous* if its elements have the same cardinality. A cartesian factorisation is *non-trivial* if it has at least two subgroups.

For a permutation group $G \leqslant \mathsf{Sym}\,\Omega$, let $\mathsf{CD}(G)$ denote the set of G-invariant cartesian decompositions of Ω and let $\mathsf{CD}_{\mathrm{tr}}(G)$ denote the set of G-invariant cartesian decompositions of Ω on which G acts transitively. Cartesian factorisations provide a way of identifying the set $\mathsf{CD}(G)$ from information internal to G.

Theorem 8.2 *Let G be an innately transitive permutation group on Ω with plinth M. Then for a fixed $\omega \in \Omega$ there is a one-to-one correspondence between the set $\mathsf{CD}(G)$ and the set of G_ω-invariant cartesian factorisations of M with respect to ω.*

Theorem 8.2 is an easy consequence of our main more detailed Theorem 8.11

8.2 Cartesian decompositions and minimal normal subgroups

The first theorem of this section shows that the class of innately transitive permutation groups is particularly suitable for studying invariant cartesian decompositions of permutation groups. We recall that a cartesian decomposition of a set Ω is a finite set of partitions that satisfies the intersection condition of Definition 1.1. In the next theorem, $G_{(\Gamma)}$ denotes the pointwise stabiliser in G of a partition Γ.

Theorem 8.3 *If G is an innately transitive group on a (not necessarily finite) set Ω with plinth M and $\mathcal{E} \in \mathsf{CD}(G)$, then each element of $\mathcal{E}$ is an M-invariant partition of Ω.*

Proof We let $\Gamma \in \mathcal{E}$ and show that each element of the G-orbit ΓG is stabilised by M. Suppose that $\Delta = \{\Gamma_1, \ldots, \Gamma_m\}$ is the G-orbit in $\mathcal{E}$ containing $\Gamma \in \mathcal{E}$ (the orbit is finite, as $\mathcal{E}$ is finite). If $m = 1$, then Γ is G-invariant, and hence M-invariant. Thus we may assume that $m \geqslant 2$. Set

$$\Sigma = \{\gamma_1 \cap \cdots \cap \gamma_m \mid \gamma_1 \in \Gamma_1, \ldots, \gamma_m \in \Gamma_m\}$$

and

$$\bar{\Gamma}_i = \{\{\sigma \in \Sigma \mid \sigma \subseteq \gamma\} \mid \gamma \in \Gamma_i\} \quad \text{for} \quad i = 1, \ldots, m.$$

By Lemma 2.16, Σ is a G-invariant partition of Ω. If $i \in \{1, \ldots, m\}$ and $\gamma \in \Gamma_i$, then the map $\gamma \mapsto \bar{\gamma}$ where $\bar{\gamma} = \{\sigma \in \Sigma \mid \sigma \subseteq \gamma\}$ is a well defined bijection $\Gamma_i \to \bar{\Gamma}_i$ and induces a bijection $\Gamma_i \mapsto \bar{\Gamma}_i$ from the G-orbit Δ to the set $\bar{\mathcal{E}} = \{\bar{\Gamma}_1, \ldots, \bar{\Gamma}_m\}$. If $\gamma \in \Gamma_i$, then $\bar{\gamma}$ is a partition of γ. Further, if $i \neq j$ and $\gamma_i \in \Gamma_i$, $\gamma_j \in \Gamma_j$, then $\gamma_i \neq \gamma_j$ and hence the partitions $\bar{\gamma}_i$ and $\bar{\gamma}_j$ are distinct, and so the sets $\bar{\Gamma}_i$ and $\bar{\Gamma}_j$ are disjoint. Thus the induced map $\Delta \to \bar{\mathcal{E}}$ is injective and hence is a bijection.

We claim that $\bar{\mathcal{E}}$ is a G-invariant cartesian decomposition of Σ. Clearly, each $\bar{\Gamma}_i$ is a non-trivial partition of Σ. Suppose that $\bar{\gamma} \in \bar{\Gamma}_i$ and $g \in G$. Then $\gamma g \in \Gamma_j$ for some j which only depends on i and not on γ. Hence

$$\bar{\gamma}g = \{\sigma g \mid \sigma \in \Sigma \text{ and } \sigma \subseteq \gamma\} = \{\sigma \in \Sigma \mid \sigma \subseteq \gamma g\}.$$

Thus $\bar{\gamma}g \in \bar{\Gamma}_j$ and similarly $\bar{\gamma}g^{-1} \in \bar{\Gamma}_i$ for each $\gamma \in \bar{\Gamma}_j$. Hence $\bar{\Gamma}_i g = \bar{\Gamma}_j$, and so $\bar{\mathcal{E}}$ is G-invariant and the G-actions on $\{\Gamma_1, \ldots, \Gamma_m\}$ and on $\{\bar{\Gamma}_1, \ldots, \bar{\Gamma}_m\}$ are equivalent. Further, if $\bar{\gamma}_i \in \bar{\Gamma}_i$ for all $i \in \{1, \ldots, m\}$ then

$$\bigcap_{i=1}^{m} \bar{\gamma}_i = \{\sigma \in \Sigma \mid \sigma \subseteq \gamma_1 \cap \cdots \cap \gamma_m\} = \{\gamma_1 \cap \cdots \cap \gamma_m\}.$$

Hence $|\bar{\gamma}_1 \cap \cdots \cap \bar{\gamma}_m| = 1$ and $\bar{\mathcal{E}}$ is a cartesian decomposition of Σ. Moreover, as above, $\bar{\gamma}g = \bar{\gamma}$ implies

$$\{\sigma \in \Sigma \mid \sigma \subseteq \gamma\} = \{\sigma \in \Sigma \mid \sigma \subseteq \gamma g\},$$

which implies $\gamma g = \gamma$. Hence $G_{(\bar{\Gamma}_i)} \leqslant G_{(\Gamma_i)}$ for all $i \leqslant m$.

Set $K = G_{(\{\bar{\Gamma}_1, \ldots, \bar{\Gamma}_m\})}$. Since K is a normal subgroup of G and M is a minimal normal subgroup of G, either $M \leqslant K$ or $M \cap K = 1$; that is, either each $\bar{\Gamma}_i$ is M-invariant or M is faithful on $\bar{\mathcal{E}}$. Suppose that $M \cap K = 1$, and so M acts on the set $\{\bar{\Gamma}_1, \ldots, \bar{\Gamma}_m\}$ faithfully. Therefore M is isomorphic to a subgroup of S_m, and, in particular, M is finite. Since M is transitive on Ω, this implies that Ω is finite and hence each $\bar{\Gamma}_i$ is finite. Note that $|\Sigma| = |\bar{\Gamma}_1|^m$, and let p be a prime dividing $|\bar{\Gamma}_1|$. Then p^m divides $|\Sigma|$. Since M is transitive on Σ, $p^m \mid |M|$. However, M is isomorphic to a subgroup of S_m, and so p^m divides $m!$, which is impossible, by Lemma 5.16. Hence $M \leqslant K$, that is, each $\bar{\Gamma}_i$ is stabilised by M, and so is each Γ_i. Hence $M \leqslant G_{(\mathcal{E})}$, that is, each $\bar{\Gamma}_i$ is M-invariant. This implies, as explained above, that each Γ_i is an M-invariant partition of Ω. □

8.3 Cartesian decompositions and affine groups

In this section we study permutation groups G with a transitive abelian minimal normal FCR-subgroup M. As M is regular, Corollary 3.13

implies that G is primitive. As explained in Section 7.1, M is a finite group and $M \cong (C_p)^k$ where C_p is a cyclic group of prime order p. Further, the O'Nan–Scott type of G is HA. As is common, in this section we identify M with the vector space $(\mathbb{F}_p)^k$ and we use additive notation in M. By Theorem 3.10, G can be embedded into $\mathsf{Hol}\, M$ such that M is embedded into its image $M\varrho$ under the right regular representation. The stabiliser G_ω of a chosen point ω is embedded into $\mathsf{Aut}(M)$. Hence we assume that $G \leqslant \mathsf{Hol}\, M$ containing $M\varrho$. Further, $\mathsf{Aut}(M)$ is naturally isomorphic to $\mathsf{GL}_k(p)$ and so the stabiliser G_0 of $0 \in M$ can be considered as a subgroup of $\mathsf{GL}_k(p)$. As M is minimal normal in G, G_0 is irreducible as a linear group.

If H is a linear group acting on a vector space V in such a way that H preserves a direct sum decomposition $V = V_1 \oplus \cdots \oplus V_\ell$ with $\ell \geqslant 2$, then we say that H is an *imprimitive linear group* (we will be careful not to confuse this concept with the concept of *imprimitive permutation groups*). If no such direct sum decomposition of V exists, we say that H is a *primitive linear group*.

The main theorem of this section connects the existence of cartesian decompositions in affine groups, such as G above, to the imprimitivity of G_0 as a linear group. Hence the task of finding G-invariant cartesian decompositions in this case is reduced to a problem in representation theory.

Recall that we denote for a permutation group G acting on Ω, the set of G-invariant transitive cartesian decompositions of Ω by $\mathsf{CD}_{\mathrm{tr}}(G)$.

Theorem 8.4 *Suppose that M is a finite elementary abelian p-group and that G is a primitive subgroup of $\mathsf{Hol}\, M$ as above.*

(i) *Given a G-invariant cartesian decomposition $\mathcal{E} = \{\Gamma_1, \ldots, \Gamma_\ell\}$ of M, let $\gamma_i \in \Gamma_i$ be the block containing 0 and let $M_i = \bigcap_{j \neq i} \gamma_j$ for all $i \in \{1, \ldots, \ell\}$. Then $M = M_1 \oplus \cdots \oplus M_\ell$ is a G_0-invariant direct sum decomposition of M, G_0 is transitive on the M_i, and $\mathcal{E} \in \mathsf{CD}_{\mathrm{tr}}(G)$.*

(ii) *If $M = M_1 \oplus \cdots \oplus M_\ell$ is a G_0-invariant direct sum decomposition, define $K_i = \bigoplus_{j \neq i} M_j$ and $\Gamma_i = \{K_i + x \mid x \in M\}$ for all $i \in \{1, \ldots, \ell\}$ and set $\mathcal{E} = \{\Gamma_1, \ldots, \Gamma_\ell\}$. Then $\mathcal{E} \in \mathsf{CD}_{\mathrm{tr}}(G)$.*

(iii) *We have $\mathsf{CD}(G) = \mathsf{CD}_{\mathrm{tr}}(G)$ and there is a one-to-one correspondence between $\mathsf{CD}(G)$ and the set of G_0-invariant direct sum decompositions of M. In particular $\mathsf{CD}_{\mathrm{tr}}(G)$ contains a non-trivial* *decomposition if and only if G_0 is an irreducible, imprimitive linear group.*

Proof (i) Let us assume that $\mathcal{E} = \{\Gamma_1, \ldots, \Gamma_\ell\}$ is a non-trivial G-invariant cartesian decomposition of M. Then by Theorem 8.3, each Γ_i is an $M\varrho$-invariant partition of M. If γ_i denotes the element of Γ_i, for all i, that contains 0, then γ_i is a subspace of M and Γ_i is the partition of M into the set of cosets of γ_i. Set, for $i \in \{1, \ldots, \ell\}$, $M_i = \bigcap_{j \neq i} \gamma_j$. Then the M_i are also subspaces of M.

We will prove that G_0 is an imprimitive linear group witnessed by the decomposition $M = M_1 \oplus \cdots \oplus M_\ell$. First we claim that $|M_i| = |\Gamma_i|$. Since $\mathcal{E} \in \mathsf{CD}(G)$, every element of M_i is an element of precisely one γ_i' in Γ_i. Further, if elements v_1, $v_2 \in M_i$ are elements of the same γ_i' then the intersection $M_i \cap \gamma_i' = \gamma_1 \cap \cdots \cap \gamma_{i-1} \cap \gamma_i' \cap \gamma_{i+1} \cap \cdots \cap \gamma_\ell$ contains v_1 and v_2, which, as $\mathcal{E}$ is a cartesian decomposition, gives that $v_1 = v_2$. Moreover, if $\gamma_i' \in \Gamma_i$ then the unique element of $M_i \cap \gamma_i'$ is an element of M_i and of γ_i'. This establishes a bijection from M_i onto Γ_i. Therefore $\prod_i |M_i| = \prod_i |\Gamma_i| = |M|$. Thus, in order to show that $M = M_1 \oplus \cdots \oplus M_\ell$, it suffices to prove that $M = M_1 + \cdots + M_\ell$. Suppose that $v \in M$ and let $v \in \gamma_1' \cap \cdots \cap \gamma_\ell'$ with $\gamma_i' \in \Gamma_i$. Note in particular that $v \in \gamma_1'$. For $i \in \{1, \ldots, \ell\}$, let v_i be the element in $\gamma_1 \cap \cdots \cap \gamma_{i-1} \cap \gamma_i' \cap \gamma_{i+1} \cap \cdots \cap \gamma_\ell$, and note that $v_1 \in \gamma_1'$, while $v_2, \ldots, v_\ell \in \gamma_1$. Thus $v_i \in M_i$ for all i. Set $v' = v_1 + \cdots + v_\ell$, so that $v' \in \gamma_1 + v_1 = \gamma_1'$. Then v and v' are in the same γ_1-coset, namely in γ_1'. Similarly, they are in the same γ_i-coset, namely in γ_i' for each i. Thus $v' \in \gamma_1' \cap \cdots \cap \gamma_\ell'$ and so $v' = v$. Therefore $M = M_1 + \cdots + M_\ell$, as claimed.

Since G_0 fixes the point 0, we find that G_0 permutes the set $\{\gamma_1, \ldots, \gamma_\ell\}$, and so also G_0 permutes $\{M_1, \ldots, M_\ell\}$. Thus the decomposition $M = M_1 \oplus \cdots \oplus M_\ell$ is G_0-invariant, and so G_0 is an imprimitive subgroup of $\mathsf{GL}_k(p)$ as claimed.

The argument in the previous paragraph also implies that the G_0-actions on $\mathcal{E}$ and on the set $\{M_i\}$ are equivalent. Since G_0 is irreducible, it is transitive on the set $\{M_i\}$, and hence it is transitive on $\mathcal{E}$, and so $\mathcal{E} \in \mathsf{CD}_{\mathrm{tr}}(G)$.

(ii) Suppose now that G_0 is an imprimitive linear group that preserves the decomposition $M = M_1 \oplus \cdots \oplus M_\ell$ with $\ell \geqslant 2$. Every element $v \in M$ can be written uniquely as $v = v_1 + \cdots + v_\ell$ where $v_i \in M_i$. For

$i \in \{1, \ldots, \ell\}$ and $w_i \in M_i$ we let

$$\gamma_i(w_i) = \{v_1 + \cdots + v_\ell \mid v_i = w_i\}$$

and set $\Gamma_i = \{\gamma_i(w_i) \mid w_i \in M_i\}$. Then each Γ_i is a partition of M. Further, $\gamma_i(0) = K_i$ where $K_i = \bigoplus_{j \neq i} M_j$ and $\Gamma_i = \{K_i + x \mid x \in M\}$. If we choose one part $\gamma_i(w_i) \in \Gamma_i$ for each i, then $\gamma_1(w_1) \cap \cdots \cap \gamma_\ell(w_\ell) = \{w_1 + \cdots + w_\ell\}$, and hence $\mathcal{E} = \{\Gamma_1, \ldots, \Gamma_\ell\}$ is a cartesian decomposition. If $i \in \{1, \ldots, \ell\}$, $w_i \in M_i$ and $g \in G_0$, then there exists j such that $M_i g = M_j$ and so $w_i g = w_j$ for some $w_j \in M_j$. Hence

$$\begin{aligned} \gamma_i(w_i)g &= \{(v_1 + \cdots + v_\ell)g \mid v_i = w_i\} \\ &= \{v_1 + \cdots + v_\ell \mid v_j = w_j\} = \gamma_j(w_j). \end{aligned}$$

Moreover, each $w_j \in M_j$ arises for some $w_i \in M_i$, since $M_i g = M_j$. Thus $\Gamma_i g = \Gamma_j$ and so $\mathcal{E}$ is a G_0-invariant cartesian decomposition of M. Since each Γ_i is invariant under M, we find that the cartesian decomposition $\mathcal{E}$ is G-invariant. Thus $\mathcal{E} \in \mathsf{CD}(G)$, and, by part (i), $\mathcal{E} \in \mathsf{CD}_{\mathrm{tr}}(G)$.

(iii) The fact that $\mathsf{CD}(G) = \mathsf{CD}_{\mathrm{tr}}(G)$ follows from part (i). For a G_0-invariant direct sum decomposition $M = M_1 \oplus \cdots \oplus M_\ell$, we let $\{M_1, \ldots, M_\ell\}\Phi$ denote the corresponding cartesian decomposition constructed in (ii). Conversely, for a G-invariant cartesian decomposition $\mathcal{E}$, let $\mathcal{E}\Psi$ denote the direct sum decomposition constructed in part (i). If $\mathcal{E} = \{\Gamma_1, \ldots, \Gamma_\ell\} \in \mathsf{CD}(G)$, then set $\mathcal{E}' = \{\Gamma_1', \ldots, \Gamma_\ell'\} = \mathcal{E}\Psi\Phi$. We claim that $\mathcal{E} = \mathcal{E}'$. Let, for all i, γ_i and γ_i' be the parts in Γ_i and Γ_i', respectively, that contain 0. As the elements of $\mathcal{E}$ and those of $\mathcal{E}'$ are M-invariant partitions and M is transitive, it suffices to prove that $\gamma_i = \gamma_i'$ for all i. However, this follows immediately from the definition of the M_i in part (i) and from the definition of the γ_i' in part (ii). Similarly, if $\mathcal{M} = \{M_1, \ldots, M_\ell\}$ is a G_0-invariant direct sum decomposition of M, then $\mathcal{M}\Phi\Psi = \mathcal{M}$. Thus the maps Φ and Ψ are bijections, as was claimed. □

Theorem 8.4 leads to examples of cartesian factorisations in elementary abelian p-groups.

Corollary 8.5 *Suppose M is a finite elementary abelian p-group and let $G \leqslant \mathsf{Hol}\, M$ a primitive permutation group containing $M\varrho$. Suppose that $M = M_1 \oplus \cdots \oplus M_\ell$ is a G_0-invariant direct sum decomposition of M and define the subgroups $K_1, \ldots, K_\ell$ as in Theorem 8.4(ii). Then*

$\mathcal{K} = \{K_1, \ldots, K_\ell\}$ *is a non-trivial* G_0*-invariant cartesian factorisation of* M.

Proof The fact that $\mathcal{K}$ is a cartesian factorisation follows from the definition of the K_i. That fact that it is G_0-invariant follows at once from the fact that $\{M_1, \ldots, M_\ell\}$ is G_0-invariant. □

8.4 Cartesian decompositions and cartesian factorisations

We are now back to considering cartesian decompositions and cartesian factorisations for general innately transitive groups.

Lemma 8.6 *Let* M *be a transitive subgroup of* $\mathsf{Sym}\,\Omega$ *and let* $\mathcal{E} \in \mathsf{CD}(M)$ *such that* $M_{(\mathcal{E})} = M$. *Suppose that* $\mathcal{E} = \{\Gamma_1, \ldots, \Gamma_\ell\}$, *let* $\omega \in \Omega$ *be a fixed element, and for* $i = 1, \ldots, \ell$ *let* $\gamma_i \in \Gamma_i$ *be such that* $\omega \in \gamma_i$. *Set* $\mathcal{K}_\omega(\mathcal{E}) = \{K_1, \ldots, K_\ell\}$ *where* $K_i = M_{\gamma_i}$ *for* $i = 1, \ldots, \ell$. *Then* $\mathcal{K}_\omega(\mathcal{E})$ *is a cartesian factorisation of* M *with respect to* ω. *Moreover, if* $\omega m = \omega'$ *for some* $m \in M$, *then* $\mathcal{K}_{\omega'}(\mathcal{E}) = \mathcal{K}_\omega(\mathcal{E})^m$.

Proof Let us prove that $\bigcap_{i=1}^{\ell} K_i = M_\omega$. Since the Γ_i are M-invariant partitions of Ω, the stabiliser of a point of Ω stabilises the block in Γ_i that contains this point. Hence $M_\omega \leqslant K_i$ for all i, and so $M_\omega \leqslant \bigcap_i K_i$. Now suppose $x \in \bigcap_i K_i$. Then x stabilises $\gamma_1, \ldots, \gamma_\ell$. Since $\mathcal{E}$ is a cartesian decomposition, $\gamma_1 \cap \cdots \cap \gamma_\ell = \{\omega\}$, and so x stabilises ω. Thus $x \in M_\omega$, and so $\bigcap_i K_i = M_\omega$.

Now we prove that (8.2) also holds. We may suppose without loss of generality that $i = 1$. Let $x \in M$, $\delta_1 = \gamma_1 x, \ldots, \delta_\ell = \gamma_\ell x$, and $\{\xi\} = \delta_1 \cap \cdots \cap \delta_\ell$. If $\{\zeta\} = \delta_1 \cap \gamma_2 \cap \cdots \cap \gamma_\ell$ then the transitivity of M on Ω implies that there exists $z \in M$ with $\xi z = \zeta$ and so $\delta_1 z = \delta_1$, $\delta_2 z = \gamma_2, \ldots, \delta_\ell z = \gamma_\ell$, whence $\gamma_j x z = \gamma_j$ for $j = 2, \ldots, \ell$ and $\gamma_1 x z x^{-1} = \gamma_1$, that is $xz \in \bigcap_{j=2}^{\ell} K_j$ and $xzx^{-1} \in K_1$. It follows that

$$x = (xzx^{-1})^{-1}(xz) \in K_1 \left(\bigcap_{j=2}^{\ell} K_j \right),$$

and we deduce that the first factorisation of (8.2) holds. The other factorisations can be proved identically. Thus $\mathcal{K}_\omega(\mathcal{E})$ is a cartesian factorisation of M with respect to ω.

If $m \in M$ and $\omega' = \omega m$ then $\{\omega'\} = \gamma_1 m \cap \cdots \cap \gamma_\ell m$ and $M_{\gamma_i m} = (M_{\gamma_i})^m$, which proves that $\mathcal{K}_{\omega'}(\mathcal{E}) = \mathcal{K}_\omega(\mathcal{E})^m$. □

Note that the cartesian factorisation in Corollary 8.5 coincides with $\mathcal{K}_0(\mathcal{E})$ where $\mathcal{E}$ is the cartesian decomposition that corresponds to the direct sum decomposition $M = M_1 \oplus \cdots \oplus M_\ell$.

If $M \leqslant \mathsf{Sym}\,\Omega$ and $\mathcal{E} \in \mathsf{CD}(M)$ such that $M_{(\mathcal{E})} = M$, then, for a fixed $\omega \in \Omega$, we define the cartesian factorisation $\mathcal{K}_\omega(\mathcal{E})$ with respect to ω as in Lemma 8.6. The last result of this section establishes one direction of the one-to-one correspondence in Theorem 8.2.

Lemma 8.7 *Let G be an innately transitive group with plinth M acting on Ω, and let $\omega \in \Omega$. If $\mathcal{E} \in \mathsf{CD}(G)$, then $M_{(\mathcal{E})} = M$. Assume that $\mathcal{K}_\omega(\mathcal{E})$ is the corresponding cartesian factorisation of M with respect to ω. Then $\mathcal{K}_\omega(\mathcal{E})$ is invariant under conjugation by G_ω, and the G_ω-actions on $\mathcal{K}_\omega(\mathcal{E})$ and on $\mathcal{E}$ are equivalent.*

Proof It follows from Theorem 8.3 that $M_{(\mathcal{E})} = M$, and so we can use Lemma 8.6 to construct $\mathcal{K}_\omega(\mathcal{E})$ for ω. Suppose that $\mathcal{E} = \{\Gamma_1, \ldots, \Gamma_\ell\}$, and let $\mathcal{K}_\omega(\mathcal{E}) = \{K_1, \ldots, K_\ell\}$ such that $K_i = M_{\gamma_i}$ where γ_i is the unique element of Γ_i containing ω. If Γ_i, $\Gamma_j \in \mathcal{E}$ and $g \in G_\omega$ such that $\Gamma_i g = \Gamma_j$ then $\omega g = \omega$, and so $\gamma_i g = \gamma_j$. Hence

$$K_i^g = (M_{\gamma_i})^g = M_{\gamma_i g} = M_{\gamma_j} = K_j,$$

and so $\mathcal{K}_\omega(\mathcal{E})$ is invariant under conjugation by G_ω. This argument also shows that the G_ω-actions on $\mathcal{E}$ and on $\mathcal{K}_\omega(\mathcal{E})$ are equivalent. □

8.5 Cartesian factorisations of abstract groups

In this section we turn our attention temporarily to factorisations of abstract groups in that we drop condition (8.1) in Definition 8.1.

Definition 8.8 A finite non-empty subset $\mathcal{K} = \{K_1, \ldots, K_\ell\}$ of proper subgroups of a group M is an *abstract cartesian factorisation* of M if (8.2) holds.

The concepts of non-triviality and homogeneity of abstract cartesian factorisations are defined as in Definition 8.1. If $\mathcal{K}$ is an abstract cartesian factorisation of M, then M has a transitive coset action on $\Omega = [M\colon M_0]$ where $M_0 = \bigcap_{K \in \mathcal{K}} K$. Thus, for $L = \mathsf{Core}_M M_0$, the set $\{K/L \mid K \in \mathcal{K}\}$ is a cartesian factorisation of the transitive permutation group M/L with respect to $M_0 \in \Omega$.

In this section we summarise the most important properties of abstract cartesian factorisations.

The following lemma is useful when working with cartesian factorisations. If $\{K_1, \ldots, K_\ell\}$ is an abstract cartesian factorisation of a group M and $I \subseteq \{1, \ldots, \ell\}$ then let K_I denote the subgroup $K_I = \bigcap_{i \in I} K_i$. We use the convention that the intersection of an empty collection of subsets of M is M. We note that each non-empty subset of an abstract cartesian factorisation of M is also an abstract cartesian factorisation of M.

Lemma 8.9 *Let $\{K_1, \ldots, K_\ell\}$ be a (possibly trivial) abstract cartesian factorisation of M, and let I, J be subsets of $\{1, \ldots, \ell\}$.*

(i) *If $x_i \in M$ for all $i \in I$, then $\bigcap_{i \in I} K_i x_i$ is a coset of K_I.*
(ii) *If M is finite, then $|M \colon K_I| = \prod_{i \in I} |M \colon K_i|$.*
(iii) *$K_I K_J = K_{I \cap J}$.*

Proof If $I = \emptyset$ then $\bigcap_{i \in I} K_i x_i = M$ by the convention stated before the lemma, and M is a coset of $K_I = M$. Thus we may assume that $I \neq \emptyset$. If an intersection of (right) cosets is non-empty then it is a (right) coset modulo the intersection of the relevant subgroups. The statement of (i) above, and the simple proof below, make use of this fact. We prove the lemma by induction on ℓ. Notice that there is nothing to prove if $\ell = 1$. Our inductive hypothesis is that $\ell > 1$ and the lemma holds for all abstract cartesian factorisations of M which consist of fewer than ℓ subgroups. Thus (i) and (ii) only have to be proved for the case $I = \{1, \ldots, \ell\}$. Set $L = \bigcap_{i > 1} K_i$, and note that $\{K_1, L\}$ is also an abstract cartesian factorisation for M (that is, $K_1 L = M$).

We also know from the inductive hypothesis that $\bigcap_{i > 1} K_i x_i$ is a coset modulo L, so for (i) it is sufficient to show that $K_1 x_1 \cap L y$ is never empty. In order to show this we choose $z \in L$ such that $K_1 z = K_1 x_1 y^{-1}$; this is possible, as $K_1 L = M$. Then $K_1 z y = K_1 x_1$, and so $zy \in K_1 x_1$, and also $zy \in Ly$. Hence $zy \in K_1 x_1 \cap Ly$, and consequently $K_1 x_1 \cap Ly$ is non-empty. This proves part (i) by induction.

To prove part (ii), assume that M is finite. If $\ell = 1$ there is nothing to prove, so assume that $\ell \geqslant 2$ and the result holds for abstract cartesian factorisations with less than ℓ subgroups. Suppose that $i \in \{2, \ldots, \ell\}$. As $M = K_{j_1} K_{j_2}$ for each $j_1 \neq j_2$, we have $|M \colon K_i| = |K_1 \colon K_1 \cap K_i|$. In the next displayed equation, we use Dedekind's Modular Law (see page 147) to show the second equality, while the third equality follows from the definition of an abstract cartesian factorisation:

$$\begin{aligned}(K_1 \cap K_i)\left(\bigcap_{j\neq 1,i}(K_j \cap K_i)\right) &= (K_1 \cap K_i)\left(\bigcap_{j\neq 1} K_j\right)\\ &= \left[K_1\left(\bigcap_{j\neq 1} K_j\right)\right] \cap K_i = K_i.\end{aligned}$$

Hence $\mathcal{K}_1 = \{K_1 \cap K_2, \ldots, K_1 \cap K_\ell\}$ is a possibly trivial abstract cartesian factorisation of K_1. Therefore the inductive hypothesis implies that

$$\left|K_1 \colon \bigcap_{i=2}^{\ell}(K_1 \cap K_i)\right| = \prod_{i=2}^{\ell} |K_1 \colon K_1 \cap K_i|.$$

Multiplying both sides by $|M \colon K_1|$, and noting that $|K_1 \colon K_1 \cap K_i| = |M \colon K_i|$, since $M = K_1K_i$, we have

$$\left|M \colon \bigcap_{i=1}^{\ell} K_i\right| = |M \colon K_1| \prod_{i=2}^{\ell} |M \colon K_i| = \prod_{i=1}^{\ell} |M \colon K_i|,$$

as required. Part (ii) is then proved by induction.

(iii) If one of I or J is empty then, by our convention, the required equality is simply $M = M$, while if $I = J$ then it is $K_I = K_I$. This proves part (iii) for $\ell = 1$, so we may assume that $\ell \geqslant 2$, and also that I, J are distinct and both are non-empty. The equations for the remaining subsets when $\ell = 2$ follow from the definition of an abstract cartesian decomposition. Thus we may assume that $\ell \geqslant 3$, and inductively we assume that the result holds for smaller cartesian factorisations. As in the previous paragraph, for $i = 1, \ldots, \ell$ the set $\mathcal{K}_i = \{K_j \cap K_i \mid j \neq i\}$ is a cartesian factorisation for the group K_i. If there exists $i \in I \cap J$, then we have

$$K_I = \bigcap_{u \in I\setminus\{i\}} (K_i \cap K_u) \quad \text{and} \quad K_J = \bigcap_{u \in J\setminus\{i\}} (K_i \cap K_u).$$

The inductive hypothesis applied to the cartesian factorisation $\mathcal{K}_i$ with the subsets $I \setminus \{i\}$ and $J \setminus \{i\}$ of $\{1, \ldots, \ell\} \setminus \{i\}$ gives that

$$\begin{aligned}K_IK_J &= \left(\bigcap_{u\in I\setminus\{i\}} (K_i \cap K_u)\right)\left(\bigcap_{u\in J\setminus\{i\}} (K_i \cap K_u)\right)\\ &= \bigcap_{u\in(I\cap J)\setminus\{i\}} (K_i \cap K_u) = \bigcap_{u\in I\cap J} K_u = K_{I\cap J}.\end{aligned}$$

Thus it remains to verify that $K_I K_J = M$ under the assumption that $I \cap J = \emptyset$. Recall that $I \neq \emptyset$ and choose $i \in I$. Then we have already proved that $K_i = K_I K_{J\cup\{i\}}$. Since $K_{J\cup\{i\}} \leqslant K_J$ and $K_J \geqslant \bigcap_{j\neq i} K_j$, we see that

$$K_I K_J = K_I K_{J\cup\{i\}} K_J = K_i K_J \geqslant K_i \left(\bigcap_{j\neq i} K_j\right) = M,$$

whence $K_I K_J = M$ as required. Assertion (iii) is now proved by induction. □

Note that, in Lemma 8.9(i), if we choose x to be any element of $\bigcap_{i\in I} K_i x_i$, then $K_i x_i = K_i x$ holds, for all $i \in I$.

8.6 Cartesian factorisations and cartesian decompositions

Recall from Lemma 2.14, that in a transitive group $M \leqslant \mathsf{Sym}\,\Omega$, a subgroup K satisfying $M_\omega \leqslant K \leqslant M$ for some $\omega \in \Omega$ determines an M-invariant partition of Ω comprising the M-translates of the K-orbit ωK.

Lemma 8.10 *Let G be an innately transitive subgroup of $\mathsf{Sym}\,\Omega$ with plinth M, and let ω be a fixed element of Ω. Suppose that $\mathcal{K} = \{K_1, \ldots, K_\ell\}$ is a G_ω-invariant cartesian factorisation of M with respect to ω, and let $\Gamma_1, \ldots, \Gamma_\ell$ be the M-invariant partitions of Ω determined by $K_1, \ldots, K_\ell$, respectively. Then $\mathcal{E} = \{\Gamma_1, \ldots, \Gamma_\ell\}$ is a G-invariant cartesian decomposition of Ω, such that $\mathcal{K}_\omega(\mathcal{E}) = \mathcal{K}$.*

Proof As $M_\omega \leqslant K_i \leqslant M$, each Γ_i is an M-invariant partition of Ω. Also, by the intersection condition of Definition 8.1, $\bigcap_i K_i = M_\omega$. For $i = 1, \ldots, \ell$ let γ_i be the unique element of Γ_i containing ω. In order to prove that $\mathcal{E}$ is a cartesian decomposition, we only have to show that

$$\left|\bigcap_{i=1}^{\ell} \delta_i\right| = 1 \quad \text{whenever} \quad \delta_1 \in \Gamma_1, \ldots, \delta_\ell \in \Gamma_\ell.$$

To do this, choose $\delta_1 \in \Gamma_1, \ldots, \delta_\ell \in \Gamma_\ell$. Now $\delta_i = \gamma_i x_i$ for some $x_i \in M$, and by Lemma 8.9(i), there exists some $x \in M$ such that $x \in \bigcap_i K_i x_i$, and so $K_i x_i = K_i x$ holds for all $i \in \underline{\ell}$. Then

$$\begin{aligned}\delta_i &= \gamma_i x_i = \{\omega k \mid k \in K_i\} x_i = \{\omega k' \mid k' \in K_i x_i\} \\ &= \{\omega k' \mid k' \in K_i x\} = \{\omega k \mid k \in K_i\} x = \gamma_i x.\end{aligned}$$

Thus

$$\bigcap_{i=1}^{\ell} \delta_i = \bigcap_{i=1}^{\ell} (\gamma_i x) = \left(\bigcap_{i=1}^{\ell} \gamma_i\right) x,$$

and therefore it is sufficient to prove that $\bigcap_{i=1}^{\ell} \gamma_i = \{\omega\}$. Suppose that $\omega' \in \gamma_1 \cap \ldots \cap \gamma_\ell$ for some $\omega' \in \Omega$. Then there is an element $x \in M$ such that $\omega x = \omega'$. Then $\gamma_i \cap \gamma_i x$ contains $\omega' = \omega x$, and so $\gamma_i x = \gamma_i$ for all i. Hence $x \in K_i$ for all $i = 1, \ldots, \ell$. Since $\bigcap_{i=1}^{\ell} K_i = M_\omega$, it follows that $x \in M_\omega$, and so $\omega' = \omega x = \omega$. Thus $\bigcap_{i=1}^{\ell} \gamma_i = \{\omega\}$, and $\mathcal{E}$ is a cartesian decomposition.

Since each Γ_i is an M-invariant partition of Ω, $\mathcal{E}$ is invariant under M. Since $\mathcal{K} = \{K_1, \ldots, K_\ell\}$ is G_ω-invariant, $\mathcal{E}$ is also G_ω-invariant, and so $\mathcal{E}$ is MG_ω-invariant. Since M is transitive, $MG_\omega = G$. Therefore $\mathcal{E}$ is G-invariant. Note that

$$\mathcal{K} = \{M_{\gamma_1}, \ldots, M_{\gamma_\ell}\} \quad \text{and} \quad K_\omega(\mathcal{E}) = \{M_{\gamma_1}, \ldots, M_{\gamma_\ell}\}.$$

Thus $\mathcal{K} = \mathcal{K}_\omega(\mathcal{E})$, as required. □

Theorem 8.11 *Let G be an innately transitive subgroup of $\mathsf{Sym}\,\Omega$ with plinth M. For $\omega \in \Omega$ the map $\mathcal{E} \mapsto \mathcal{K}_\omega(\mathcal{E})$ is a bijection between the set $\mathsf{CD}(G)$ and the set of G_ω-invariant cartesian factorisations of M with respect to ω.*

Proof Let $\mathcal{C}$ denote the set of G_ω-invariant cartesian factorisations of M with respect to ω. In Lemma 8.6, we explicitly constructed a map $\Psi\colon \mathsf{CD}(G) \to \mathcal{C}$ for which $\Psi(\mathcal{E}) = \mathcal{K}_\omega(\mathcal{E})$. We claim that Ψ is a bijection. Let $\mathcal{K} \in \mathcal{C}$, let $\Gamma_1, \ldots, \Gamma_\ell$ be the M-invariant partitions determined by the elements $K_1, \ldots, K_\ell$ of $\mathcal{K}$, and let $\mathcal{E} = \{\Gamma_1, \ldots, \Gamma_\ell\}$. We proved in Lemma 8.10 that $\mathcal{E}$ is a G-invariant cartesian decomposition of Ω such that $\mathcal{K}_\omega(\mathcal{E}) = \mathcal{K}$. Hence Ψ is surjective.

Suppose now that $\mathcal{E}_1$, $\mathcal{E}_2 \in \mathsf{CD}(G)$ are such that $\Psi(\mathcal{E}_1) = \Psi(\mathcal{E}_2)$ and let $\mathcal{K}$ denote this common cartesian factorisation. Let $\mathcal{E}$ be the set of M-invariant partitions determined by the elements of $\mathcal{K}$. Then, by the definition of $\Psi(\mathcal{E}_i)$ in Lemma 8.6, $\mathcal{E}_1 = \mathcal{E}$ and $\mathcal{E}_2 = \mathcal{E}$. Thus Ψ is injective, and so Ψ is a bijection. □

Theorem 8.2 is an immediate consequence of the previous result.

8.7 Factorisations of groups

The connection between cartesian decompositions and cartesian factorisations in Theorem 8.2 shows that understanding factorisations of the plinths of innately transitive groups is crucial in our investigation of cartesian decompositions. As these plinths are characteristically simple groups (Lemma 3.14), we need detailed information about related factorisations of characteristically simple groups.

In this section we prove some elementary results about factorisations of groups that will be used in the descriptions of the cartesian factorisations. In order to give detailed descriptions of cartesian factorisations, such as in Theorem 8.17, we need results that go well beyond the ones presented in this section and require detailed knowledge of the subgroup structure of non-abelian finite simple groups. These results are summarised in Lemmas 8.14 and 8.15 and in the Appendix.

If X and Y are subgroups in a group G, then we denote $X^Y = \{X^y \mid y \in Y\}$. Thus X^Y is the set of Y-conjugates of X.

Lemma 8.12 *Let G be a group and let A, B be subgroups of G.*

(i) *Assertions (a)–(c) below are equivalent:*
   (a) *$G = AB$;*
   (b) *B is a transitive subgroup of G in its action on the right cosets of A;*
   (c) *A is a transitive subgroup of G in its action on the right cosets of B.*

(ii) *If G is finite, then $G = AB$ if and only if $|G| = |A||B|/|A \cap B|$.*

(iii) *If $G = AB$ with A, $B \neq G$, then A and B are not conjugate in G.*

(iv) *If $G = AB$ and there is an automorphism of G that swaps the conjugacy classes A^G and B^G, then there is an automorphism of G swapping the subgroups A and B.*

Proof (i) Suppose first that $G = AB$ and let us prove that assertion (b) holds. It suffices to show that for each $g \in G$ there is some $b \in B$ such that the right cosets Ag and Ab coincide. Indeed, as $G = AB$, there are $a \in A$ and $b \in B$ such that $g = ab$. Therefore $Ag = Aab = Ab$, as claimed. Interchanging the roles of A and B and noting that $G = AB$ implies $G = BA$, we obtain that assertion (a) implies assertion (c). Suppose now that B is transitive by conjugation on the set of right cosets $[G\colon A]$. If $g \in G$, then, as B is transitive on $[G\colon A]$, there is some $b \in B$ such that $Ag = Ab$, and so $gb^{-1} = a$ with some $a \in A$. Therefore $g = ab$, and it follows that $G = AB$ and so (b) implies (a). The same argument shows that (c) implies (a).

(ii) It is well known that $|AB| = |A||B|/|A \cap B|$ (see (Robinson 1996, 1.3.11)). Therefore $G = AB$ if and only if $|G| = |A||B|/|A \cap B|$, and hence the statement follows.

(iii) Suppose that $G = AB$ and $A,\ B \neq G$. Then, by part (a), B is transitive by right multiplication on the right cosets of A. The subgroup A is a point stabiliser for this G-action. If B were conjugate to A, then B would be another point stabiliser (Lemma 2.1), and so it could not be transitive. Thus B is not conjugate to A.

(iv) If $G = AB$, then $A^G = A^{AB} = A^B$, and so the conjugation action of B is transitive on the conjugacy class A^G. As $G = AB$ implies $G = BA$, we also obtain, in this case, that the conjugation action of A is transitive on B^G. Let σ be an automorphism of G swapping A^G and B^G. Then $A\sigma = B^g$ and $B\sigma = A^h$ for some $g,\ h \in G$. Since $B^G = B^A$, there exists $a \in A$ such that for the inner automorphism ι_a of G induced by a, $A\sigma\iota_a = B$. Similarly, since $A^G = A^B$ and $B\sigma\iota_a = A^{ha} \in A^G$, there exists $b \in B$ such that $B\sigma\iota_a\iota_b = A$. Then also $A\sigma\iota_a\iota_b = B^b = B$, and so $\sigma\iota_a\iota_b$ is a suitable automorphism. □

The following lemma simplifies the proof that certain factorisations with three subgroups are cartesian.

Lemma 8.13 *Suppose that G is a group and $A,\ B,\ C \leqslant G$ such that $G = AB = AC = BC = A(B \cap C)$. Then $G = B(A \cap C) = C(A \cap B)$. In particular, $\{A, B, C\}$ is an abstract cartesian factorisation of G.*

Proof As $G = A(B \cap C)$,

$$|G| = \frac{|A||B \cap C|}{|A \cap B \cap C|} = \frac{|A||B||C|}{|G||A \cap B \cap C|} = \frac{|B||A \cap C|}{|A \cap B \cap C|}.$$

Hence $G = B(A \cap C)$ and $G = C(A \cap B)$ is proved similarly. □

Factorisations of finite simple and almost simple groups have been widely studied. The starting point for the investigation of the factorisations of finite almost simple groups is the classification of their maximal factorisations, proved in (Liebeck *et al.* 1990). A factorisation $G = AB$ of a finite simple group G is said to be a *full factorisation* if for each prime p dividing $|G|$, we have that p divides both $|A|$ and $|B|$. Though the definition of a full factorisation would make sense for an arbitrary finite group, we only define it for finite simple groups, since we want to define this concept for finite characteristically simple groups slightly differently (namely, coordinatewise); see Section A.2. A set $\{A_1, \ldots, A_k\}$ of proper subgroups of a (possibly infinite) simple group G is said to be a *strong multiple factorisation* if $G = A_i(A_j \cap A_l)$ whenever $|\{i, j, l\}| = 3$. Hence a set of proper subgroups of G is a strong multiple factorisation if and only if all three-element subsets are abstract cartesian factorisations. We will extend this concept for characteristically simple FCR-groups in Section A.2. Although the concept of a strong multiple factorisation appears to be more general than that of cartesian factorisations, Theorem A.1(B) shows that every strong multiple factorisation of a finite simple group is a cartesian factorisation with three subgroups.

The factorisations of finite simple groups that are most relevant to our work are the full factorisations and strong multiple factorisations. These factorisations were classified in (Baddeley & Praeger 1998) and a summary is given in Theorem A.1 of the Appendix. The following two lemmas extract several facts from the works cited above concerning factorisations of finite simple groups that we will use in our study of cartesian decompositions. A weaker version of Lemma 8.14(iii) appeared in related works of Baumeister; see (Baumeister 1997, Theorem 1.1(ii)) and (Baumeister 2004) and also in a work of Pálfy and Saxl (see (Pálfy & Saxl 1990)). For the notation concerning finite simple groups, we recommend that the reader should consult (Wilson 2009). For a set $S = \{A_i \mid i \in I\}$ of subgroups of a group G, we define the *setwise normaliser* $\mathsf{N}_G(S)$ as

$$\mathsf{N}_G(S) = \{g \in G \mid A_i^g \in S \text{ for all } i \in I\}.$$

Lemma 8.14 *Let T be a non-abelian finite simple group. Then the following hold.*

(i) *A cartesian factorisation of T has at most three subgroups.*

(ii) *If $\{A, B, C\}$ is a cartesian factorisation of T with three subgroups, then the groups T, A, B and C are as in one of the rows of Table 8.1. In particular, A, B and C are pairwise non-isomorphic.*

(iii) *If A and B are proper subgroups of T such that $T = AB$ and $|A| = |B|$, then T, A, B are as in one of the rows of Table 8.2. Further, the following are valid.*

   (a) *A and B are isomorphic maximal subgroups of T;*

	T	A	B	C
1	$\mathsf{Sp}_{4a}(2)$, $a \geqslant 2$	$\mathsf{Sp}_{2a}(4) \cdot 2$	$\mathsf{O}^-_{4a}(2)$	$\mathsf{O}^+_{4a}(2)$
2	$\mathsf{P}\Omega^+_8(3)$	$\Omega_7(3)$	$\mathbb{Z}_3^6 \rtimes \mathsf{PSL}_4(3)$	$\mathsf{P}\Omega^+_8(2)$
3	$\mathsf{Sp}_6(2)$	$\mathsf{G}_2(2)$	$\mathsf{O}^-_6(2)$	$\mathsf{O}^+_6(2)$
		$\mathsf{G}_2(2)'$	$\mathsf{O}^-_6(2)$	$\mathsf{O}^+_6(2)$
		$\mathsf{G}_2(2)$	$\mathsf{O}^-_6(2)'$	$\mathsf{O}^+_6(2)$
		$\mathsf{G}_2(2)$	$\mathsf{O}^-_6(2)$	$\mathsf{O}^+_6(2)'$

Table 8.1. *Cartesian factorisations of finite simple groups with three subgroups*

(b) *there exists an automorphism* $\vartheta \in \mathsf{Aut}(T)$ *such that* $A\vartheta = B$ *and* $B\vartheta = A$*;*
(c) $A \cap B$ *is self-normalising in* T*;*
(d) *if* T *is as in row* 1, 2, *or* 4 *of Table 8.2, then*

$$\mathsf{N}_{\mathsf{Aut}(T)}(A \cap B) = \mathsf{N}_{\mathsf{Aut}(T)}(\{A, B\}) = N,$$

say, and moreover $TN = \mathsf{Aut}(T)$*;*
(e) $\mathsf{N}_T(A' \cap B') = \mathsf{N}_T(A \cap B) = A \cap B$ *and* $\mathsf{C}_T(A' \cap B') = \mathsf{C}_T(A \cap B) = 1$.

	T	A, B
1	A_6	A_5
2	M_{12}	M_{11}
3	$\mathsf{P}\Omega^+_8(q)$	$\Omega_7(q)$
4	$\mathsf{Sp}_4(q)$, $q \geqslant 4$ even	$\mathsf{Sp}_2(q^2).2$

Table 8.2. *Factorisations of finite simple groups in Lemma 8.14(iii)*

Proof If $\{A_1, \ldots, A_k\}$ is a cartesian factorisation of T, then the set $\{A_1, \ldots, A_k\}$ is a strong multiple factorisation. By Theorem A.1(B), a strong multiple factorisation of T contains at most 3 subgroups, and this implies part (i). The characterisation of strong multiple factorisation of finite simple groups with three subgroups in Theorem A.1(B) implies part (ii).

Let us prove part (iii). Suppose that $T = AB$ as in the assumptions of the lemma. Then $|T| = |A||B|/|A \cap B|$, by Lemma 8.12(ii). Thus, if p is a prime that does not divide the order $|A| = |B|$, then p does not divide $|T|$. Therefore the factorisation $T = AB$ is a full factorisation. Full factorisations of finite simple groups are characterised in Theorem A.1(A). By inspection of Table A.1, we obtain that the only possibilities that satisfy the condition $|A| = |B|$ are the ones presented in Table 8.2. In particular, A and B are maximal subgroups of T, as claimed. For the proof of parts (b)–(d), see (Baddeley *et al.* 2004b, Lemma 5.2), and for part (e) consult (Baddeley *et al.* 2008, Lemma 4.2). □

Let $T = \mathsf{P}\Omega_8^+(q)$. Then, as shown in (Kleidman 1987, pp. 181–182), $\mathsf{Aut}(T) = \Theta \rtimes \Phi$, where Φ is the group of field automorphisms of T, and Θ is a certain subgroup of $\mathsf{Aut}(T)$ containing the commutator subgroup $\mathsf{Aut}(T)'$. We also have $\mathsf{Out}(T) = \mathsf{Aut}(T)/T = \Theta/T \times \Phi T/T$, and $\Theta/T \cong \mathsf{S}_m$ where $m = 3$ for even q, and $m = 4$ for odd q. Let $\pi\colon \Theta \to \mathsf{S}_m$ denote the natural epimorphism. The following lemma derives the information about $\mathsf{P}\Omega_8^+(q)$ similar to that in Lemma 8.14(iii)(d)–(e); the proof can be found in (Baddeley *et al.* 2004b, Lemma 5.3).

Lemma 8.15 *Let $T = \mathsf{P}\Omega_8^+(q)$, let A, B be subgroups of T such that A, $B \cong \Omega_7(q)$ and $AB = T$, and set $C = A \cap B$. Then the following hold:*

(i) $\Phi \leqslant \mathsf{N}_{\mathsf{Aut}(T)}(A) \cap \mathsf{N}_{\mathsf{Aut}(T)}(B)$;
(ii) $(\mathsf{N}_{\mathsf{Aut}(T)}(A) \cap \Theta)\pi \cong (\mathsf{N}_{\mathsf{Aut}(T)}(B) \cap \Theta)\pi \cong \mathbb{Z}_2 \times \mathbb{Z}_2$;
(iii) $(\mathsf{N}_{\mathsf{Aut}(T)}(C) \cap \Theta)\pi \cong \mathsf{S}_3$ *and* $|\mathsf{N}_{\mathsf{Aut}(T)}(C) : \mathsf{N}_{\mathsf{Aut}(T)}(\{A, B\})| = 3$;
(iv) $T\mathsf{N}_{\mathsf{Aut}(T)}(\{A, B\}) = T\Phi\langle\vartheta\rangle$, *where ϑ is as in Lemma 8.14(iii), so that* $(\mathsf{N}_{\mathsf{Aut}(T)}(\{A, B\}) \cap \Theta)\pi \cong \mathbb{Z}_2$.

The following result shows that a factorisation of a direct product that involves a full strip leads to cartesian factorisations of a factor.

Lemma 8.16 *Suppose that $T_1, \ldots, T_k$ are pairwise isomorphic groups and let $M = T_1 \times \cdots \times T_k$. Suppose that $\alpha_i\colon T_1 \to T_i$ is an isomorphism for $i = 2, \ldots, k$, and K is a subgroup of M such that $K\sigma_i \neq T_i$ for $i = 1, \ldots, k$ and $K = K_1\sigma_1 \times \cdots \times K\sigma_k$ where $\sigma_i\colon M \to T_i$ is the i-th coordinate projection. Then*

$$\{(t, t\alpha_2, \ldots, t\alpha_k) \mid t \in T_1\}K = M, \tag{8.3}$$

if and only if

$$t_1(K\sigma_1) \cap t_2(K\sigma_2\alpha_2^{-1}) \cap \cdots ! \cap t_k(K\sigma_k\alpha_k^{-1}) \neq \emptyset \quad \textit{for all } t_1, \ldots, t_k \in T_1. \tag{8.4}$$

Moreover, if either (8.3) or (8.4) holds, then

$$\{K\sigma_1, K\sigma_2\alpha_2^{-1}, \ldots, K\sigma_k\alpha_k^{-1}\}$$

is an abstract cartesian factorisation of T_1, and, in particular, $T_1 = (K\sigma_1)(K\sigma_i\alpha_i^{-1})$ holds for all $i \geqslant 2$. Further, if T_i is a finite non-abelian simple group, then $k \leqslant 3$.

Proof If claim (8.3) holds, then, for all $(t_1, \ldots, t_k) \in M$, there exist $t \in T_1$ and $(a_1, \ldots, a_k) \in K$ such that

$$(t_1, \ldots, t_k) = (t, t\alpha_2, \ldots, t\alpha_k)(a_1, \ldots, a_k).$$

We obtain, taking $\alpha_1 = \mathrm{id}_{T_1}$, that $t_i = t\alpha_i a_i$ for all i; that is, $t = (t_i a_i^{-1})\alpha_i^{-1}$, and hence t is an element of the left coset $t_i\alpha_i^{-1}(K\sigma_i\alpha_i^{-1})$ for all i, and so t is an element of the intersection of these cosets. Since the t_i are chosen arbitrarily, we obtain that (8.4) holds. Reversing this argument shows that (8.4) implies (8.3).

Let us now show the second statement. Since (8.3) and (8.4) are equivalent, it suffices to assume (8.3). For each $t \in T_1$ there are elements $x \in T_1$ and $(a_1, \ldots, a_k) \in K$ such that

$$(t, 1, \ldots, 1) = (x, x\alpha_2, \ldots, x\alpha_k)(a_1, \ldots, a_k).$$

Then $t = xa_1$ and $x = a_i^{-1}\alpha_i^{-1}$ for $i = 2, \ldots, k$, and so $x \in K\sigma_2\alpha_2^{-1} \cap \cdots \cap K\sigma_k\alpha_k^{-1}$. So

$$T_1 = K\sigma_1(K\sigma_2\alpha_2^{-1} \cap \cdots \cap K\sigma_k\alpha_k^{-1}),$$

and, setting $\alpha_1 = \mathrm{id}_{T_1}$, we obtain similarly that the factorisations

$$T_1 = K\sigma_i\alpha_i^{-1} \bigcap_{j \neq i} K\sigma_j\alpha_j^{-1} \tag{8.5}$$

are valid for all i. Thus

$$\{K\sigma_1, K\sigma_2\alpha_2^{-1}, \ldots, K\sigma_k\alpha_k^{-1}\} \tag{8.6}$$

is an abstract cartesian factorisation of T_1. In particular, we have, for $i \geqslant 2$, that $T_1 = (K\sigma_1)(K\sigma_i\alpha_i^{-1})$. If the T_i are finite, non-abelian simple groups, then $k \leqslant 3$ follows from Lemma 8.14(i). □

8.8 Cartesian decompositions preserved by finite simple groups

Since little is known about factorisations of infinite simple groups, we will concentrate in this section on finite groups and use available results about factorisations of finite simple groups. We combine the results of Sections 8.1–8.6 with the information about factorisations of finite simple groups given in Lemmas 8.14 and 8.15 to obtain a full classification of cartesian decompositions that are preserved by a transitive action of a non-abelian finite simple group.

Theorem 8.17 *Let G be a finite innately transitive permutation group on Ω with a non-abelian, simple plinth T, let $\omega \in \Omega$, $\mathcal{E} \in \mathsf{CD}(G)$, and let W be the stabiliser of $\mathcal{E}$ in $\mathsf{Sym}\,\Omega$. Then $|\mathcal{E}| \leqslant 3$ and the following hold.*

(i) *Suppose that $\mathcal{E}$ is homogeneous. Then $|\mathcal{E}| = 2$, and G, T, W, the subgroups $K \in \mathcal{K}_\omega(\mathcal{E})$, and $|\Omega|$ are as in Table 8.3. In particular, the set $\mathcal{K}_\omega(\mathcal{E})$ contains two isomorphic subgroups. Moreover, the group G is quasiprimitive and T is the unique minimal normal subgroup of G, and exactly one of the following holds:*

  (a) $|\mathsf{CD}(G)| = 1$;

  (b) $|\mathsf{CD}(G)| = 3$, *T is as in row 3 of Table 8.3, $G \leqslant T\Phi$ where Φ is the group of field automorphisms of T.*

(ii) *Suppose that $|\mathcal{E}| = 3$. If W is the stabiliser in $\mathsf{Sym}\,\Omega$ of $\mathcal{E}$, then T, W, the elements of $\mathcal{K}_\omega(\mathcal{E})$, and $|\Omega|$ are as in Table 8.4.*

Proof Suppose that $\mathcal{E} \in \mathsf{CD}(G)$. Then Theorem 8.3 implies that $T_{(\mathcal{E})} = T$. Let ℓ be the index of $\mathcal{E}$, and let $\mathcal{K}_\omega(\mathcal{E}) = \{K_1, \ldots, K_\ell\}$ be the corresponding cartesian factorisation for T. Then we obtain from Lemma 8.14(i) that $\ell \leqslant 3$.

(i) If $\ell = 3$ then Table 8.1 shows that K_1, K_2, K_3 have different sizes. Thus if $\mathcal{E}$ is homogeneous then $\ell = 2$ and the factorisation $T = K_1K_2$ is as in Lemma 8.14(iii). Hence T, K_1, K_2, and $|\Omega|$ are as in Table 8.3.

Let us now prove that G is quasiprimitive. As T is transitive on Ω, we have $\mathsf{C}_{\mathsf{Sym}\,\Omega}(T) \cong \mathsf{N}_T(T_\omega)/T_\omega$ by Theorem 3.2(i). On the other hand $T_\omega = K_1 \cap K_2$, and Lemma 8.14(iii) shows that $\mathsf{N}_T(K_1 \cap K_2) = K_1 \cap K_2 = T_\omega$. Hence $\mathsf{C}_{\mathsf{Sym}\,\Omega}(T) = 1$, and so T is the unique minimal normal subgroup of G. Hence G is an almost simple quasiprimitive group acting on Ω.

	G	T	W	K	$\|\Omega\|$
1	$\mathsf{A}_6 \leqslant G \leqslant \mathsf{P\Gamma L}_2(9)$	A_6	$\mathsf{S}_6 \wr \mathsf{S}_2$	A_5	36
2	$\mathsf{M}_{12} \leqslant G \leqslant \mathsf{Aut}(\mathsf{M}_{12})$	M_{12}	$\mathsf{S}_{12} \wr \mathsf{S}_2$	M_{11}	144
3	$\mathsf{P}\Omega_8^+(q) \leqslant G \leqslant \mathsf{P}\Omega_8^+(q)\Phi\langle\vartheta\rangle$ Φ: field automorphisms ϑ is as in Lemma 8.14(iii)	$\mathsf{P}\Omega_8^+(q)$	$\mathsf{S}_{(d/2)q^3(q^4-1)} \wr \mathsf{S}_2$ $d = (4, q^4 - 1)$	$\Omega_7(q)$	$(d^2/4)q^6(q^4-1)^2$
4	$\mathsf{Sp}_4(q) \leqslant G \leqslant \mathsf{Aut}(\mathsf{Sp}_4(q))$	$\mathsf{Sp}_4(q)$, $q \geqslant 4$ even	$\mathsf{S}_{q^2(q^2-1)} \wr \mathsf{S}_2$	$\mathsf{Sp}_2(q^2).2$	$q^4(q^2-1)^2$

Table 8.3. *Homogeneous cartesian decompositions preserved by almost simple groups*

	T	W	$\mathcal{K}_\omega(\mathcal{E})$	$\|\Omega\|$
1	$\mathsf{Sp}_{4a}(2)$, $a \geqslant 2$	$\mathsf{S}_{n_1} \times \mathsf{S}_{n_2} \times \mathsf{S}_{n_3}$ $n_1 = \|\mathsf{Sp}_{4a}(2) : \mathsf{Sp}_{2a}(4) \cdot 2\|$ $n_2 = \|\mathsf{Sp}_{4a}(2) : \mathsf{O}_{4a}^-(2)\|$ $n_3 = \|\mathsf{Sp}_{4a}(2) : \mathsf{O}_{4a}^+(2)\|$	$\mathsf{Sp}_{2a}(4) \cdot 2$, $\mathsf{O}_{4a}^-(2)$, $\mathsf{O}_{4a}^+(2)$	$n_1 \cdot n_2 \cdot n_3$
2	$\mathsf{P}\Omega_8^+(3)$	$\mathsf{S}_{1080} \times \mathsf{S}_{1120} \times \mathsf{S}_{28431}$	$\Omega_7(3)$, $\mathbb{Z}_3^6 \rtimes \mathsf{PSL}_4(3)$, $\mathsf{P}\Omega_8^+(2)$	34,390,137,600
3	$\mathsf{Sp}_6(2)$	$\mathsf{S}_{120} \times \mathsf{S}_{28} \times \mathsf{S}_{36}$ $\mathsf{S}_{240} \times \mathsf{S}_{28} \times \mathsf{S}_{36}$ $\mathsf{S}_{120} \times \mathsf{S}_{56} \times \mathsf{S}_{36}$ $\mathsf{S}_{120} \times \mathsf{S}_{28} \times \mathsf{S}_{72}$	$\mathsf{G}_2(2)$, $\mathsf{O}_6^-(2)$, $\mathsf{O}_6^+(2)$ $\mathsf{G}_2(2)'$, $\mathsf{O}_6^-(2)$, $\mathsf{O}_6^+(2)$ $\mathsf{G}_2(2)$, $\mathsf{O}_6^-(2)'$, $\mathsf{O}_6^+(2)$ $\mathsf{G}_2(2)$, $\mathsf{O}_6^-(2)$, $\mathsf{O}_6^+(2)'$	120,960 241,920 241,920 241,920

Table 8.4. *Cartesian decompositions with index 3 preserved by almost simple groups*

Now we prove that the information given in the G-column of Table 8.3 is correct. Since T is the unique minimal normal subgroup of G, we have that G is an almost simple group and $T \leqslant G \leqslant \mathsf{Aut}(T)$. Let $N = \mathsf{N}_{\mathsf{Aut}(T)}(\{K_1, K_2\})$. Note that $G = TG_\omega$ and $G_\omega \leqslant N$. On the other hand, N has the property that, since A and B are not conjugate in T,

$$T \cap N = \mathsf{N}_T(K_1) \cap \mathsf{N}_T(K_2) = K_1 \cap K_2 = T_\omega,$$

and so the T-action on Ω can be extended to TN with point stabiliser N. Thus $G \leqslant TN$. By Lemmas 8.14(iii) and 8.15(iv), for $T \cong \mathsf{A}_6$, M_{12}, $\mathsf{P\Omega}_8^+(q)$, and $\mathsf{Sp}_4(q)$, we have $TN = \mathsf{P\Gamma L}_2(9)$, $\mathsf{Aut}(\mathsf{M}_{12})$, $\mathsf{P\Omega}_8^+(q)\Phi\langle\vartheta\rangle$ (where Φ is the group of field automorphisms and ϑ is as in Lemma 8.14(iii)), and $\mathsf{Aut}(\mathsf{Sp}_4(q))$, respectively. Hence the assertion follows.

Finally we prove the claim about $|\mathsf{CD}(G)|$. Suppose that $L_1,\ L_2 < T$ is such that $|L_1| = |L_2|$, $L_1L_2 = T$ and $L_1 \cap L_2 = T_\omega$. Inspection of Table 8.2, shows that there is an element $\alpha \in \mathsf{Aut}(T)$ such that $\{K_1\alpha, K_2\alpha\} = \{L_1, L_2\}$, and so $\alpha \in \mathsf{N}_{\mathsf{Aut}(T)}(T_\omega) = \mathsf{N}_{\mathsf{Aut}(T)}(K_1 \cap K_2)$. Lemma 8.14(iii) implies that, if T is as in row 1, 2, or 4 of Table 8.3, then

$$\mathsf{N}_{\mathsf{Aut}(T)}(\{K_1, K_2\}) = \mathsf{N}_{\mathsf{Aut}(T)}(T_\omega)$$

and so $\{L_1, L_2\} = \{K_1\alpha, K_2\alpha\} = \{K_1, K_2\}$. Thus $|\mathsf{CD}(G)| = 1$ in these cases, as asserted.

Suppose now that $T \cong \mathsf{P\Omega}_8^+(q)$ for some q. Then we obtain from Lemma 8.15(iii) that $|\mathsf{N}_{\mathsf{Aut}(T)}(T_\omega) : \mathsf{N}_{\mathsf{Aut}(T)}(\{K_1, K_2\})| = 3$, and so the $\mathsf{N}_{\mathsf{Aut}(T)}(T_\omega)$-orbit containing $\{K_1, K_2\}$ has 3 elements, which gives rise to 3 different choices of cartesian factorisations with respect to ω. Let $\mathcal{E}_1$, $\mathcal{E}_2$, and $\mathcal{E}_3$ denote the corresponding cartesian decompositions of Ω, such that $\mathcal{E} = \mathcal{E}_1$. We computed above that $\mathsf{C}_{\mathsf{Sym}\,\Omega}(T) = 1$, and this implies that $\mathsf{N}_{\mathsf{Sym}\,\Omega}(T) = \mathsf{Aut}(T) \cap \mathsf{Sym}\,\Omega$. In other words, $N = \mathsf{N}_{\mathsf{Sym}\,\Omega}(T)$ is the largest subgroup of $\mathsf{Aut}(T)$ that extends the T-action on Ω. Since T is a transitive subgroup of N, we have $N = TN_\omega$. As T_ω is a normal subgroup of N_ω, it follows that $N \leqslant T\mathsf{N}_{\mathsf{Aut}(T)}(T_\omega)$. On the other hand

$$\begin{aligned}|T\mathsf{N}_{\mathsf{Aut}(T)}(T_\omega) : \mathsf{N}_{\mathsf{Aut}(T)}(T_\omega)|\\ &= |T : T \cap \mathsf{N}_{\mathsf{Aut}(T)}(T_\omega)| = |T : \mathsf{N}_T(T_\omega)| = |T : T_\omega|,\end{aligned}$$

by Lemma 8.14(iii). This shows that the T-action on Ω can be extended to $T\mathsf{N}_{\mathsf{Aut}(T)}(T_\omega)$ with point stabiliser $\mathsf{N}_{\mathsf{Aut}(T)}(T_\omega)$. In other words $T\mathsf{N}_{\mathsf{Aut}(T)}(T_\omega)$ is the largest subgroup of $\mathsf{Aut}(T)$ that extends the T-action on Ω. The stabiliser of $\mathcal{E}_1$ in $T\mathsf{N}_{\mathsf{Aut}(T)}(T_\omega)$ is the subgroup $T\mathsf{N}_{\mathsf{Aut}(T)}(\{K_1, K_2\})$. Hence if $G \leqslant \mathsf{Aut}(T)$ is such that $T \leqslant G$ and G leaves the cartesian decomposition $\mathcal{E}_1$ invariant, then $G \leqslant T\mathsf{N}_{\mathsf{Aut}(T)}(\{K_1, K_2\}) = T\Phi\langle\vartheta\rangle$, by Lemma 8.15(iv). If $\mathsf{CD}(G) \neq \{\mathcal{E}\}$, then G leaves $\mathcal{E}_1$, $\mathcal{E}_2$, and $\mathcal{E}_3$ invariant. Therefore G lies in the kernel of the action of $T\mathsf{N}_{\mathsf{Aut}(T)}(T_\omega)$ on $\{\mathcal{E}_1, \mathcal{E}_2, \mathcal{E}_3\}$. Hence $G \leqslant T\Phi$, as required.

(ii) Suppose that $|\mathcal{E}| = 3$. Then $\{K_1, K_2, K_3\}$ is an abstract cartesian factorisation of T. Therefore the K_i are as in Table 8.1. Thus we obtain that T, K_1, K_2, K_3, and the degree $|\Omega| = |T : K_1 \cap K_2 \cap K_3|$ of G are as in Table 8.4. □

The main result of this section is the following theorem.

Theorem 8.18 *Let Ω be a finite set, let $T < W < \mathsf{Sym}\,\Omega$ such that T is a finite simple group, and W is permutationally isomorphic to a wreath product $\mathsf{Sym}\,\Gamma \wr \mathsf{S}_\ell$ in product action. Then either T is intransitive, or T, W, and $|\Omega|$ are as in Table 8.5. Moreover, if T is transitive, then $\mathsf{N}_{\mathsf{Sym}\,\Omega}(T)$ is an almost simple group and T is not regular.*

	T	W	$\|\Omega\|$
1	A_6	$\mathsf{S}_6 \wr \mathsf{S}_2$	36
2	M_{12}	$\mathsf{S}_{12} \wr \mathsf{S}_2$	144
3	$\mathsf{P}\Omega_8^+(q)$	$\mathsf{S}_{(d/2)q^3(q^4-1)} \wr \mathsf{S}_2$	$(d^2/4)q^6(q^4-1)^2$ $d = (4, q^4 - 1)$
4	$\mathsf{Sp}_4(q)$, $q \geqslant 4, q$ even	$\mathsf{S}_{q^2(q^2-1)} \wr \mathsf{S}_2$	$q^4(q^2-1)^2$

Table 8.5. *Transitive simple subgroups of wreath products*

Proof The proof of Theorem 8.18 is now easy, because, if T is transitive, then Theorem 8.17 implies that $\mathsf{C}_{\mathsf{Sym}\,\Omega}(T) = 1$, and so $\mathsf{N}_{\mathsf{Sym}\,\Omega}(T)$ is an almost simple group with socle T. Theorem 8.17 implies that T, W, and Ω are as in Table 8.3. The possibilities for simple T are tabulated in Table 8.5. The fact that T is non-regular can be shown by observing that, in all rows of Table 8.3, we have $|T| \neq |\Omega|$. □

By Theorem 8.18, if a finite transitive simple group T preserves a non-trivial cartesian decomposition, then T is not regular. This observation gives the following corollary.

Corollary 8.19 *Suppose that G is a finite permutation group on Ω that contains a simple regular subgroup T. Then G does not preserve a non-trivial cartesian decomposition of Ω.*

The fact that T is non-regular in Theorem 8.18 implies that certain groups do not preserve cartesian decompositions. See Chapter 7 (in particular, Theorem 7.8 and Table 7.1) for the descriptions of the quasiprimitive and primitive groups that appear in the following result.

Corollary 8.20 *If G is a finite permutation group such that one of the following cases is valid, then G does not preserve a non-trivial cartesian decomposition of Ω:*

(i) *G is quasiprimitive of type* $\mathrm{AS}_{\mathrm{reg}}$*;*
(ii) *G is quasiprimitive of type* SD *with socle isomorphic to T^2;*
(iii) *G is primitive of type* HS.

Proof The claim follows from the fact that if one of (i)–(iii) is valid, then G contains a regular simple subgroup. In fact, if G is of type $\mathrm{AS}_{\mathrm{reg}}$ or HS, then it has a regular simple plinth. If G is of type SD with socle isomorphic to T^2, then a simple direct factor of T^2 is simple and regular. □

We will show in Theorem 9.19 that possibly infinite quasiprimitive groups of type SD do not preserve non-trivial cartesian decompositions independently of the number of simple factors of the socle.

We conclude this chapter with a result about cartesian decompositions preserved by finite primitive groups of AS type. The result rephrases (Liebeck *et al.* 1987, Lemma 3.3). Theorem 8.21 also provides examples of transitive simple permutation groups T acting on Ω such that $\mathsf{N}_{\mathsf{Sym}\,\Omega}(T)$ is primitive, but $\mathsf{N}_{\mathsf{Sym}\,\Omega}(T)$ is not maximal in $\mathsf{Sym}\,\Omega$. Hence a group H that satisfies Theorem 7.11(vi) may not be maximal.

Theorem 8.21 *Suppose that G is a finite primitive permutation group acting on Ω preserving a non-trivial cartesian decomposition $\mathcal{E}$ of Ω. If G has a non-abelian simple minimal normal subgroup T, then the following hold.*

(i) *G is transitive on $\mathcal{E}$ and $|\mathcal{E}| = 2$.*
(ii) *T must be as in one of the rows 1, 2, or 4 of Table 8.5.*

In particular T is not regular and the O'Nan–Scott type of G is AS.

Proof The fact that G is transitive on $\mathcal{E}$ follows from Corollary 2.17. In particular, $\mathcal{E}$ is homogeneous and Theorem 8.18 implies that $|\mathcal{E}| = 2$ and T must be as in one of the rows of Table 8.5. The fact that line 3 is impossible is non-trivial and requires an understanding of the subgroup structure of $\mathsf{P}\Omega_8^+(q)$. The main idea is that if $T \cong \mathsf{P}\Omega_8^+(q)$, then T_α is isomorphic to the exceptional group of Lie type $\mathsf{G}_2(q)$ and G_α is not maximal unless G involves a triality automorphism of T, that is, a graph automorphism of order 3; see the proof of (Liebeck *et al.* 1987, Lemma 4.3) and (Kleidman 1987). On the other hand, the first column of the corresponding row of Table 8.3 shows that G cannot involve such an automorphism.

By Theorem 8.18, T is not regular and the O'Nan–Scott type of the primitive group G must be AS (see Section 7.2). □

9

Transitive cartesian decompositions for innately transitive groups

The aim of this chapter is to extend the results of Chapter 8 on cartesian factorisations of simple groups for characteristically simple groups. This will enable us to study G-invariant cartesian decompositions on which G acts transitively. Such cartesian decompositions are called *transitive*. The initial study of transitive cartesian decompositions will be carried out in the context of innately transitive groups with a non-abelian FCR-plinth. To progress further, we will require in Corollary 9.3, towards the end of Section 9.3, and again in Section 9.7, knowledge on the factorisations of finite simple groups that is not yet available for infinite simple groups, and so we will restrict our attention to finite innately transitive groups. The main theorem of this chapter is Theorem 9.7 which is a characterisation of transitive cartesian decompositions preserved by finite innately transitive groups.

9.1 Abstract cartesian factorisations involving strips

Cartesian factorisations were introduced in Chapter 8 to characterise cartesian decompositions preserved by innately transitive groups. In this section we shall investigate cartesian factorisations in characteristically simple FCR-groups whose subgroups involve strips. Let us introduce notation for this section. Let $M = T_1 \times \cdots \times T_k$ be a non-abelian, characteristically simple FCR-group with simple normal subgroups $T_1, \ldots, T_k$, and let G_0 be a subgroup of $\mathsf{Aut}(M)$ such that the natural action of G_0 on $\{T_1, \ldots, T_k\}$ is transitive. Suppose, in addition, that $\mathcal{K} = \{K_1, \ldots, K_\ell\}$ is an abstract G_0-invariant cartesian factorisation of M. Set $M_0 = \bigcap_i K_i$.

Recall that if I is a subset of $\{T_1, \ldots, T_k\}$ or a subset of $\{1, \ldots, k\}$, then σ_I denotes the corresponding coordinate projection from M onto either $\prod_{T_i \in I} T_i$ or onto $\prod_{i \in I} T_i$. Strips were introduced in Section 4.4. We recommend that the reader should consult this section to revise the basic definitions and properties. According to the definition given in Section 4.6, a strip X is involved in a subgroup K of M if $K = X \times K\sigma_{\mathcal{T} \setminus \mathrm{Supp}\, X}$ where $\mathcal{T} = \{T_1, \ldots, T_k\}$. We say that a strip X is involved in a cartesian factorisation $\mathcal{K}$ if X is involved in a member $K \in \mathcal{K}$. In this case, (8.2) implies that X is involved in a unique member of $\mathcal{K}$. Uniform automorphisms were introduced in Section 4.8.

In this section we prove the following theorem.

Theorem 9.1 *Suppose that $\mathcal{K} = \{K_1, \ldots, K_\ell\}$, $M = T_1 \times \cdots \times T_k$, M_0, and G_0 are as above and let X_1 and X_2 be two non-trivial full strips involved in $\mathcal{K}$ such that $\mathrm{Supp}\, X_1 \cap \mathrm{Supp}\, X_2 \neq \emptyset$. Then the following both hold:*

(i) *the T_i admit uniform automorphisms;*
(ii) *M_0 is not a subdirect subgroup of M.*

Proof The proof of this theorem is quite involved, and so we split it into a series of claims. Assume that the hypotheses of Theorem 9.1 hold. Assume, moreover, that either T_1 does not admit a uniform automorphism, or that M_0 is a subdirect subgroup of M.

Claim 1. $|\mathrm{Supp}\, X_1 \cap \mathrm{Supp}\, X_2| = 1$.

Proof of Claim 1. By the definition of 'being involved' for strips, if distinct strips are involved in the same K_i then they are disjoint as strips. Thus we may assume without loss of generality that X_1 is involved in K_1 and X_2 is involved in K_2. Suppose to the contrary that $T_1,\ T_2 \in \mathrm{Supp}\, X_1 \cap \mathrm{Supp}\, X_2$ and set $\sigma = \sigma_{\{1,2\}}$. Then $Y_1 = K_1\sigma$ and $Y_2 = K_2\sigma$ are non-trivial full strips in $T_1 \times T_2$ such that $Y_1 Y_2 = M\sigma = T_1 \times T_2$. By Lemma 4.28, T_1 admits a uniform automorphism. Thus, by our own assumption, M_0 is a subdirect subgroup of M. Then $K_1 \cap K_2$ is also a subdirect subgroup of M, and so $(K_1 \cap K_2)\sigma$ is a subdirect subgroup in $T_1 \times T_2$. Now $Y_1 = \{(t, t\alpha) \mid t \in T_1\}$ and $Y_2 = \{(t, t\beta) \mid t \in T_1\}$ where $\alpha,\ \beta \colon T_1 \to T_2$ are isomorphisms. Then

$$(K_1 \cap K_2)\sigma \leqslant Y_1 \cap Y_2 = \{(t, t\alpha) \mid t \in T_1 \text{ such that } t\alpha = t\beta\}.$$

Now the fact that $(K_1 \cap K_2)\sigma$ is subdirect in $T_1 \times T_2$ implies that $t\alpha = t\beta$ for all $t \in T_1$, and hence $\alpha = \beta$. However, this implies that $Y_1 = Y_2$, yielding $Y_1 Y_2 \neq T_1 \times T_2$, which is a contradiction proving the claim. □

Claim 2. There exists a sequence of strips $X_1,\ldots,X_a$, with $a \geqslant 3$, involved in $\mathcal{K}$ such that

(a) X_1 and X_j are disjoint for $j \notin \{a,2\}$;
(b) for $i = 2,\ldots,a-1$ and $j \notin \{i-1,i+1\}$, the strips X_i and X_j are disjoint;
(c) X_a and X_j are disjoint for $j \notin \{a-1,1\}$;
(d) and finally,

$$|\operatorname{Supp} X_1 \cap \operatorname{Supp} X_2| = |\operatorname{Supp} X_2 \cap \operatorname{Supp} X_3| = \cdots$$
$$= |\operatorname{Supp} X_{a-1} \cap \operatorname{Supp} X_a| = |\operatorname{Supp} X_a \cap \operatorname{Supp} X_1| = 1.$$

Proof of Claim 2. By Claim 1, $\operatorname{Supp} X_1 \cap \operatorname{Supp} X_2 = \{T_t\}$ for some $t \leqslant k$. Choose $g \in G_0$ such that $T_t^g \in \operatorname{Supp} X_2 \backslash \operatorname{Supp} X_1$; such an element g exists since G_0 is transitive on $\mathcal{T} = \{T_1,\ldots,T_k\}$ and $\operatorname{Supp} X_2 \backslash \operatorname{Supp} X_1$ is non-empty. Now G_0 acts by conjugation on the set of full strips involved in $\mathcal{K}$, and so both X_1^g and X_2^g are full strips involved in $\mathcal{K}$. As T_t^g is in both $\operatorname{Supp} X_1^g$ and $\operatorname{Supp} X_2^g$, but is not in $\operatorname{Supp} X_1$ we deduce that there exists a non-trivial strip X_3 in $\mathcal{K}$ distinct from X_1, X_2 such that $\operatorname{Supp} X_3 \cap (\operatorname{Supp} X_2 \setminus \operatorname{Supp} X_1) \neq \emptyset$ (namely, we can take X_3 to be one of X_1^g or X_2^g as at least one of these is distinct from X_1 and X_2). Proceeding in this way we construct a sequence $X_1, X_2, \ldots$ of distinct, non-trivial strips in $\mathcal{K}$ such that $\operatorname{Supp} X_{d+1} \cap (\operatorname{Supp} X_d \setminus \operatorname{Supp} X_{d-1}) \neq \emptyset$ for each $d \geqslant 2$. Since k is finite, there exists a such that

$$\operatorname{Supp} X_a \cap (\operatorname{Supp} X_1 \cup \cdots \cup \operatorname{Supp} X_{a-2}) \neq \emptyset.$$

Let a be the least integer such that this property holds. The conditions imposed on X_1 and X_2 imply that $a \geqslant 3$. By removing some initial segment of the sequence and relabelling the X_i if necessary, we may assume that the intersection $\operatorname{Supp} X_a \cap \operatorname{Supp} X_1$ is non-empty, while $\operatorname{Supp} X_a \cap \operatorname{Supp} X_d = \emptyset$ if $2 \leqslant d \leqslant a-2$ for some $a \geqslant 3$. Now, applying Claim 1 a number of times, the sequence $X_1,\ldots,X_a$ is as required. □

Now assume that the conditions of Claim 2 are valid, X_1 and X_2 are non-disjoint strips involved in $\mathcal{K}$ and select $X_1,\ldots,X_a$ as in the proof of Claim 2. By relabelling the K_i we may assume that X_1 is involved in K_1. Let $1 = i_1 < i_2 < \cdots < i_d < a$ be such that among the X_i the strips $X_{i_1},\ldots,X_{i_d}$ are precisely the ones that are involved in K_1. Note that X_a is not involved in K_1 since $\operatorname{Supp} X_a$ and $\operatorname{Supp} X_1$ are not disjoint. Also, $i_{j+1} \geqslant i_j + 2$ for all $j = 1,\ldots,d-1$ since $\operatorname{Supp} X_{i_j}$

and $\operatorname{Supp} X_{i_j+1}$ are not disjoint. We may also relabel the T_i so that

$$\{T_1\} = \operatorname{Supp} X_a \cap \operatorname{Supp} X_1 \quad \text{and} \quad \{T_2\} = \operatorname{Supp} X_1 \cap \operatorname{Supp} X_2,$$

and so that for $j = 2, \ldots, d$,

$$\begin{aligned} \{T_{2j-1}\} &= \operatorname{Supp} X_{i_j-1} \cap \operatorname{Supp} X_{i_j}; \\ \{T_{2j}\} &= \operatorname{Supp} X_{i_j} \cap \operatorname{Supp} X_{i_j+1}. \end{aligned}$$

It follows from Claim 2, that $T_1, \ldots, T_{2d}$ are pairwise distinct. Define the projection map

$$\sigma\colon M \to T_1 \times \cdots \times T_{2d} \quad \text{and} \quad \widehat{K_1} = \bigcap_{i \neq 1} K_i. \tag{9.1}$$

Claim 3. Using the notation introduced above, the following hold.

(a) $K_1\sigma$ is a direct product $Y_1 \times \cdots \times Y_d$ such that each Y_i is a non-trivial full strip in $T_{2i-1} \times T_{2i}$ and $Y_i = \{(t, t\alpha_i) \mid t \in T_{2i-1}\}$ for some isomorphism $\alpha_i\colon T_{2i-1} \to T_{2i}$.

(b) $\widehat{K_1} \leqslant Z_1 \times \cdots \times Z_d$, such that, for $i = 1, \ldots, d-1$, $Z_i = \{(t, t\beta_i) \mid t \in T_{2i}\}$ is a non-trivial full strip in $T_{2i} \times T_{2i+1}$ where $\beta_i\colon T_{2i} \to T_{2i+1}$ is an isomorphism. Further $Z_d = \{(t\beta_d, t) \mid t \in T_{2d}\}$ is a non-trivial full strip in $T_{2d} \times T_1$ where $\beta_d\colon T_{2d} \to T_1$ is an isomorphism.

Proof of Claim 3. Assertion (a) follows from the observation that

$$K_1\sigma = (X_{i_1} \times \cdots \times X_{i_d})\sigma = Y_1 \times \cdots \times Y_d \quad \text{with} \quad Y_j = X_{i_j}\sigma \text{ for each } j.$$

Let us prove assertion (b). It suffices to show that $\widehat{K_1}\sigma_{\{2i,2i+1\}} \leqslant Z_i$ for $i = 1, \ldots, d-1$ and $\widehat{K_1}\sigma_{\{1,2d\}} \leqslant Z_d$. We prove the claim for $i = 1$, that is, for $\widehat{K_1}\sigma_{\{2,3\}}$, noting that the proof for the other projections is identical. Set $r = i_2 - 1$. Then the strips $X_2, \ldots, X_r$ are 'between' X_1 and X_{i_2} in the sequence of the X_i and they are not involved in K_1. Choose $T_{m_1}, \ldots, T_{m_r}$ such that $\{T_{m_i}\} = \operatorname{Supp} X_i \cap \operatorname{Supp} X_{i+1}$. By the choice made earlier, we have $T_{m_1} = T_2$ and $T_{m_r} = T_3$. Let σ' denote the projection onto $T_{m_1} \times \cdots \times T_{m_r}$. By Claim 2, the indices $m_1, \ldots, m_r$ are pairwise distinct. Suppose that $x = (t_{m_1}, \ldots, t_{m_r})$ is an element of $\widehat{K_1}\sigma'$ with $t_i \in T_{m_i}$. The strip X_2 is involved in K_m for some $m \neq 1$. Recall that X_2 covers $T_{m_1} = T_2$ and T_{m_2}. Then $x \in K_m\sigma'$ and hence $(t_{m_1}, t_{m_2}) \in X_2\sigma_{\{m_1,m_2\}}$. Thus there exists an isomorphism $\gamma_{m_1}\colon T_{m_1} \to T_{m_2}$ such that $X_2\sigma_{\{m_1,m_2\}} = \{(t, t\gamma_{m_1}) \mid t \in T_{m_1}\}$. Using the same argument, we find that there exist isomorphisms $\gamma_{m_i}\colon T_{m_i} \to$

$T_{m_{i+1}}$, for $i = 1, \ldots, r-1$, such that $X_{i+1}\sigma_{\{m_i,m_{i+1}\}} = \{(t, t\gamma_{m_2}) \mid t \in T_{m_i}\}$. Thus $t_{m_r} = t_{m_1}\gamma_{m_1}\cdots\gamma_{m_{r-1}}$. Set $\beta_1 = \gamma_{m_1}\cdots\gamma_{m_{r-1}}$. This argument shows that

$$\widehat{K_1}\sigma_{\{2,3\}} \leqslant \{(t, t\beta_1) \mid t \in T_2\} := Z_1.$$

Therefore $\widehat{K_1}\sigma_{\{2,3\}} \leqslant Z_1$. The proof for the other projections is identical. This shows that assertion (b) holds. □

Claim 4. Use the notation of Claim 3, and set $\alpha = \alpha_1\beta_1\alpha_2\cdots\alpha_d\beta_d$. Then the following hold.

(a) $\alpha \in \mathsf{Aut}(T_1)$ and α is uniform.

(b) M_0 is not a subdirect subgroup of M where $M_0 = \bigcap_i K_i$.

Proof of Claim 4. (a) It follows from Claim 3 that $\alpha \in \mathsf{Aut}(T_1)$. Since $K_1\widehat{K_1} = M$, we have $(K_1\sigma)(\widehat{K_1}\sigma) = T_1 \times \cdots \times T_{2d}$ with σ and $\widehat{K}_1$ as in (9.1). Therefore

$$(Y_1 \times \cdots \times Y_d)(Z_1 \times \cdots \times Z_d) = T_1 \times \cdots \times T_{2d}. \qquad (9.2)$$

Since the factorisation in (9.2) is as in Lemma 4.29, it follows from Lemma 4.29 that α is uniform.

(b) By definition, $K_1 \cap \widehat{K_1} = M_0$. Suppose that M_0 is subdirect in M. Then $M_0\sigma$ is also a subdirect subgroup of $T_1 \times \cdots \times T_{2d}$. Suppose that $x = (t_1, t_2, \ldots, t_{2d}) \in M_0\sigma$. Then $x \in K_1\sigma$ and $x \in \widehat{K_1}\sigma$, and so $t_{2i} = t_{2i-1}\alpha_i$, for $i = 1, \ldots, d$, and also $t_{2i+1} = t_{2i}\beta_i$ for $i = 1, \ldots, d-1$, and $t_1 = t_d\beta_d$. Thus $t_1 = t_1\alpha_1\beta_1\cdots\alpha_d\beta_d = t_1\alpha$. Since $M_0\sigma$ is subdirect, this has to hold for all $t_1 \in T_1$, and hence $\alpha = 1$. However, by part (a), α is uniform, which is a contradiction, as the identity automorphism is not uniform. □

Now Theorem 9.1 follows at once from Claim 4. □

9.2 The structure of transitive cartesian factorisations

Suppose that G is an innately transitive permutation group acting on Ω with a non-abelian FCR-plinth M. The group M has the form $M = T_1 \times \cdots \times T_k$ where the T_i are non-abelian, simple groups. Recall from Chapter 8 that $\mathsf{CD}(G)$ denotes the set of G-invariant cartesian decompositions of Ω. Let $\mathcal{E} \in \mathsf{CD}(G)$ and let $\mathcal{K}_\omega(\mathcal{E})$ be a corresponding cartesian factorisation $\{K_1, \ldots, K_\ell\}$ for M with respect to some $\omega \in \Omega$, as defined in Definition 8.1. Suppose that, for $i \in \{1, \ldots, k\}$,

$\sigma_i\colon M \to T_i$ denotes the i-th coordinate projection. Then equation (8.2) implies that, for all $i \leqslant k$ and $j \leqslant \ell$,

$$K_j\sigma_i\left(\bigcap_{j'\neq j} K_{j'}\sigma_i\right) = T_i. \qquad (9.3)$$

In particular this means that if $K_j\sigma_i$ is a proper subgroup of T_i then $K_{j'}\sigma_i \neq K_j\sigma_i$ for all $j' \in \{1,\ldots,\ell\}\setminus\{j\}$. It is thus important to understand the following sets of subgroups:

$$\mathcal{F}_i(\mathcal{E},M,\omega) = \{K_j\sigma_i \mid j = 1,\ldots,\ell,\ K_j\sigma_i \neq T_i\}. \qquad (9.4)$$

Note that $|\mathcal{F}_i(\mathcal{E},M,\omega)|$ is the number of indices j such that $K_j\sigma_i \neq T_i$. The set $\mathcal{F}_i(\mathcal{E},M,\omega)$ is independent of i up to isomorphism, in the sense that if $i_1,\ i_2 \in \{1,\ldots,k\}$ and $g \in G_\omega$ are such that $T_{i_1}^g = T_{i_2}$ then $\mathcal{F}_{i_1}(\mathcal{E},M,\omega)^g = \{L^g \mid L \in \mathcal{F}_{i_1}(\mathcal{E},M,\omega)\} = \mathcal{F}_{i_2}(\mathcal{E},M,\omega)$. This argument also shows that the elements of $\mathcal{F}_{i_1}(\mathcal{E},M,\omega)$ are actually G_ω-conjugate to the elements of $\mathcal{F}_{i_2}(\mathcal{E},M,\omega)$.

If $G \leqslant \mathsf{Sym}\,\Omega$, and $\mathcal{E} \in \mathsf{CD}(G)$, then $\mathcal{E}$ is said to be a transitive decomposition, if G acts transitively on $\mathcal{E}$. Recall that $\mathsf{CD}_{\mathrm{tr}}(G)$ denotes the set of transitive G-invariant cartesian decompositions of Ω.

Theorem 9.2 *Suppose that G is an innately transitive permutation group with a non-abelian* FCR-*plinth $M = T_1\times\cdots\times T_k$ where $T_1,\ldots,T_k$ are pairwise isomorphic simple groups and $k \geqslant 1$. Let $\mathcal{E} \in \mathsf{CD}_{\mathrm{tr}}(G)$ with a corresponding cartesian factorisation $\mathcal{K}$ of M with respect to $\omega \in \Omega$. For $i = 1,\ldots,k$, let $\mathcal{F}_i = \mathcal{F}_i(\mathcal{E},M,\omega)$ be defined as in (9.4). Then the number $|\mathcal{F}_i|$ is independent of i and if $|\mathcal{F}_i| \geqslant 1$, then $\mathcal{F}_i$ is an abstract cartesian factorisation of T_i. Further, if T_i does not admit a uniform automorphism, then the following hold.*

(i) *Set $\mathcal{P} = \{\mathrm{Supp}\,X \mid X$ is a non-trivial, full strip involved in $\mathcal{K}\}$. If $\mathcal{P} \neq \emptyset$ then $\mathcal{P}$ is a G-invariant partition of $\{T_1,\ldots,T_k\}$.*

(ii) *Suppose that a non-trivial full strip X of length s is involved in $\mathcal{K}$ and $|\mathcal{F}_i| \geqslant 1$. Then T_1 admits a cartesian factorisation $\{A_1,\ldots,A_s\}$ such that the A_i are proper subgroups that are members of the same $\mathsf{Aut}(T_1)$-orbit.*

Proof Suppose that $i_1,\ i_2 \in \{1,\ldots,k\}$ and $g \in G_\omega$ such that $T_{i_1}^g = T_{i_2}$. Then it was observed after equation (9.4) that $\mathcal{F}_{i_1}^g = \mathcal{F}_{i_2}$, and so $|\mathcal{F}_i|$ is independent of i. The definition of $\mathcal{F}_i$ and (9.3) imply that $\mathcal{F}_i$ is an abstract cartesian factorisation of T_i.

For the rest of the proof, we assume that the T_i do not admit uniform automorphisms.

(i) Let $\mathcal{P} = \{\operatorname{Supp} X \mid X \text{ is a non-trivial, full strip involved in } \mathcal{K}\}$. By Theorem 9.1, either $\mathcal{P} = \emptyset$ or $\mathcal{P}$ is a partition of $\{T_1, \ldots, T_k\}$. Moreover, if $\operatorname{Supp} X \in \mathcal{P}$ and $g \in G_\omega$, then there exists $K \in \mathcal{K}$ such that X is involved in K. Therefore $K^g \in \mathcal{K}$ and X^g is involved in K^g. Thus X^g is involved in $\mathcal{K}$, and so $(\operatorname{Supp} X)^g = \operatorname{Supp}(X^g) \in \mathcal{P}$. Hence $\mathcal{P}$ is G_ω-invariant. Since M acts trivially on $\{T_1, \ldots, T_k\}$ by conjugation and $G = MG_\omega$, we have that $\mathcal{P}$ is a G-invariant partition of $\{T_1, \ldots, T_k\}$.

(ii) Suppose without loss of generality that

$$X = \{(t, t\alpha_2, \ldots, t\alpha_s) \mid t \in T_1\}$$

is a non-trivial, full strip involved in K_1 covering $T_1, \ldots, T_s$ where, for $i = 2, \ldots, s$, $\alpha_i \colon T_1 \to T_i$ is an isomorphism. Set $\widehat{K}_1 = \bigcap_{i \geqslant 2} K_i$. Let i, j be distinct elements of $\{1, \ldots, s\}$ and let $g \in G_\omega$ such that $T_i^g = T_j$. Then X^g is a strip involved in $\mathcal{K}$ such that $T_j \in \operatorname{Supp} X \cap \operatorname{Supp} X^g$. Therefore by Theorem 9.1, $X = X^g$, and so $K_1^g = K_1$, and also $\widehat{K}_1^g = \widehat{K}_1$. Since $T_i^g = T_j$ it follows that $(\widehat{K}_1\sigma_i)^g = \widehat{K}_1\sigma_j$. In particular, there exists, for each $i = 2, \ldots, s$, an isomorphism $\beta_i \colon T_1 \to T_i$, namely the conjugation induced by some $g_i \in G_\omega$ with $T_1^{g_i} = T_i$, such that $\widehat{K}_1\sigma_i = \widehat{K}_1\sigma_1\beta_i$. Now, as $|\mathcal{F}_i| \geqslant 1$, there exists t such that $K_t\sigma_i < T_i$, and, as $K_1\sigma_i = T_i$, we have $t \neq 1$. Hence $\widehat{K}_1\sigma_i < T_i$ holds for all $i = 1, \ldots, s$. In particular, the projections $\widehat{K}_1\sigma_i$ are all proper. Thus the factorisation $M\sigma_{\operatorname{Supp} X} = (K_1\sigma_{\operatorname{Supp} X})(\widehat{K}_1\sigma_{\operatorname{Supp} X})$ is as in Lemma 8.16. Therefore the set

$$\{\widehat{K}_1\sigma_1, \widehat{K}_1\sigma_2\alpha_2^{-1}, \ldots, \widehat{K}_1\sigma_s\alpha_s^{-1}\} \\ = \{\widehat{K}_1\sigma_1, \widehat{K}_1\sigma_1\beta_2\alpha_2^{-1}, \ldots, \widehat{K}_1\sigma_1\beta_s\alpha_s^{-1}\}$$

is an abstract cartesian factorisation of T_1 whose members are all elements of the $\mathsf{Aut}(T_1)$-orbit of $\widehat{K}_1\sigma_1$. □

If, in the previous theorem, the plinth M is finite, then we obtain even stronger assertions using the available knowledge of the factorisations of finite simple groups. In particular, an abstract cartesian factorisation of a finite simple group can contain at most three subgroups. Further, such an abstract cartesian factorisation with pairwise isomorphic subgroups contains two subgroups; see Lemma 8.14(ii).

Corollary 9.3 *Suppose that the conditions of Theorem 9.2 hold and that M is finite. Then the following are valid.*

(i) $|\mathcal{F}_i| \leqslant 3$.
(ii) *Suppose that a non-trivial full strip X is involved in $\mathcal{K}$. Then $|\mathcal{F}_i| \leqslant 1$. Further, if $|\mathcal{F}_i| = 1$, then $|\operatorname{Supp} X| = 2$.*

Proof (i) By Theorem 9.2, the set $\mathcal{F}_i$ is an abstract cartesian factorisation of T_i. By Lemma 8.14(i), $|\mathcal{F}_i| \leqslant 3$.

(ii) Suppose, as in the proof of Theorem 9.2, that K_1 involves a non-trivial full strip X covering $T_1, \ldots, T_s$. If $|\mathcal{F}_i| = 0$, then there is nothing to prove, so assume that $|\mathcal{F}_i| \geqslant 1$. Set $\widehat{K}_1 = \bigcap_{i \geqslant 2} K_i$. By Theorem 9.2, the finite simple group T_1 admits an abstract cartesian factorisation with s pairwise isomorphic proper subgroups. By Lemma 8.14(ii), $s = 2$. In particular $X = \{(t, t\alpha) \mid t \in T_1\}$ with some isomorphism $\alpha \colon T_1 \to T_2$ and $\{\widehat{K}_1\sigma_1, \widehat{K}_1\sigma_2\alpha^{-1}\}$ is a factorisation of T_1 with isomorphic proper subgroups. Now Lemma 8.14(iii) implies that $\widehat{K}_1\sigma_1$ and $\widehat{K}_1\sigma_2\alpha^{-1}$ are maximal in T_1. If $|\mathcal{F}_1| \geqslant 2$, then $K_{j_1}\sigma_1 < T_1$ and $K_{j_2}\sigma_1 < T_1$ for distinct j_1, $j_2 \in \{2, \ldots, \ell\}$, and, as noted above, $K_{j_1}\sigma_1 \neq K_{j_2}\sigma_1$. On the other hand, $\widehat{K}_1\sigma_1 < K_{j_1}\sigma_1 \cap K_{j_2}\sigma_1 < K_{j_1}\sigma_1 < T_1$, which contradicts the maximality of $\widehat{K}_1\sigma_1$ in T_1. Therefore $|\mathcal{F}_i| \leqslant 1$. □

9.3 A classification of transitive cartesian decompositions

If G is an innately transitive group acting on Ω with a non-abelian FCR-plinth M, then the set $\mathsf{CD}_{\mathrm{tr}}(G)$ of G-invariant transitive cartesian decompositions of Ω is further subdivided according to the structure of the subgroups in the corresponding cartesian factorisations for M as follows. Assume that $M = T_1 \times \cdots \times T_k$ where the T_i are simple groups. For $\mathcal{E} \in \mathsf{CD}_{\mathrm{tr}}(G)$ and $\omega \in \Omega$ the sets $\mathcal{F}_i = \mathcal{F}_i(\mathcal{E}, M, \omega)$ are as defined in (9.4).

Proposition 9.4 *Suppose that G, Ω, ω, $M = T_1 \times \cdots \times T_k$ and $\mathcal{E}$ are as above, and let $i \in \underline{k}$. Set $r = |\mathcal{F}_i(\mathcal{E}, M, \omega)|$ and set s to be the number of non-trivial full strips that are involved in $\mathcal{K}_\omega(\mathcal{E})$ and that cover T_i. Then the following hold.*

(i) *The numbers r and s are invariants of the G-invariant cartesian decomposition $\mathcal{E}$; that is, they are independent of ω, M and i. Moreover, $r + s \geqslant 1$.*
(ii) *If T_i does not admit a uniform automorphism, then $s \leqslant 1$.*

(iii) *If T_i is finite, then*

$$(r, s) \in \{(1,0), (0,1), (1,1), (2,0), (3,0)\},$$

and in particular, $r \leqslant 3$ and $s \leqslant 1$.

Proof (i) In order to prove that the above invariants of cartesian decompositions are well defined, we need to show that the properties used in their definitions are independent of the choice of the plinth M, the point ω, and the index i. For a fixed transitive minimal normal subgroup M and $\omega \in \Omega$, we obtain from Theorem 9.2 that r is independent of i. Similarly, if $X_1, \ldots, X_s$ are the non-trivial full strips involved in $\mathcal{K}(\mathcal{E}, M, \omega)$ that cover T_1 and $g \in G_\omega$ is such that $T_1^g = T_i$, then $X_1^g, \ldots, X_s^g$ are the non-trivial full strips involved in $\mathcal{K}(\mathcal{E}, M, \omega)$ that cover T_i. Hence s is also independent of i.

Next we verify that, given M, the numbers r and s are independent of ω. Let $\omega_1, \omega_2 \in \Omega$, and let $\mathcal{K}_1 = \mathcal{K}_{\omega_1}(\mathcal{E})$, $\mathcal{K}_2 = \mathcal{K}_{\omega_2}(\mathcal{E})$ be the corresponding cartesian factorisations for M. As M is a transitive subgroup of G, there is some $m \in M$ such that $\omega_1 m = \omega_2$ and Lemma 8.6 implies that $\mathcal{K}_1^m = \mathcal{K}_2$. Since m normalises T_1, $\mathcal{K}_1$ involves the non-trivial full strips $X_1 \ldots, X_s$ covering T_1, if and only if $\mathcal{K}_2 = \mathcal{K}_1^m$ involves the non-trivial full strips $X_1^m, \ldots, X_s^m$ also covering T_1. Thus the number of non-trivial full strips covering T_1 involved in a cartesian factorisation corresponding to $\mathcal{E}$ is independent of the choice of ω. Moreover, $K\sigma_i < T_i$ holds for some $K \in \mathcal{K}_1$ if and only if $K^m\sigma_i < T_i$ holds. Thus the number $|\mathcal{F}_i(\mathcal{E}, M, \omega)|$ is independent of the choice ω.

Next we show that r is independent of the plinth M. If M is the unique transitive, minimal normal subgroup of G, then there is nothing to prove. If this is not the case, then, by Theorem 3.6, G has exactly two transitive, minimal normal subgroups M and $\widehat{M}$, they are isomorphic, they both are regular on Ω, and $\mathsf{C}_G(M) = \widehat{M}$, $\mathsf{C}_G(\widehat{M}) = M$ also hold. Moreover, there is an involution $\alpha \in \mathsf{Sym}\,\Omega$ stabilising ω and centralising G_ω that interchanges M and $\widehat{M}$ (Corollary 3.11). Further, identifying Ω with M and ω with 1, α can be taken to be the map $x \mapsto x^{-1}$. Let $\widehat{T}_1 = T_1^\alpha, \ldots, \widehat{T}_k = T_k^\alpha$. Then $\widehat{M} = \widehat{T}_1 \times \cdots \times \widehat{T}_k$. Let $\widehat{\sigma}_i \colon \widehat{M} \to \widehat{T}_i$ be the natural projection map, and define $\mathcal{F}_1(\mathcal{E}, M, \omega)$, $\mathcal{F}_1(\mathcal{E}, \widehat{M}, \omega)$ as in (9.4). Let $\mathcal{K}$ and $\widehat{\mathcal{K}}$ be the cartesian factorisations for M and $\widehat{M}$, respectively, with respect to ω. For $i = 1, \ldots, \ell$, let $\gamma_i \in \Gamma_i$ such that $\omega \in \gamma_i$. Note that the identification of Ω with M and ω with 1 implies that the γ_i are subgroups of M, and so α stabilises each γ_i setwise. As $\omega\alpha = \omega$, we obtain that $\widehat{M}_{\gamma_i} = (M_{\gamma_i})^\alpha$, and so, by the definition of $\mathcal{K}$ and $\widehat{\mathcal{K}}$, we have $\widehat{\mathcal{K}} = \mathcal{K}^\alpha = \{K^\alpha \mid K \in \mathcal{K}\}$. Let $(t_1, \ldots, t_k) \in M$ with $t_1 \in T_1, \ldots, t_k \in T_k$. Then $(t_1^\alpha, \ldots, t_k^\alpha) \in \widehat{M}$ with $t_1^\alpha \in \widehat{T}_1, \ldots, t_k^\alpha \in \widehat{T}_k$. Thus $((t_1, \ldots, t_k)\sigma_i)^\alpha = (t_1, \ldots, t_k)^\alpha \widehat{\sigma}_i$ for all $i \in \{1, \ldots, k\}$. Therefore

$$K^\alpha \widehat{\sigma}_1 = (K\sigma_1)^\alpha \quad \text{for all} \quad K \in \mathcal{K}, \tag{9.5}$$

and so $K\sigma_1 \neq T_1$ if and only if $K^\alpha \widehat{\sigma}_1 \neq \widehat{T}_1$. Hence $|\mathcal{F}_1(\mathcal{E}, M, \omega)| = |\mathcal{F}_1(\mathcal{E}, \widehat{M}, \omega)|$. It also follows that if $X_1, \ldots, X_s$ are the non-trivial full strips involved in $\mathcal{K}$ covering T_1, then $X_1^\alpha, \ldots, X_s^\alpha$ are the non-trivial full strips involved in $\mathcal{K}^\alpha$ covering $\widehat{T}_1$. This shows that the definitions of r and s do not depend on the choice of the plinth.

(ii) Suppose that T_i does not admit a uniform automorphism. Then Theorem 9.1 implies that two distinct non-trivial, full strips X_1 and X_2 that are involved in $\mathcal{K}_\omega(\mathcal{E})$ must be disjoint. Thus there may exist at most one non-trivial, full strip involved in $\mathcal{K}_\omega(\mathcal{E})$ that covers T_i. Thus $s \leqslant 1$.

(iii) By Lemma 4.26, if T_i is finite then T_i does not admit a uniform automorphism, and so by part (ii), $s \leqslant 1$. The possible values for (r, s) now follow from Corollary 9.3. □

Recall the O'Nan–Scott type of a quasiprimitive permutation group with an FCR–plinth as defined in Chapter 7. Groups of type SD and CD were introduced in Section 7.4.

Theorem 9.5 *Suppose that G is an innately transitive permutation group on Ω with a non-abelian* FCR-*plinth M. If $\mathcal{E}$ is a G-invariant cartesian decomposition of Ω such that $r = 0$ and $s = 1$, with r and s as in Proposition 9.4, then G is a quasiprimitive group with type* CD.

Proof Assume that $M = T_1 \times \cdots \times T_k$ where the T_i are simple groups. Suppose that $\mathcal{E} \in \mathsf{CD}(G)$ such that $r = 0$ and $s = 1$. Let ω be a fixed element of Ω and set $\mathcal{K} = \mathcal{K}_\omega(\mathcal{E})$. Since $s = 1$, every T_i is covered by precisely one non-trivial full strip that is involved in some element of $\mathcal{K}$. For $K \in \mathcal{K}$, we define

$$\mathcal{X}_K = \{X \mid X \text{ is a non-trivial full strip involved in } K\}$$

and

$$I_K = \bigcup_{X \in \mathcal{X}_K} \operatorname{Supp} X.$$

Since $r = 0$, the elements of $\mathcal{K}$ are subdirect subgroups of M, and so $\mathcal{X}_K$ and I_K are non-empty by Theorem 4.16(iii). In particular,

$$K = \prod_{X \in \mathcal{X}_K} X \times \prod_{T_i \notin I_K} T_i \quad \text{for all} \quad K \in \mathcal{K}.$$

Set $\mathcal{X} = \bigcup_{K \in \mathcal{K}} \mathcal{X}_K$. Since the stabiliser G_ω is transitive on the set $\{T_1, \ldots, T_k\}$ and also acts on the set $\mathcal{X}$ by conjugation, each T_i is covered by a strip in $\mathcal{X}$. Also since $s = 1$, each T_i is covered by precisely one element of $\mathcal{X}$. Therefore

$$M_\omega = \bigcap_{K \in \mathcal{K}} K = \prod_{K \in \mathcal{K}} \left(\prod_{X \in \mathcal{X}_K} X \right) = \prod_{X \in \mathcal{X}} X.$$

Thus M_ω is a subdirect subgroup of M, and hence the type of G is either SD or CD. The group G is quasiprimitive by Theorem 7.3. Since there are at least $|\mathcal{K}| \geqslant 2$ strips X involved in $\mathcal{K}$, we have that $M_\omega \not\cong T_1$; therefore G is of type CD. □

When M in Proposition 9.4 is a finite group, then Proposition 9.4(iii) gives that there are only five possibilities for the pair (r, s).

Definition 9.6 Let G be a not necessarily finite innately transitive permutation group on Ω with a non-abelian FCR-plinth $M = T_1 \times \cdots \times T_k$ where the T_i are simple groups and let $\omega \in \Omega$. We recall that, for $\mathcal{E} \in \mathsf{CD}_{\mathrm{tr}}(G)$, s is the number of non-trivial full strips involved in $\mathcal{K}_\omega(\mathcal{E})$ that cover T_1 and r is the number of subgroups K in $\mathcal{K}_\omega(\mathcal{E})$ such that $K\sigma_1 < T_1$. We define the following families of transitive cartesian decompositions:

$$\begin{aligned}
\mathsf{CD}_{\mathrm{S}}(G) &= \{\mathcal{E} \in \mathsf{CD}_{\mathrm{tr}}(G) \mid r = 0 \text{ and } s = 1\};\\
\mathsf{CD}_{1}(G) &= \{\mathcal{E} \in \mathsf{CD}_{\mathrm{tr}}(G) \mid r = 1 \text{ and } s = 0\};\\
\mathsf{CD}_{1\mathrm{S}}(G) &= \{\mathcal{E} \in \mathsf{CD}_{\mathrm{tr}}(G) \mid s = r = 1\};\\
\mathsf{CD}_{2\sim}(G) &= \{\mathcal{E} \in \mathsf{CD}_{\mathrm{tr}}(G) \mid r = 2,\ s = 0 \text{ and}\\
&\qquad \text{the subgroups of } \mathcal{F}_i(\mathcal{E}, M, \omega) \text{ are isomorphic}\};\\
\mathsf{CD}_{2\not\sim}(G) &= \{\mathcal{E} \in \mathsf{CD}_{\mathrm{tr}}(G) \mid r = 2,\ s = 0 \text{ and}\\
&\qquad \text{the subgroups in } \mathcal{F}_i(\mathcal{E}, M, \omega) \text{ are not isomorphic}\};\\
\mathsf{CD}_{3}(G) &= \{\mathcal{E} \in \mathsf{CD}_{\mathrm{tr}}(G) \mid r = 3 \text{ and } s = 0\}.
\end{aligned}$$

As mentioned above, in Definition 9.6, we do not assume that the group G is finite. In Section 9.4, we show that for each class defined above, there exists an innately transitive permutation group admitting a cartesian decomposition that belongs to that class. In fact, for each of these classes, there exists a *finite* innately transitive permutation group G for which the class is non-empty.

Theorem 9.7 (6-class Theorem) *If G is an innately transitive permutation group with a non-abelian* FCR*-plinth M, then the classes $\mathsf{CD}_1(G)$, $\mathsf{CD}_{\mathrm{S}}(G)$, $\mathsf{CD}_{1\mathrm{S}}(G)$, $\mathsf{CD}_{2\sim}(G)$, $\mathsf{CD}_{2\not\sim}(G)$, and $\mathsf{CD}_3(G)$ are subsets of $\mathsf{CD}_{\mathrm{tr}}(G)$ and are independent of the choices of ω, i, and M. If G is finite, then these classes form a partition of $\mathsf{CD}_{\mathrm{tr}}(G)$. Moreover, if M is a finite simple group, then $\mathsf{CD}_{\mathrm{tr}}(G) = \mathsf{CD}_{2\sim}(G)$.*

Proof By Proposition 9.4, we only need to show that the definitions of the classes $\mathsf{CD}_{2\sim}(G)$ and $\mathsf{CD}_{2\not\sim}(G)$ are independent of ω, of i, and of M. The simple argument given after (9.4) shows that, for $j = 1,\ 2$, the subgroups in $\mathcal{F}_i(\mathcal{E}, M, \omega_j)$ are isomorphic for some i if and only if they are isomorphic for all i. Suppose that the subgroups in $\mathcal{F}_1(\mathcal{E}, M, \omega_1)$ are isomorphic. Let $\omega_2 \in \Omega$ and $m \in M$ such that $\omega_1 m = \omega_2$. If $\mathcal{F}_1(\mathcal{E}, M, \omega_1) = \{A, B\}$, then $\mathcal{F}_1(\mathcal{E}, M, \omega_1)^m = \mathcal{F}_1(\mathcal{E}, M, \omega_2)$, and hence the subgroups A^m and B^m in $\mathcal{F}_1(\mathcal{E}, M, \omega_2)$ are isomorphic. Hence the definitions of $\mathsf{CD}_{2\sim}(G)$ and $\mathsf{CD}_{2\not\sim}(G)$ are independent of the choices of ω and i.

To show that the definitions of these classes are independent of the plinth M, let us use the notation introduced in the last paragraph of the proof of Proposition 9.4 and assume that $\widehat{M}$ is another transitive minimal normal subgroup of G. Suppose that $\mathcal{F}_1(\mathcal{E}, M, \omega) = \{A, B\}$ and $\mathcal{F}_1(\mathcal{E}, \widehat{M}, \omega) = \{\widehat{A}, \widehat{B}\}$. Let α be an involution of $\mathsf{Sym}\,\Omega$ that centralises G_ω and interchanges M and $\widehat{M}$ (see Corollary 3.11). Equation (9.5) implies that we may assume without loss of generality that $\widehat{A} = A^\alpha$ and $\widehat{B} = B^\alpha$. Hence subgroups A and B are isomorphic, if and only if $\widehat{A}$ and $\widehat{B}$ are isomorphic.

When M is finite, then Corollary 9.3 implies that transitive G-invariant cartesian decompositions can be classified into one of these six classes.

If M is a finite simple group, then the map σ_1 is the identity map $M \to M$, and so $\mathcal{K}_\omega(\mathcal{E}) = \mathcal{F}_1(\mathcal{E}, M, \omega)$. Theorem 8.17 implies that $|\mathcal{E}| = 2$ and hence that $|\mathcal{F}_1(\mathcal{E}, M, \omega)| = 2$. As $\mathcal{E} \in \mathsf{CD}_{\mathrm{tr}}(G)$, the stabiliser G_ω

is transitive on $\mathcal{K}_\omega(\mathcal{E})$, and hence on $\mathcal{F}_1(\mathcal{E}, M, \omega)$. Thus $\mathcal{E} \in \mathsf{CD}_{2\sim}(G)$. □

9.4 Examples of cartesian decompositions

Next we give several examples to show that each of the classes of cartesian decompositions defined before Theorem 9.7 is non-empty for certain groups. In the following constructions, the simple group T need not be finite. Based on the general Example 9.9, one can construct more explicit examples with T finite using the material in Lemmas 8.14 and 8.15 and in the Appendix.

Example 9.8 Cartesian decompositions in $\mathsf{CD}_1(G)$ and $\mathsf{CD}_\mathrm{S}(G)$ are easily constructed using the product action of wreath products introduced in Section 5.2.2. Suppose that H is a quasiprimitive group with non-abelian FCR-socle $M = T_1 \times \cdots \times T_k$ acting on Γ and let K be a transitive subgroup of S_ℓ. Then by Theorem 5.18, $G = H \wr K$ is a quasiprimitive group with socle M^ℓ acting on Γ^ℓ preserving the natural cartesian decomposition $\mathcal{E}$ of Γ^ℓ (see Section 5.2.2). Now $\mathcal{E} \in \mathsf{CD}_1(G)$ if, for $\gamma \in \Gamma$, the stabiliser M_γ is not a subdirect subgroup of M, while if M_γ is subdirect, then $\mathcal{E} \in \mathsf{CD}_\mathrm{S}(G)$.

Example 9.9 Let T be a non-abelian simple group and let $A_1, \ldots, A_m$ be an abstract cartesian factorisation of T with proper subgroups. Let $K_1 = A_1 \times \cdots \times A_m$. Let C_m denote the cyclic group of order m with generator x. Then C_m acts on T^m permuting the factors. More precisely, $(t_1, \ldots, t_m)^x = (t_m, t_1, \ldots, t_{m-1})$. Set, for $i = 1, \ldots, m$, $K_i = (K_1)^{x^i}$. Consider T as a permutation group acting on the right coset space $\Gamma = [T\colon A_1 \cap \cdots \cap A_m]$, let $\gamma_0 = A_1 \cap \cdots \cap A_m \in \Gamma$, let $G = T \wr C_m$ and consider G as a permutation group acting on Γ^m (see Section 5.2.2). In this way $G = T \wr C_m$ is a quasiprimitive group with socle T^m and $\mathcal{K} = \{K_1, \ldots, K_m\}$ is an $(A_1 \cap \cdots \cap A_m)^m \rtimes C_m$-invariant cartesian factorisation of T^m with respect to $\omega = (\gamma_0, \ldots, \gamma_0)$. Hence if $\Omega = \Gamma^m$, then G acts on Ω preserving the cartesian decomposition $\mathcal{E} = \mathcal{E}(\{K_1, \ldots, K_m\})$. Further, if s and r are defined as in Proposition 9.4, then $r = m$ and $s = 0$. Choosing T to be a suitable finite group, we may construct cartesian decompositions in the classes $\mathsf{CD}_{2\not\sim}(G)$, $\mathsf{CD}_{2\sim}(G)$, and $\mathsf{CD}_3(G)$ using the cartesian factorisations given in Lemma 8.14.

Finally we construct examples of cartesian decompositions in the class $\mathsf{CD}_{1\mathrm{S}}(G)$.

Example 9.10 Let T be a simple group and let A, B be proper isomorphic subgroups of T such that $T = AB$ and there exists $\vartheta \in \mathsf{Aut}(T)$ that interchanges A and B. For T finite, the possibilities for T, A, and B can be found in Table 8.2. Define the subgroups K_1 and K_2 of T^4 by

$$\begin{aligned} K_1 &= \{(t_1, t_2, t_3, t_4) \mid t_1 \in A,\ t_2 \in B,\ t_3 = t_4\}; \\ K_2 &= \{(t_1, t_2, t_3, t_4) \mid t_1 = t_2,\ t_3 \in A,\ t_4 \in B\}. \end{aligned}$$

One can check using Lemma 8.12(ii) that $(A \times B)\{(t,t) \mid t \in T\} = T \times T$, and hence $K_1K_2 = T^4$. Suppose that $C \cong C_2 \times C_2$ is a group generated by π_1, π_2 and C acts on T^4 as follows:

$$(t_1, t_2, t_3, t_4)^{\pi_1} = (t_2\vartheta, t_1\vartheta, t_4\vartheta, t_3\vartheta) \text{ and } (t_1, t_2, t_3, t_4)^{\pi_2} = (t_3, t_4, t_1, t_2).$$

It is easy to see that this defines an action of C on T^4 and we let $G = T^4 \rtimes C$. Then G is a group with socle T^4. Further, π_1 normalises both K_1 and K_2 while π_2 swaps K_1 and K_2. The group G acts quasiprimitively on the right coset space $[G\colon G_0]$ where $G_0 = (K_1 \cap K_2)C$. If ω denotes the trivial coset $(K_1 \cap K_2)C$ then $M_\omega = K_1 \cap K_2$. Thus $\{K_1, K_2\}$ is a G_ω-invariant cartesian factorisation of M and the corresponding cartesian decomposition is an element of $\mathsf{CD}_{1\mathrm{S}}(G)$.

9.5 Simple group factorisations and cartesian decompositions

Theorem 9.2 shows that the existence of a G-invariant cartesian decomposition in a particular class may pose severe restriction on the structure of G. This is made more explicit in the following theorem. Recall that an innately transitive group G with plinth M has type CD if a point stabiliser M_ω is a subdirect subgroup of M and is not simple; see Section 7.4.

Theorem 9.11 *Suppose that G is an innately transitive permutation group with a non-abelian* FCR*-plinth M, and let T be the isomorphism type of a simple direct factor of M. Then the following hold.*

(i) *Suppose that either $\mathsf{CD}_{2\sim}(G) \neq \emptyset$ or $\mathsf{CD}_{1\mathrm{S}}(G) \neq \emptyset$ and T does not admit a uniform automorphism. Then T admits a factorisation with two proper, isomorphic subgroups. If T is finite, then it is isomorphic to one of the groups A_6, M_{12}, $\mathsf{P}\Omega_8^+(q)$, or $\mathsf{Sp}_4(2^a)$ with $a \geqslant 2$.*

(ii) *If* $\mathsf{CD}_{2\not\sim}(G) \neq \emptyset$ *then* T *admits a factorisation with proper subgroups.*

(iii) *If* $\mathsf{CD}_3(G) \neq \emptyset$ *then* T *admits a cartesian factorisation with three subgroups. If* T *is finite, then it is isomorphic to one of the groups* $\mathsf{Sp}_{4a}(2)$ *with* $a \geqslant 2$, $\mathsf{P}\Omega_8^+(3)$, *or* $\mathsf{Sp}_6(2)$.

Moreover, for each $\mathrm{x} \in \{\mathrm{S},\ 1,\ 1\mathrm{S},\ 2\sim,\ 2\not\sim,\ 3\}$ *there is some* G *as above such that* $\mathsf{CD}_{\mathrm{x}}(G) \neq \emptyset$.

Proof It follows from Theorem 9.2 and the definitions of $\mathsf{CD}_{2\sim}(G)$ and $\mathsf{CD}_{2\not\sim}(G)$ that the claimed factorisations are admitted by T. Factorisations of finite simple groups with isomorphic subgroups were listed in Table 8.2, while cartesian factorisations with three subgroups of finite simple groups can be found in Table 8.1. These results imply that the isomorphism type of T in parts (i) and (iii) is as claimed.

The final assertion follows from Examples 9.8–9.10. □

There are finite simple groups that admit no factorisations with proper subgroups: for example, $\mathsf{PSL}_2(q)$ with $q \equiv 1 \pmod 4$ and $q \notin \{5, 9, 29\}$, $\mathsf{PSU}_{2m+1}(q)$ (with a finite number of exceptions) and some sporadic groups (see the tables in (Liebeck *et al.* 1990)). Therefore Theorem 9.11(ii) also restricts the isomorphism type of T.

9.6 The cartesian decompositions in $\mathsf{CD}_{\mathrm{S}}(G)$ and in $\mathsf{CD}_1(G)$

In this section we describe the elements of $\mathsf{CD}_{\mathrm{S}}(G)$ and $\mathsf{CD}_1(G)$ for an innately transitive permutation group G. First we characterise normal cartesian decompositions, which were defined in Section 4.7, in terms of the parameters r and s introduced in Proposition 9.4. By Theorem 4.22, a normal cartesian decomposition for an innately transitive group is transitive.

Theorem 9.12 *Let* G *be an innately transitive group on* Ω *with a non-abelian* FCR*-plinth* M. *A cartesian decomposition* $\mathcal{E} \in \mathsf{CD}_{\mathrm{tr}}(G)$ *is normal if and only if* $r = 1$ *and* $s = 0$, *or* $r = 0$ *and* $s = 1$.

Proof First we note that Theorem 4.22 implies that any G-invariant normal cartesian decomposition is transitive and M-normal. Suppose that $M = T_1 \times \cdots \times T_k$ where the T_i are non-abelian simple groups. The definition of normal cartesian decompositions implies that a decomposition $\mathcal{E}$ is M-normal if and only if for all T_i there is a unique $\Gamma \in \mathcal{E}$

such that T_i is not in the kernel of the M-action on Γ. In other words, $\mathcal{E}$ is normal if and only if for all T_i there is a unique $K \in \mathcal{K}(\mathcal{E})$ such that $T_i \not\leqslant K$. For this unique K, we must have that either $K\sigma_i$ is a proper subgroup of T_i or K contains a non-trivial full strip that covers T_i. Thus $r = 1$ and $s = 0$, or $r = 0$ and $s = 1$. □

Using the above characterisation of normal cartesian decompositions, we can describe the class of cartesian decompositions that are preserved by quasiprimitive groups of type SD or CD as defined in Section 7.4.

Theorem 9.13 *Let G be a quasiprimitive group with* FCR*-plinth $M = T_1 \times \cdots \times T_k$.*

(i) *If the O'Nan–Scott type of G is* SD*, then $\mathsf{CD}(G) = \emptyset$.*
(ii) *The O'Nan–Scott type of G is* CD *if and only if $\mathsf{CD}_{\mathrm{S}}(G) \neq \emptyset$. Further, in this case, $\mathsf{CD}(G) = \mathsf{CD}_{\mathrm{S}}(G)$.*

Proof Let T denote the common isomorphism type of the T_i. Suppose that M_ω is a subdirect subgroup of M. Suppose that $|\mathsf{CD}(G)| \geqslant 1$ and that $\mathcal{E}$ is a non-trivial element in $\mathsf{CD}(G)$. For each $K \in \mathcal{K}_\omega(\mathcal{E})$ we have $M_\omega \leqslant K$, and so all elements of $\mathcal{K}_\omega(\mathcal{E})$ are subdirect subgroups of M, and so $r = 0$. Let $K_1,\ K_2 \in \mathcal{K}_\omega(\mathcal{E})$ be two distinct subgroups. Then $K_1,\ K_2 \neq M$, and so, by Theorem 4.16, $K_1,\ K_2$ involve non-trivial full strips X_1 and X_2, say, and by (8.2), $X_1 \neq X_2$. By Theorem 9.1, X_1 and X_2 are disjoint strips, and so $s = 1$. Thus $\mathcal{E} \in \mathsf{CD}_{\mathrm{S}}(G)$. Suppose that $T_{i_1} \in \operatorname{Supp} X_1$ and $T_{i_2} \in \operatorname{Supp} X_2$. Then $T_{i_1}^g = T_{i_2}$ for some $g \in G_\omega$, and so Theorem 9.1 implies that $X_1^g = X_2$. This, in turn, yields $K_1^g = K_2$. Thus $\mathcal{E} \in \mathsf{CD}_{\mathrm{tr}}(G)$. Note that our argument also implies that if a non-trivial, full strip X is involved in $\mathcal{K}$, then it is also involved in M_ω. The supports of the full strips involved in M_ω form a G-invariant partition $\mathcal{P}$ of $\{T_1, \ldots, T_k\}$ such that $M_\omega = \prod_{P \in \mathcal{P}} M_\omega \sigma_P$. Since $|\mathcal{P}| \geqslant 2$, M_ω is not simple, and so the type of G is CD (Table 7.1). This shows item (i) and the second statement of item (ii).

The two directions of the first statement of item (ii) follow from Theorem 9.5 and from the construction in Example 9.8. □

Remark 9.14 Suppose that G is a primitive permutation group with type HC (see Section 7.3). Hence G has two minimal normal subgroups M and N. Let π be the involution in Corollary 3.11 that swaps M and N and set L to be the group generated by G and π. Then L is a primitive permutation group of type CD and its unique minimal normal subgroup is $M \times N$. Further L preserves a cartesian decomposition $\mathcal{E}$ such that $\mathcal{E} \in \mathsf{CD}_{\mathrm{S}}(L)$ by Theorem 9.13. The cartesian decomposition $\mathcal{E}$ is invariant also under the smaller permutation group G. However, to determine its type relative to G, we need to consider a corresponding cartesian factorisation in a minimal normal subgroup of G, for instance, in M. Assuming that $M = T_1 \times \cdots \times T_k$ where the T_i are simple groups, a cartesian factorisation of M that corresponds to this cartesian decomposition is

$$\{T_1 \times \cdots \times T_{i-1} \times T_{i+1} \times \cdots \times T_k \mid i \in \underline{k}\}.$$

Hence the cartesian decomposition $\mathcal{E}$ is a member of $\mathsf{CD}_1(G)$.

This example shows that if $\mathcal{E}$ is a cartesian decomposition preserved by two innately transitive groups G and L, then $\mathcal{E}$ may belong to different classes relative to G and L.

9.7 The structure of finite innately transitive permutation groups preserving a transitive cartesian decomposition

In this section, we further investigate the structure of *finite* innately transitive groups that preserve cartesian decompositions. Concentrating on the classes $\mathsf{CD}_{1\mathrm{S}}(G)$, $\mathsf{CD}_{2\sim}(G)$, $\mathsf{CD}_3(G)$, the information obtained from the corresponding cartesian factorisation is so strong that we will be able to restrict the isomorphism type of the plinth and the permutational isomorphism type of G.

Throughout the section we assume that G is a finite innately transitive group acting on Ω with a non-abelian plinth $M = T_1 \times \cdots \times T_k$ where the T_i are the simple direct factors of M all isomorphic to a finite simple group T. Further, we suppose that $\mathcal{E} = \{\Gamma_1, \ldots, \Gamma_\ell\}$ is a G-invariant cartesian decomposition of Ω. As usual we denote by σ_i the i-th coordinate projection $M \to T_i$. We fix $\omega \in \Omega$ and set $\mathcal{K} = \mathcal{K}_\omega(\mathcal{E}) = \{K_1, \ldots, K_\ell\}$ to denote the cartesian factorisation of M that corresponds to $\mathcal{E}$ with respect to ω, as in Lemma 8.6.

We analyse the decompositions type by type. For the types $\mathsf{CD}_{1\mathrm{S}}(G)$, $\mathsf{CD}_{2\sim}(G)$, $\mathsf{CD}_3(G)$ of cartesian decompositions, the cartesian factorisation $\mathcal{K}$ of M and, in turn, the resulting factorisation of T lead to strong restrictions on the structure of G. The factorisations of T that are obtained from cartesian decompositions of type $\mathsf{CD}_{2\not\sim}(G)$ are too generic to draw strong conclusions. The main results of the section are summarised in the following theorem.

Theorem 9.15 *Suppose that G is a finite innately transitive group with non-abelian plinth $M \cong T^k$ where T is a finite simple group with $k \geqslant 2$ and let $\mathcal{E} \in \mathsf{CD}_{\mathrm{tr}}(G)$. Define $\mathcal{F}_i(\mathcal{E}, M, \omega)$ as in (9.4).*

(i) *If $\mathcal{E} \in \mathsf{CD}_{1\mathrm{S}}(G)$ or $\mathcal{E} \in \mathsf{CD}_{2\sim}(G)$, then the isomorphism type of T is as in one of the rows of Table 8.2. Further, the subgroups contained in the $\mathcal{F}_i(\mathcal{E}, M, \omega)$ are all isomorphic to the group in the third column of the corresponding row of Table 8.2.*

(ii) *If $\mathcal{E} \in \mathsf{CD}_3(G)$, then the isomorphism type of T and those of the subgroups in $\mathcal{F}(\mathcal{E}, M, \omega)$ are as in one of the rows of Table 8.1.*

Further if $\mathcal{E}$ is as in (i)–(ii) and T is not as in the 4th row of Table 8.2 or in the 3rd row of Table 8.1, then G preserves a normal cartesian decomposition of Ω. The index of this normal decomposition is $k/2$ if $\mathcal{E} \in \mathsf{CD}_{1\mathrm{S}}(G)$, while it is k in the other cases.

The proof of the theorem will be given in Section 9.7.4.

9.7.1 Cartesian decompositions in $\mathsf{CD}_{1\mathrm{S}}(G)$

In this section we assume that G is a finite innately transitive group with a non-abelian plinth $M = T_1 \times \cdots \times T_k = T^k$, that $\mathcal{E}$ is a G-transitive cartesian decomposition and that $\mathcal{E} \in \mathsf{CD}_{1\mathrm{S}}(G)$ (see Definition 9.6). For $K \in \mathcal{K}$, let $\mathcal{X}_K$ denote the set of non-trivial, full strips involved in K, and set $\mathcal{X} = \mathcal{X}_{K_1} \cup \cdots \cup \mathcal{X}_{K_\ell}$. By Theorem 9.2 and Corollary 9.3, $\mathcal{X}$ contains $k/2$ pairwise disjoint, full strips, each of length 2. By the definition of $\mathsf{CD}_{1\mathrm{S}}(G)$, for each $i \in \{1, \ldots, k\}$ there exists precisely one K_j such that $K_j\sigma_i < T_i$.

Recall that G_ω is transitive on $\mathcal{K}_\omega(\mathcal{E})$ since $\mathcal{E} \in \mathsf{CD}_{\mathrm{tr}}(G)$. For $K \in \mathcal{K}$, set

$$\mathcal{T}_K = \{T_i \mid T_i \notin \operatorname{Supp} X \text{ for all } X \in \mathcal{X}_K\}.$$

The set $\mathcal{T}_K$ is non-empty, by the definition of the class $\mathsf{CD}_{1\mathrm{S}}(G)$.

Proposition 9.16 *Let A denote the isomorphism type of the group in $\mathcal{F}_i(\mathcal{E}, M, \omega)$ (which is independent of i). Then T and A are as in one of the rows of Table 8.2. If one of the rows 1–3 of Table 8.2 is valid, then*

$$K = \prod_{X \in \mathcal{X}_K} X \times \prod_{T_m \in \mathcal{T}_K} K\sigma_m \quad \text{for all} \quad K \in \mathcal{K}, \tag{9.6}$$

while if row 4 is valid, then

$$\prod_{X \in \mathcal{X}_K} X \times \prod_{T_m \in \mathcal{T}_K} (K\sigma_m)' \leqslant K \tag{9.7}$$

for all $K \in \mathcal{K}$.

Proof For $i = 1, \ldots, \ell$ let $\widehat{K}_i$ denote $\bigcap_{j \neq i} K_j$. By Corollary 9.3, each non-trivial, full strip involved in any of the K_i has length 2. Suppose without loss of generality that X is a non-trivial full strip involved in K_1 covering T_1 and T_2. Thus by (8.2), $T_1 \times T_2 = (K_1\sigma_{\{1,2\}})(\widehat{K}_1\sigma_{\{1,2\}})$. Suppose that $X = \{(t, t\alpha) \mid t \in T_1\}$ for some isomorphism $\alpha\colon T_1 \to T_2$. As was shown in the proof of Theorem 9.2(ii), $T_1 = (\widehat{K}_1\sigma_1)(\widehat{K}_1\sigma_2\alpha^{-1})$ is a factorisation with proper, isomorphic subgroups. Therefore Lemma 8.14(iii) implies that $T_1 \cong T$ is as in Table 8.2 and the isomorphism types of $\widehat{K}_1\sigma_1$ and $\widehat{K}_1\sigma_2$ are as in the third column of the corresponding row of Table 8.2.

Suppose that $K_j\sigma_i \neq T_i$ for some i and j. Then there is a non-trivial full strip $X \in \mathcal{X}$ covering T_i; assume that $X \in \mathcal{X}_{K_m}$ for some $m \in \{1, \ldots, \ell\} \setminus \{j\}$ and that $\operatorname{Supp} X = \{T_i, T_{i'}\}$. Then as shown above, $\widehat{K}_m\sigma_i$ is as in the third column of Table 8.2. In particular, $\widehat{K}_m\sigma_i$ is a maximal subgroup of T_i. Since $\widehat{K}_m \leqslant K_j$, we obtain that $\widehat{K}_m\sigma_i \leqslant K_j\sigma_i < T_i$, and so $\widehat{K}_m\sigma_i = K_j\sigma_i$. Therefore $K_j\sigma_i$ is also as in the third column of the corresponding row of the table.

We have proved so far that for all i and j either $K_j\sigma_i = T_i$ or $K_j\sigma_i \cong A$ where A is as in the third column of Table 8.2. In particular A is a maximal subgroup of T, A is almost simple, and if T is as in rows 1–3 of Table 8.2 then A is simple. Suppose by contradiction that (9.7) fails to hold for some K_j. Set $S = \mathcal{T}_{K_j}$ and $\overline{S} = \{T_1, \ldots, T_k\} \backslash \mathcal{T}_{K_j}$, and write $\sigma_S, \sigma_{\overline{S}}$ for the projection of M onto $\prod_{s \in S} T_s$ and $\prod_{s \in \overline{S}} T_s$ respectively. Then it follows from the definition of $\mathcal{X}_{K_j}$ that

$$K_j = K_j\sigma_S \times K_j\sigma_{\overline{S}}.$$

As (9.7) fails for K_j we must have that

$$\prod_{m \in S} (K_j\sigma_m)' \nleqslant K_j\sigma_S.$$

Thus it follows from Lemma 4.10 that there are distinct elements i_1, i_2 of S such that

$$(K_j\sigma_{i_1})' \times (K_j\sigma_{i_2})' \nleqslant K_j\sigma_{\{i_1,i_2\}}. \tag{9.8}$$

If $K_j\sigma_{i_1} = T_{i_1}$ then, by Theorem 4.16(ii), there is a full strip X involved in K_j covering T_{i_1}. By the definition of S, we must have that $X = T_{i_1}$, and so $K_j\sigma_{i_1} \leqslant K_j$. Hence $K_j\sigma_{i_1} \leqslant K_j\sigma_{\{i_1,i_2\}}$, and also $K_j\sigma_{i_2} \leqslant K_j\sigma_{\{i_1,i_2\}}$. Hence $K_j\sigma_{i_1} \times K_j\sigma_{i_2} = K_j\sigma_{\{i_1,i_2\}}$, contradicting (9.8). Thus $K_j\sigma_{i_1}$ is a proper subgroup of T_{i_1}, and also $K_j\sigma_{i_2}$ is a proper subgroup of T_{i_2}.

By Theorem 9.2, G_ω is transitive on $\mathcal{X}$, and so there are (not necessarily distinct) strips X_1 and X_2 in $\mathcal{X}$ such that X_1 covers T_{i_1} and X_2 covers T_{i_2}. Suppose that $X_1 = X_2$. Then $\operatorname{Supp} X_1 = \{T_{i_1}, T_{i_2}\}$, and let $j_1 \in \{1, \ldots, \ell\} \setminus \{j\}$ be such that $X_1 \in \mathcal{X}_{K_{j_1}}$. Then, as verified above, $\widehat{K}_{j_1}\sigma_{i_1}$ and $\widehat{K}_{j_1}\sigma_{i_2}$ are maximal subgroups of T_{i_1} and T_{i_2}, respectively, and, in addition, $\widehat{K}_{j_1}\sigma_{i_1} \cong \widehat{K}_{j_1}\sigma_{i_2}$. Thus the factorisation

$$X_1(\widehat{K}_{j_1}\sigma_{\{i_1,i_2\}}) = (K_{j_1}\sigma_{\{i_1,i_2\}})(\widehat{K}_{j_1}\sigma_{\{i_1,i_2\}}) = T_{i_1} \times T_{i_2}$$

is a full strip factorisation (see Section A.2). Hence Theorem A.6 implies that

$$(\widehat{K}_{j_1}\sigma_{i_1})' \times (\widehat{K}_{j_1}\sigma_{i_2})' \leqslant \widehat{K}_{j_1}\sigma_{\{i_1,i_2\}}.$$

Note that $j \neq j_1$, and so $\widehat{K}_{j_1}\sigma_{\{i_1,i_2\}} \leqslant K_j\sigma_{\{i_1,i_2\}}$. Moreover, $\widehat{K}_{j_1}\sigma_{i_1}$ is a maximal subgroup of T_{i_1} and so is $K_j\sigma_{i_1}$. As $\widehat{K}_{j_1}\sigma_{i_1} \leqslant K_j\sigma_{i_1}$, we obtain that $\widehat{K}_{j_1}\sigma_{i_1} = K_j\sigma_{i_1}$, and, similarly, $K_{j_1}\sigma_{i_2} = K_j\sigma_{i_2}$. Therefore

$$(K_j\sigma_{i_1})' \times (K_j\sigma_{i_2})' \leqslant K_j\sigma_{\{i_1,i_2\}},$$

which is a contradiction. Hence $X_1 \neq X_2$.

Suppose that X_1 is involved in K_{j_1} and X_2 is involved in K_{j_2}, where j_1 and j_2 are not necessarily distinct elements of $\{1, \ldots, \ell\} \setminus \{j\}$. Let $I = \operatorname{Supp} X_1 \cup \operatorname{Supp} X_2$ and set $\widehat{K}_{j_1,j_2} = \bigcap_{m \neq j_1,j_2} K_m$. Then, by Lemma 8.9(iii), $(K_{j_1} \cap K_{j_2})\widehat{K}_{j_1,j_2} = M$, and so

$$M\sigma_I = ((K_{j_1} \cap K_{j_2})\sigma_I)(\widehat{K}_{j_1,j_2}\sigma_I).$$

Suppose that $n \in \operatorname{Supp} X_1 \cup \operatorname{Supp} X_2$; in fact suppose without loss of generality that $n \in \operatorname{Supp} X_1$. Then the argument above shows that $\widehat{K}_{j_1}\sigma_n \cong A$ and also $K_{j'}\sigma_n \cong A$ where A is as in the third column of Table 8.2 and $j' \in \{1, \ldots, \ell\}$ is such that $K_{j'}\sigma_n < T_n$. Since,

$$\widehat{K}_{j_1}\sigma_n \leqslant \widehat{K}_{j_1,j_2}\sigma_n \leqslant K_{j'}\sigma_n,$$

we obtain that $\widehat{K}_{j_1,j_2}\sigma_n \cong A$ holds for all $n \in \operatorname{Supp} X_1 \cup \operatorname{Supp} X_2$.

Clearly $(K_{j_1} \cap K_{j_2})\sigma_I \leqslant X_1 \times X_2$, and so

$$M\sigma_I = (X_1 \times X_2)(\widehat{K}_{j_1,j_2}\sigma_I).$$

Then it follows from Theorem A.6 that

$$\begin{aligned}(\widehat{K}_{j_1,j_2}\sigma_{\min X_1})' \quad &\times \quad (\widehat{K}_{j_1,j_2}\sigma_{\max X_1})' \\ &\times \quad (\widehat{K}_{j_1,j_2}\sigma_{\min X_2})' \times (\widehat{K}_{j_1,j_2}\sigma_{\max X_2})' \leqslant \widehat{K}_{j_1,j_2}\sigma_I.\end{aligned}$$

As $i_1,\ i_2 \in I$, we obtain that

$$(\widehat{K}_{j_1,j_2}\sigma_{i_1})' \times (\widehat{K}_{j_1,j_2}\sigma_{i_2})' \leqslant \widehat{K}_{j_1,j_2}\sigma_{\{i_1,i_2\}} \leqslant K_j\sigma_{\{i_1,i_2\}}.$$

Since $(\widehat{K}_{j_1,j_2}\sigma_{i_1})' = (K_j\sigma_{i_1})'$ and $(\widehat{K}_{j_1,j_2}\sigma_{i_2})' = (K_j\sigma_{i_2})'$, this is a contradiction. Hence (9.7) holds. If T is as in rows 1–3, then $K_j\sigma_i$ is simple, and hence perfect. This proves (9.6). □

9.7.2 Cartesian decompositions in $\mathsf{CD}_{2\sim}(G)$

In this section we assume that G is a finite innately transitive group with a non-abelian plinth $M = T_1 \times \cdots \times T_k = T^k$, that $\mathcal{E}$ is G-transitive and that $\mathcal{E} \in \mathsf{CD}_{2\sim}(G)$ (Definition 9.6). Recall that the sets $\mathcal{F}_i(\mathcal{E}, M, \omega)$ were defined in (9.4). By definition, each $\mathcal{F}_i(\mathcal{E}, M, \omega)$ contains two isomorphic subgroups of T_i.

Proposition 9.17 *Suppose that G, M, T, ω, $\mathcal{E}$, $\mathcal{F}_i(\mathcal{E}, M, \omega)$ are as above. Then the isomorphism type of T and that of the subgroups in $\mathcal{F}_i(\mathcal{E}, M, \omega)$ are as in one of the rows of Table 8.2. If one of the rows 1–3 of Table 8.2 is valid then*

$$K = K\sigma_1 \times \cdots \times K\sigma_k \quad \text{for} \quad K \in \mathcal{K}, \tag{9.9}$$

while if row 4 is valid then

$$(K\sigma_1)' \times \cdots \times (K\sigma_k)' \leqslant K \quad \text{and} \quad \frac{K}{(K\sigma_1)' \times \cdots \times (K\sigma_k)'} \leqslant \mathbb{Z}_2^k \tag{9.10}$$

for $K \in \mathcal{K}$.

Proof For $i = 1, \ldots, k$, we have $\mathcal{F}_i(\mathcal{E}, M, \omega) = \{A_i, B_i\}$, as above. Since G acts transitively on $\mathcal{T}$ by conjugation, and since, by the definition of $\mathsf{CD}_{2\sim}(G)$, A_i and B_i are isomorphic, the subgroups $A_1, \ldots, A_k$ and $B_1, \ldots, B_k$ are pairwise isomorphic. Also, since $T_1 = A_1B_1$ is a factorisation of a finite simple group with two isomorphic subgroups, it follows from Lemma 8.14 that T and $\mathcal{F}_1(\mathcal{E}, M, \omega)$ are as in Table 8.2. Suppose that $(K_j\sigma_1)' \times \cdots \times (K_j\sigma_k)' \not\leqslant K_j$, for some j. Then it follows from Lemma 4.10 that there are $i_1,\ i_2 \in \{1,\ldots,k\}$ such that

$$(K_j\sigma_{i_1})' \times (K_j\sigma_{i_2})' \not\leqslant K_j\sigma_{\{i_1,i_2\}}. \tag{9.11}$$

Suppose first that $(K_j\sigma_{i_1}) = T_{i_1}$. Then Theorem 4.16 implies that K_j involves a full strip X covering T_{i_1}. However, by Corollary 9.3, X cannot be a non-trivial strip since $\mathcal{E} \in \mathsf{CD}_{2\sim}(G)$. Thus $X = T_{i_1}$, and so $T_{i_1} \leqslant K_j$. This, however, implies that $K_j\sigma_{i_1} \leqslant K_j\sigma_{\{i_1,i_2\}}$, and in this case we must also have $K_j\sigma_{i_2} \leqslant K_j\sigma_{\{i_1,i_2\}}$. Therefore $K_j\sigma_{i_1} \times K_j\sigma_{i_2} = K_j\sigma_{\{i_1,i_2\}}$ contradicting (9.11). Hence $K_j\sigma_{i_1} < T_{i_1}$, and the same argument shows that $K_j\sigma_{i_2} < T_{i_2}$.

Since $\mathcal{E} \in \mathsf{CD}_{2\sim}(G)$, there exist $j_1,\ j_2 \in \{1,\ldots,\ell\} \setminus \{j\}$ such that $K_{j_1}\sigma_{i_1} < T_{i_1}$ and $K_{j_2}\sigma_{i_2} < T_{i_2}$. It follows from (8.2) that $K_j(K_{j_1} \cap K_{j_2}) = M$ (where possibly $j_1 = j_2$) and so

$$K_j\sigma_{\{i_1,i_2\}}\left(K_{j_1}\sigma_{\{i_1,i_2\}} \cap K_{j_2}\sigma_{\{i_1,i_2\}}\right) = T_{i_1} \times T_{i_2}.$$

Note that

$$K_{j_1}\sigma_{\{i_1,i_2\}} \cap K_{j_2}\sigma_{\{i_1,i_2\}} \leqslant K_{j_1}\sigma_{i_1} \times K_{j_2}\sigma_{i_2}$$

and hence

$$K_j\sigma_{\{i_1,i_2\}}\left(K_{j_1}\sigma_{i_1} \times K_{j_2}\sigma_{i_2}\right) = T_{i_1} \times T_{i_2}.$$

By an observation made at the beginning of this proof, $K_j\sigma_{i_1}$, $K_j\sigma_{i_2}$, $K_{j_1}\sigma_{i_1}$, $K_{j_2}\sigma_{i_2}$ are pairwise isomorphic. Therefore the factorisation in the previous displayed equation is a full factorisation (see Section A.2). On the other hand (9.11) holds, and this contradicts Theorem A.2. Hence the first inequality of (9.10) holds for all $K \in \mathcal{K}_\omega(\mathcal{E})$. If T is not as in row 4 of Table 8.2 then the elements of the $\mathcal{F}_i$ are finite simple groups, and the stronger equation (9.9) also follows.

Finally if T is as in row 4 of Table 8.2 then $K_j\sigma_i/(K_j\sigma_i)' \cong \mathbb{Z}_2$, and hence

$$\frac{K_j}{(K_j\sigma_1)' \times \cdots \times (K_j\sigma_k)'} \leqslant \frac{K_j\sigma_1 \times \cdots \times K_j\sigma_k}{(K_j\sigma_1)' \times \cdots \times (K_j\sigma_k)'} \cong \mathbb{Z}_2^k.$$

□

9.7.3 Cartesian decompositions in $\mathsf{CD}_3(G)$

In this section we assume that G is a finite innately transitive group with a non-abelian plinth $M = T_1 \times \cdots \times T_k = T^k$, that $\mathcal{E}$ is G-transitive and that $\mathcal{E} \in \mathsf{CD}_3(G)$ (Definition 9.6). By the definition of the class $\mathsf{CD}_3(G)$, the sets $\mathcal{F}_i(\mathcal{E}, M, \omega)$ contain three subgroups.

Proposition 9.18 *The following hold.*

(i) *The isomorphism type of the simple direct factor T of M and those of the subgroups A, B, and C in $\mathcal{F}_i(\mathcal{E}, M, \omega)$ are as in Table 8.1.*

(ii) *For $j = 1, \ldots, \ell$, $(K_j\sigma_1)' \times \cdots \times (K_j\sigma_k)' \leqslant K_j$ and if T is as in row 1 or row 2 of Table 8.1 then $K_j\sigma_1 \times \cdots \times K_j\sigma_k = K_j$.*

Proof (i) By Theorem 9.2, the set $\{A, B, C\}$ is a cartesian factorisation of T, and we obtain from Lemma 8.14(ii) that T, A, B, and C must be as in Table 8.1.

(ii) Suppose that $\mathcal{F}_1(\mathcal{E}, M, \omega) = \{A, B, C\}$ for some subgroups A, B, and C of T. Then A', B', and C' are perfect groups. Moreover for all $i \in \{1, \ldots, k\}$ we have that either $K_j\sigma_i = T$, or $K_j\sigma_i$ is isomorphic to one of A, B, or C. We claim that

$$(K_i\sigma_1)' \times \cdots \times (K_i\sigma_k)' \leqslant K_i$$

for $i = 1, \ldots, \ell$. Since $(K_j\sigma_i)'$ is a perfect group, for all i and j, it follows from Lemma 4.10 that we only have to prove that

$$(K_j\sigma_{i_1})' \times (K_j\sigma_{i_2})' \leqslant K_j\sigma_{\{i_1,i_2\}} \tag{9.12}$$

for $j = 1, \ldots, \ell$, i_1, $i_2 \in \{1, \ldots, k\}$.

Suppose that $j \in \{1, \ldots, \ell\}$ and i_1, $i_2 \in \{1, \ldots, k\}$ are such that (9.12) does not hold. If $K_j\sigma_{i_1} \cong T$ or $K_j\sigma_{i_2} \cong T$ then Theorem 4.16 implies that $K_j\sigma_{\{i_1,i_2\}}$ is a diagonal subgroup of $T \times T$ isomorphic to T. However, this implies that a non-trivial full strip is involved in K_j, which is not the case by Corollary 9.3. This is a contradiction, and so each of the $K_j\sigma_{i_1}$, $K_j\sigma_{i_2}$ is isomorphic to one of A, B, or C.

Suppose without loss of generality that $K_j\sigma_{i_1} \cong A$. The isomorphism types of A, B, C and Theorem 4.8 imply that if $K_j\sigma_{\{i_1,i_2\}}$ were a subgroup of $A \times B$ or $A \times C$ then $K_j\sigma_{\{i_1,i_2\}}$ would have to contain $A' \times B'$ or $A' \times C'$, respectively. Hence the only possibility is that $K_j\sigma_{i_2} \cong A$. Then Theorem 4.8 implies that either $K_j\sigma_{\{i_1,i_2\}} \cong A$ or $A \cong \mathbb{Z}_3^6 \rtimes \mathsf{PSL}_4(3)$ and $K_j\sigma_{\{i_1,i_2\}}$ contains a normal subgroup isomorphic to $(\mathbb{Z}_3^6)^2$ and the quotient is isomorphic to $\mathsf{PSL}_4(3)$. There are j_1, j_2, j_3, $j_4 \in \{1, \ldots, \ell\} \setminus \{j\}$ not necessarily pairwise distinct, but satisfying $j_1 \neq j_3$ and $j_2 \neq j_4$, such that

$K_{j_1}\sigma_{i_1}$, $K_{j_2}\sigma_{i_2} \cong B$ and $K_{j_3}\sigma_{i_1}$, $K_{j_4}\sigma_{i_2} \cong C$. The defining properties of strong multiple factorisations imply that $\{K_j, K_{j_1} \cap K_{j_2}, K_{j_3} \cap K_{j_4}\}$ is a strong multiple factorisation of M, as defined in Section A.2, and hence $\{K_j\sigma_{\{i_1,i_2\}}, (K_{j_1} \cap K_{j_2})\sigma_{\{i_1,i_2\}}, (K_{j_3} \cap K_{j_4})\sigma_{\{i_1,i_2\}}\}$ is a strong multiple factorisation of $T \times T$. Considering that $(K_{j_1} \cap K_{j_2})\sigma_{\{i_1,i_2\}} \leqslant B \times B$ and $(K_{j_3} \cap K_{j_4})\sigma_{\{i_1,i_2\}} \leqslant C \times C$, we find that $\{K_j\sigma_{\{i_1,i_2\}}, B \times B, C \times C\}$ is a strong multiple factorisation of $T \times T$. However, (9.12) does not hold, which is a contradiction by Theorem A.4. Thus $(K_i\sigma_1)' \times \cdots \times (K_i\sigma_k)' \leqslant K_i$.

If T is as in row 2 then $K_j\sigma_i$ is a perfect group, for all $i \in \{1, \ldots, k\}$ and $j \in \{1, \ldots, \ell\}$, and so $K_i = K_i\sigma_1 \times \cdots \times K_i\sigma_k$. Let us now suppose that T is as in row 1 and set $\bar{K}_i = K_i\sigma_1 \times \cdots \times K_i\sigma_k$ for all i. Since $\bigcap_i K_i = \bigcap_i \bar{K}_i$ (see (Baddeley & Praeger 1998) page 181), it follows that $\bar{\mathcal{K}} = \{\bar{K}_1, \ldots, \bar{K}_i\}$ is a cartesian factorisation of subgroups for M. Therefore

$$|\Omega| = \prod_{i=1}^{\ell} |M \colon K_i| = \prod_{i=1}^{\ell} |M \colon \bar{K}_i|,$$

which forces $|M \colon K_i| = |M \colon \bar{K}_i|$, and hence $K_i = \bar{K}_i$ for all i. □

9.7.4 The proof of Theorem 9.15

Assertions (i)–(ii) follow from Propositions 9.16–9.18. The existence of the normal cartesian decomposition follows from the fact that in these cases M admits a G_ω-invariant direct decomposition $M = M_1 \times \cdots \times M_\ell$ such that $M_\omega = \prod_i (M_i \cap M_\omega)$. If $\mathcal{E} \in \mathsf{CD}_{2\sim}(G) \cup \mathsf{CD}_3(G)$ then this direct decomposition coincides with the direct decomposition $M = \prod T_i$. If $\mathcal{E} \in \mathsf{CD}_{1\mathrm{S}}(G)$ then $\ell = k/2$ and, for each $i \leqslant k/2$, we have $M_i = \prod_{T_j \in \mathsf{Supp}\, X_i} T_j$ where $X_1, \ldots, X_{k/2}$ are the strips involved in $\mathcal{K}$. Now the assertion follows by applying Theorem 4.24.

9.8 Cartesian decompositions preserved by finite primitive groups

The following theorem characterises cartesian decompositions preserved by finite primitive permutation groups with non-abelian socle. The case of primitive groups with abelian socle (that is, of type HA) is completely resolved in Theorem 8.4. As was noticed in Remark 9.14, the type of a G-invariant cartesian decomposition depends on the group G. The same cartesian decomposition may have a different type when considered for a different permutation group L.

Theorem 9.19 *Suppose that G is a finite primitive permutation group acting on Ω preserving a non-trivial cartesian decomposition $\mathcal{E}$.*

(i) *If the O'Nan–Scott type of G is* AS, *then $\mathcal{E} \in \mathsf{CD}_{2\sim}(G)$.*
(ii) *If the O'Nan–Scott type of G is* CD, *then $\mathcal{E} \in \mathsf{CD}_{\mathrm{S}}(G)$.*
(iii) *If the O'Nan–Scott type of G is* HC *or* TW, *then $\mathcal{E} \in \mathsf{CD}_1(G)$.*
(iv) *If the type of G is* PA, *then $\mathcal{E} \in \mathsf{CD}_1(G) \cup \mathsf{CD}_{2\sim}(G)$. Further, if $\mathcal{E} \in \mathsf{CD}_{2\sim}(G)$, then the simple composition factor of the unique minimal normal subgroup of G is isomorphic to A_6, M_{12}, or to $\mathsf{Sp}_4(2^a)$ with $a \geqslant 2$.*
(v) *The O'Nan–Scott type of G is not* SD *or* HS.

Proof First note that G acts transitively on $\mathcal{E}$ by Corollary 2.17. Let M be a minimal normal subgroup of G and assume that $M = T_1 \times \cdots \times T_k$ where the T_i are finite pairwise isomorphic non-abelian simple groups. If G has type AS, then the statement of the theorem follows from Theorem 9.7. Part (ii) follows from Theorem 9.13.

To prove parts (iii) and (iv), suppose that $k \geqslant 2$ and that the O'Nan–Scott type of G is HC, TW, or PA. Let $\mathcal{K} = \{K_1, \ldots, K_\ell\}$ denote the cartesian factorisation $\mathcal{K}_\omega(\mathcal{E})$ with some $\omega \in \Omega$. We claim that $\mathcal{E} \notin \mathsf{CD}_{\mathrm{S}}(G) \cup \mathsf{CD}_{1\mathrm{S}}(G) \cup \mathsf{CD}_{2\nsim}(G) \cup \mathsf{CD}_3(G)$. The fact that $\mathcal{E} \notin \mathsf{CD}_{\mathrm{S}}(G)$ follows from Theorem 9.13, since G is not of CD type. Assume with the aim of obtaining a contradiction that $\mathcal{E} \in \mathsf{CD}_{1\mathrm{S}}(G) \cup \mathsf{CD}_{2\nsim}(G) \cup \mathsf{CD}_3(G)$. Define, for $i \leqslant k$, the factorisation $\mathcal{F}_i = \mathcal{F}_i(\mathcal{E}, M, \omega)$ of T_i as in (9.4) and let $A \in \mathcal{F}_1$. For each i there exists a unique subgroup $A_i \in \mathcal{F}_i$ such that $A_i \cong A$ and let $X = A_1 \times \cdots \times A_k$. The subgroup X is a proper subgroup of M that properly contains M_ω.

We claim that X is normalised by G_ω. Let $g \in G_\omega$ and $i \in \{1, \ldots, k\}$. Then $A_i = K_j\sigma_i$ with some j, and so $(A_i)^g = (K_j\sigma_i)^g = (K_j)^g\sigma_{i'}$ where i' is such that $(T_i)^g = T_{i'}$. Since $(K_j)^g \in \mathcal{K}$, $(A_i)^g \in \mathcal{F}_{i'}$. Since $A_{i'}$ is the unique member of $\mathcal{F}_{i'}$ that is isomorphic to A (and hence to A_i), we find that $(A_i)^g = A_{i'}$. Therefore G_ω permutes the A_i, which gives that X is normalised by G_ω. Thus XG_ω is a subgroup of G containing G_ω. Since $M \cap (XG_\omega) = X > M_\omega$, $G_\omega < XG_\omega < G$, which contradicts the primitivity of G.

Thus $\mathcal{E} \in \mathsf{CD}_1(G) \cup \mathsf{CD}_{2\sim}(G)$. If $\mathcal{E} \in \mathsf{CD}_{2\sim}(G)$, then $M_\omega \neq 1$ by Proposition 9.17. This proves claim (iii). Turning to claim (iv), it follows from Theorem 9.11 that the T_i are isomorphic to one of the groups A_6, M_{12}, $\mathsf{P}\Omega_8^+(q)$, or $\mathsf{Sp}_4(2^a)$ with $a \geqslant 2$. That fact that $T_i \not\cong \mathsf{P}\Omega_8^+(q)$ can be verified as in the proof of Theorem 8.21.

(v) If G has type SD or HS, then $\mathsf{CD}(G) = \emptyset$ by Theorem 9.13 and Corollary 8.20. $\square$

10

Intransitive cartesian decompositions

In this chapter we consider cartesian decompositions $\mathcal{E}$ that are invariant under the action of an innately transitive group G with a non-abelian FCR-plinth, such that G is intransitive on $\mathcal{E}$. Obtaining a good description of this case requires a detailed knowledge about the factorisations of simple groups that is only available, at the moment, for finite simple groups. Thus, while most of the results in Sections 10.1 and 10.2 apply for general groups, those from Section 10.3 on apply only for finite groups. We prove that such a finite innately transitive group has at most 3 orbits on a cartesian decomposition (Theorem 10.7) and when the cartesian decomposition is homogeneous, then the number of orbits is at most 2 (Theorem 10.11).

10.1 Intransitive cartesian decompositions

Let us introduce notation that we will keep throughout this chapter. Suppose that G is an innately transitive permutation group acting on Ω with FCR-plinth M and let $\mathcal{E} = \{\Gamma_1, \ldots, \Gamma_\ell\}$ be a G-invariant cartesian decomposition of Ω on which G acts intransitively. By Theorem 4.22, $\mathcal{E}$ is not normal, so we expect to find rather different behaviour and different examples from those we have seen in Chapter 9. It also follows from Theorem 8.4 that M is non-abelian. Thus $M = T_1 \times \cdots \times T_k$ where each of the T_i is isomorphic to the same non-abelian simple group T. For $i \geqslant 1$, we let $\sigma_i \colon M \to T_i$ be the i-th coordinate projection. Suppose that $\Xi_1, \ldots, \Xi_s$ are the G-orbits on $\mathcal{E}$, and that $\mathcal{K}_\omega(\mathcal{E}) = \{L_1, \ldots, L_\ell\}$ is the corresponding cartesian factorisation of M with respect to some fixed $\omega \in \Omega$. Let $\underline{s} = \{1, \ldots, s\}$.

For $i \in \underline{s}$ set

$$K_i = \bigcap_{\Gamma_j \in \Xi_i} L_j \quad \text{and} \quad \mathcal{K} = \{K_1, \ldots, K_s\}$$

and

$$\Omega_i = \left\{ \bigcap \gamma_j \mid \gamma_j \in \Gamma_j,\ \Gamma_j \in \Xi_i \right\}.$$

Let $\bar{\mathcal{E}} = \{\Omega_1, \ldots, \Omega_s\}$. By Lemma 2.16, each Ω_i is a G-invariant partition of Ω. If $\gamma \in \Gamma_j$ for some $\Gamma_j \in \Xi_i$ then γ is a union of blocks from Ω_i and we set

$$\begin{aligned} \bar{\gamma} &= \{\sigma \mid \sigma \in \Omega_i,\ \sigma \subseteq \gamma\}; \\ \bar{\Gamma}_j &= \{\bar{\gamma} \mid \gamma \in \Gamma_j\}; \\ \bar{\Xi}_i &= \{\bar{\Gamma}_j \mid \Gamma_j \in \Xi_i\}. \end{aligned}$$

Using induction, Lemma 4.4 shows that $\bar{\mathcal{E}}$ is a cartesian decomposition of Ω acted upon trivially by G. By the same lemma, each G-invariant partition Ω_i in $\bar{\mathcal{E}}$ admits a G-invariant, transitive cartesian decomposition, namely $\bar{\Xi}_i$. Thus the study of the original intransitive decomposition $\mathcal{E}$ can be carried out via the study of a G-trivial decomposition, and the study of several transitive cartesian decompositions.

The first result in this chapter gives a general description of intransitive cartesian decompositions.

Proposition 10.1 *The set $\bar{\mathcal{E}} = \{\Omega_1, \ldots, \Omega_s\}$ is a G-invariant cartesian decomposition of Ω such that the cartesian factorisation $\mathcal{K}_\omega(\bar{\mathcal{E}})$ coincides with $\mathcal{K} = \{K_1, \ldots, K_s\}$. Moreover, for each i, Ω_i is a G-invariant partition of Ω, K_i is normalised by G_ω, $\bar{\Xi}_i \in \mathsf{CD}_{\mathrm{tr}}(G^{\Omega_i})$, and M acts faithfully on Ω_i. Further, if $\overline{\omega}$ is the block of Ω_i that contains ω, then $\mathcal{K}_{\overline{\omega}}(\bar{\Xi}_i) = \{L_j \mid \Gamma_j \in \Xi_i\}$.*

Proof It follows from Lemma 2.16 that Ω_i is a G-invariant partition of Ω for each i. Lemma 4.4(i) implies by induction that $\bar{\mathcal{E}} = \{\Omega_1, \ldots, \Omega_s\}$ is a cartesian decomposition of Ω on which G acts trivially. Further, Lemma 4.4(ii) shows that $\bar{\Xi}_i$ is a G-invariant cartesian decomposition of Ω_i and that the actions on $\bar{\Xi}_i$ and Ξ_i are equivalent. As Ξ_i is a G-orbit, $\bar{\Xi}_i \in \mathsf{CD}_{\mathrm{tr}}(G^{\Omega_i})$.

Let us now prove that $\mathcal{K} = \mathcal{K}_\omega(\bar{\mathcal{E}})$. Assume without loss of generality that $\Xi_1 = \{\Gamma_1, \ldots, \Gamma_m\}$. For $i \in \underline{m}$, let γ_i be the element of Γ_i that contains ω. We claim that K_1 is the stabiliser in M of the element

$\gamma_1 \cap \cdots \cap \gamma_m$ in Ω_1. Now $K_1 = L_1 \cap \cdots \cap L_m$ where L_j is the stabiliser in M of γ_j. Hence K_1 stabilises $\gamma_1 \cap \cdots \cap \gamma_m$. Now suppose that some element $g \in M$ stabilises $\gamma_1 \cap \cdots \cap \gamma_m$. The definition of a cartesian factorisation implies that $\gamma_1 \cap \cdots \cap \gamma_m$ is non-empty. As $\gamma_1, \ldots, \gamma_m$ are blocks of imprimitivity for the M-actions on $\Gamma_1, \ldots, \Gamma_m$, respectively, it follows that g fixes each of $\gamma_1, \ldots, \gamma_m$ setwise. Thus $g \in L_1 \cap \cdots \cap L_m$. Therefore K_1 is the stabiliser in M of $\gamma_1 \cap \cdots \cap \gamma_m$. As Ω_1 is a G-invariant partition of Ω, K_1 is normalised by G_ω. Moreover, since $K_1 \neq M$ it follows that $\bigcap_{g \in G} K_1^g$ is a normal subgroup of G properly contained in M. As M is a minimal normal subgroup of G, this implies that $\bigcap_{g \in G} K_1^g = 1$, and so M acts faithfully on Ω_1. The last assertion concerning $\mathcal{K}_{\overline{\omega}}(\bar{\Xi}_i)$ follows from the fact that an element $g \in M$ preserves the block $\gamma \in \Gamma_j$ that contains ω if and only if the same element preserves the corresponding $\bar{\gamma} \in \bar{\Gamma}_j$ that contains $\bar{\omega}$. □

10.2 Cartesian factorisations involving non-trivial strips

First we investigate the case when one of the K_i, as defined in Section 10.1, is a subdirect subgroup of M; that is, K_i is the direct product of pairwise disjoint full strips. Let us start with a motivating example.

Example 10.2 Let T be a simple group, let A be a proper subgroup of T and let τ be an automorphism of T with finite order r where $r \geqslant 2$ such that the intersection $t_1 A \cap t_2(A\tau) \cap \cdots \cap t_r(A\tau^{r-1})$ is non-empty for all $t_1, \ldots, t_r \in T$. By Lemma 8.16, the family $A, A\tau, \ldots, A\tau^{r-1}$ is an abstract cartesian factorisation of T with isomorphic subgroups. Hence if T is a finite simple group, then $r = 2$ and the possibilities for T and A are listed in Table 8.2.

Suppose that $k \geqslant 1$ and set $M = (T^r)^k = T^{rk}$. Let

$$D = \{(t, \ldots, t) \mid t \in T\}$$

be the straight diagonal subgroup in T^r and consider the following two subgroups of M:

$$K_1 = D^k \qquad \text{and} \qquad K_2 = (A \times A\tau \cdots \times A\tau^{r-1})^k.$$

We obtain from Lemma 8.16 that $M = K_1 K_2$. Every automorphism of M can be written in the form $(\beta_1, \ldots, \beta_{rk})\pi$ where $\beta_i \in \mathsf{Aut}(T)$ and π is a permutation in S_{rk}. Consider the following two automorphisms of M:

$$\begin{aligned}\sigma_1 &= (\tau,\ldots,\tau,\mathrm{id},\ldots,\mathrm{id})(1,\ldots,r); \quad \text{and}\\ \sigma_2 &= (\mathrm{id},\ldots,\mathrm{id})(1,r+1,\ldots,(k-1)r+1)\\ &\quad \times(2,r+2,\ldots,(k-1)r+2)\cdots(r,2r,\ldots,kr)\end{aligned}$$

where in the first component of σ_1, the first r coordinates are non-trivial. The group $H = \langle\sigma_1,\sigma_2\rangle$ permutes transitively the rk simple direct factors of M and normalises K_1 and K_2. Set $G = M \rtimes H$, $G_0 = (K_1 \cap K_2)H$, $\Omega = [G\colon G_0]$, and consider the right coset action of G on Ω. Let ω be the trivial coset G_0. Then $M_\omega = K_1 \cap K_2$ and the set $\mathcal{K} = \{K_1, K_2\}$ is a G_ω-invariant cartesian factorisation of M such that G_ω acts trivially by conjugation on $\mathcal{K}$. Thus $\mathcal{E}(\mathcal{K})$ is an intransitive cartesian decomposition of Ω such that one member of the corresponding cartesian factorisation is a subdirect subgroup.

Lemma 10.3 *Assume the notation introduced in Section 10.1 and let $K_j \in \mathcal{K}$. Then for each $i \geqslant 1$, there exists $g_i \in G_\omega$ such that $K_j\sigma_i = (K_j\sigma_1)^{g_i}$. Further, if $K_j\sigma_i = T_i$, for some i, then K_j is a subdirect subgroup of M and it is a direct product of pairwise disjoint, non-trivial, full strips in M.*

Proof By Proposition 10.1, K_j is normalised by G_ω. If $i \in \underline{k}$ then there is an element $g_i \in G_\omega$ such that $(T_1)^{g_i} = T_i$, and so $K_j\sigma_i = K_j^{g_i}\sigma_i = (K_j\sigma_1)^{g_i}$. Thus the projections $K_j\sigma_i$ are pairwise isomorphic for all i. Thus either K_j is a subdirect subgroup of M, and hence is a direct product of full strips by Theorem 4.16, or $K_j\sigma_i < T_i$ for all i. Assume now that K_j is a subdirect subgroup. If T_i is contained in K_j, then, since G_ω normalises K_j and is transitive by conjugation on $\{T_1,\ldots,T_k\}$, we obtain that K_j must be equal to M, which is impossible. Hence if K_j is a subdirect subgroup of M, then it is the direct product of non-trivial full strips. □

Uniform automorphisms were defined in Section 4.8 in order to aid the study of factorisations with subgroups that involve strips.

Theorem 10.4 *Assume the notation of Section 10.1, and suppose that T_1 does not admit a uniform automorphism. Then at most one member of $\mathcal{K}$ is a subdirect subgroup of M. Further, if K_j is a subdirect subgroup and K_j involves a strip X of length r, then T_1 contains proper subgroups $A_1,\ldots,A_r$ that belong to the same $\mathsf{Aut}(T_1)$-orbit such that*

$t_1A_1 \cap \cdots \cap t_rA_r \neq \emptyset$ for all $t_1, \ldots, t_r \in T_1$ and the set $\{A_1, \ldots, A_r\}$ is an abstract cartesian factorisation of T_1.

Proof Suppose without loss of generality that K_1 is a subdirect subgroup of M, and hence, by Lemma 10.3, K_1 is a direct product of pairwise disjoint, non-trivial, full strips. If K_i, with $i \neq 1$, is a subdirect subgroup of M, then K_i too is a direct product of pairwise disjoint, non-trivial full strips. Since K_1 and K_i are distinct members of a cartesian factorisation, $K_1K_i = M$, which is a contradiction by Theorem 4.30. Thus K_1 is the unique member of $\mathcal{K}$ that is a subdirect subgroup, and this proves the first claim.

Let us now prove the second claim of the lemma. Set $\overline{K}_1 = \bigcap_{j \geqslant 2} K_j$ and note that $K_1\overline{K}_1 = M$. Assume without loss of generality that

$$X = \{(t, t\alpha_2, \ldots, t\alpha_r) \mid t \in T_1\}$$

is a full strip involved in K_1 that covers $T_1, \ldots, T_r$ where $\alpha_i \colon T_1 \to T_i$ is an isomorphism for $i \geqslant 2$. As above, $\sigma_i \colon M \to T_i$ denotes the i-th coordinate projection and let σ denote the projection from M onto $T_1 \times \cdots \times T_r$. Since, for $j \geqslant 2$ and $i \geqslant 1$, the projection $K_j\sigma_i$ is a proper subgroup of T_i, we find that $\overline{K}_1\sigma_i < T_i$ for all i. Since, by Proposition 10.1, $\overline{K}_1$ is normalised by G_ω and G_ω permutes transitively by conjugation the set $\{T_1, \ldots, T_k\}$, for all $i \geqslant 1$ there exists some $g_i \in G_\omega$ such that $(\overline{K}_1\sigma_1)^{g_i} = \overline{K}_1\sigma_i$. Applying the projection σ to the factorisation $K_1\overline{K}_1 = M$, we obtain that

$$X(\overline{K}_1\sigma) = T_1 \times \cdots \times T_r.$$

Now it follows from Lemma 8.16 that the subgroups $A_1 = \overline{K}_1\sigma_1$, $A_i = \overline{K}_1\sigma_i\alpha_i^{-1}$ (for $i \geqslant 2$) satisfy the intersection property claimed in the lemma, and so they form an abstract cartesian factorisation. Further, $A_i = A_1^{g_i}\alpha_i^{-1}$, for all i, and so the subgroups $A_1, \ldots, A_r$ belong to the same $\mathsf{Aut}(T_1)$-orbit. □

When G is a finite group, we obtain a stronger theorem.

Theorem 10.5 *Using the notation of Section 10.1, assume that G is finite and that K_1 is a subdirect subgroup of M. Then $s = 2$ and (M, K_1, K_2) is a full strip factorisation. In particular, the isomorphism types of T and $K_2\sigma_i$ are as in Table 8.2. Further, K_1 is the direct product of full strips of length 2, $K_2' = (K_2\sigma_1)' \times \cdots \times (K_2\sigma_k)'$, and if T is not as in row 4 of Table 8.2 then $K_2 = K_2\sigma_1 \times \cdots \times K_2\sigma_k$.*

Proof By Lemma 10.3, K_1 is a direct product of pairwise disjoint non-trivial full strips. Since a non-abelian finite simple group does not admit a uniform automorphism (Lemma 4.26), Theorem 10.4 implies that K_1 is the unique member of $\mathcal{K}$ that is a subdirect subgroup and hence $K_2\sigma_i < T_i$ for all i. Since, by Proposition 10.1, G_ω normalises K_2, $K_2\sigma_i \cong K_2\sigma_j$ for all i and j. Thus (M, K_1, K_2) is a full strip factorisation. Lemma A.5 implies that all strips involved in K_1 have length 2.

We now show that $s = 2$. Suppose on the contrary that $s \geqslant 3$. Let X be a strip in K_1 whose support is, without loss of generality, $\{T_1, T_2\}$. Then $X = \{(t, t\alpha) \mid t \in T_1\}$ for some isomorphism $\alpha\colon T_1 \to T_2$. For $i \geqslant 2$, $K_i\sigma_1$ and $K_i\sigma_2$ are proper subgroups of T_1 and T_2, respectively, and it follows from Lemma 8.16 that $(T_1, \{K_i\sigma_1, K_i\sigma_2\alpha^{-1}\})$ is a factorisation with isomorphic subgroups. As K_2 and K_3 are normalised by G_ω, so is their intersection $K_2 \cap K_3$. Hence $(K_2 \cap K_3)\sigma_1 \cong (K_2 \cap K_3)\sigma_2$. Since $K_1(K_2 \cap K_3) = M$ and $K_1\sigma_{\{1,2\}}$ is the full strip X, we obtain from Lemma 8.16 that $(T_1, \{(K_2 \cap K_3)\sigma_1, (K_2 \cap K_3)\sigma_2\alpha^{-1}\})$ is also a factorisation with isomorphic subgroups. In such factorisations the subgroups involved are maximal subgroups of T_1 (see Table 8.2), and so $(K_2 \cap K_3)\sigma_1$ and $(K_2 \cap K_3)\sigma_2\alpha^{-1}$ are maximal subgroups of T_1. However, $(K_2 \cap K_3)\sigma_1 \leqslant K_2\sigma_1 \cap K_3\sigma_1$, which, as $K_2\sigma_1$ and $K_3\sigma_1$ are proper subgroups of T_1, implies that $(K_2 \cap K_3)\sigma_1$, $K_2\sigma_1$, and $K_3\sigma_1$ coincide. Hence $(K_2K_3)\sigma_1 = (K_2\sigma_1)(K_3\sigma_1) < T_1$ which is a contradiction, as $K_2K_3 = M$. Thus $s = 2$. The rest of the theorem follows from Theorem A.6 and from the fact that the subgroups A and B in rows 1–3 of Table 8.2 are perfect. □

If G is finite and K_1 is a subdirect subgroup of M, then we prove that $\mathsf{C}_G(M)$ is small, in fact, in most cases $\mathsf{C}_G(M) = 1$ and G is quasiprimitive.

Proposition 10.6 *Let G, M, and $\mathcal{K}$ be as in Section 10.1 and assume that G is finite and that that $K_1\sigma_1 = T_1$. If the group T is as in rows 1–3 of Table 8.2, then $\mathsf{C}_{\mathsf{Sym}\,\Omega}(M) = 1$, and in particular G is quasiprimitive of type* PA. *If T is as in row 4 then $\mathsf{N}_M(M_\omega) = K_1 \cap \mathsf{N}_M(K_2)$, and*

$$\mathsf{C}_{\mathsf{Sym}\,\Omega}(M) \cong (K_1 \cap \mathsf{N}_M(K_2))/(K_1 \cap K_2) \cong \mathsf{N}_M(K_2)/K_2.$$

In particular $\mathsf{C}_{\mathsf{Sym}\,\Omega}(M)$ is an elementary abelian 2-group of rank at most $k/2$, and all minimal normal subgroups of G different from M are elementary abelian 2-groups.

Proof By Theorem 10.5, $s = 2$, and so, $M_\omega = K_1 \cap K_2$. Note that by Theorem 3.2

$$\mathsf{C}_{\mathsf{Sym}\,\Omega}(M) \cong \mathsf{N}_M(M_\omega)/M_\omega = \mathsf{N}_M(K_1 \cap K_2)/(K_1 \cap K_2).$$

If T is as in one of the rows 1–3 of Table 8.2, then Proposition A.7 implies that $K_1 \cap K_2$ is self-normalising in M, and hence $\mathsf{C}_{\mathsf{Sym}\,\Omega}(M) = 1$. This implies that M is the unique minimal normal subgroup in G, and so G is quasiprimitive. As K_1 involves a non-trivial full strip, $k \geqslant 2$. Moreover, it follows from Table 8.2 that $M_\omega \neq 1$ and M_ω is not a subdirect subgroup of M. Thus G has quasiprimitive type PA. If T is as in row 4 of Table 8.2, then, again by Proposition A.7, we only have to prove that $(K_1 \cap \mathsf{N}_M(K_2))/(K_1 \cap K_2)$ and $\mathsf{N}_M(K_2)/K_2$ are isomorphic. As $K_1K_2 = M$, we have $\mathsf{N}_M(K_2) = (K_1 \cap \mathsf{N}_M(K_2))K_2$. By the second isomorphism theorem, $\mathsf{N}_M(K_2)/K_2 \cong (K_1 \cap \mathsf{N}_M(K_2))/(K_1 \cap K_2)$. □

10.3 Bounding the number of orbits in an intransitive cartesian factorisation of a finite group

Suppose that the conditions of Section 10.1 hold. We can define, similarly to (9.4), the sets

$$\mathcal{F}_i = \{K_j\sigma_i \mid j \in \underline{s},\ K_j\sigma_i \neq T_i\}.$$

Then, as in Chapter 9, the set $\mathcal{F}_i$ is independent of i, up to conjugacy by G_ω, and $\mathcal{F}_i$ is an abstract cartesian factorisation of T_i. Further, if T_i does not admit a uniform automorphism, then $K_j\sigma_i = T_i$ for at most one j (see Theorem 10.4). Thus using information about the isomorphism type of T_i, it is possible to bound the number s of orbits that G has on the original cartesian decomposition $\mathcal{E}$.

For infinite simple groups we do not have the necessary information available to prove a general upper bound for s. Thus we concentrate on finite groups. By Proposition 10.1, $\mathcal{K} = \{K_1, \ldots, K_s\}$ is an abstract cartesian factorisation of M. If, in addition, $\sigma_i(K_j) \neq T_i$ holds for all i and j, then this set is also a strong multiple factorisation of M (see Section A.2 for a definition).

Theorem 10.7 *Suppose that G is finite. The index s of the cartesian factorisation $\mathcal{K}$ in Section 10.1 is at most 3. Further, if $s = 3$ then $(M, \{K_1, K_2, K_3\})$ is a strong multiple factorisation. Hence, in* *this case, for all i, the factorisation $(T_i, \{K_1\sigma_i, K_2\sigma_i, K_3\sigma_i\})$ is in Table 8.1. Moreover if T is as in the first two rows then*

$$K_i = K_i\sigma_1 \times \cdots \times K_i\sigma_k \quad \text{for} \quad i = 1,\ 2,\ 3,$$

while if T is as in the third row then

$$(K_i\sigma_1)' \times \cdots \times (K_i\sigma_k)' \leqslant K_i \leqslant K_i\sigma_1 \times \cdots \times K_i\sigma_k \quad \text{for} \quad i = 1,\ 2,\ 3.$$

Proof If $K_j\sigma_i = T_i$ for some i and j then, by Theorem 10.5, we have $s = 2$. Therefore we may assume without loss of generality that all projections $K_j\sigma_i$ are proper in T_i. Then $K_1, \ldots, K_s$ form a strong multiple factorisation of M. Thus, by Theorem A.4, $s \leqslant 3$.

If $s = 3$, then, by Theorem 10.5, $K_j\sigma_i < T_i$ for all i and j. Thus, by (1.2), $\{K_1, K_2, K_3\}$ is a strong multiple factorisation of M, and T_i, $K_1\sigma_i$, $K_2\sigma_i$, $K_3\sigma_i$ are as in Table 8.1. The assertions about the K_i follow from Theorem A.4. □

A generic example with $s = 3$ can easily be constructed as follows.

Example 10.8 Let A, B, C be maximal subgroups of a non-abelian simple group T forming a cartesian factorisation of T, and let $K_1 = A^k$, $K_2 = B^k$, $K_3 = C^k$ be the corresponding subgroups of $M = T^k$. Then $(T, \{A, B, C\})$ and $(M, \{K_1, K_2, K_3\})$ are cartesian factorisations. Identify M with $\mathsf{Inn}(M)$ in $\mathsf{Aut}(M)$, and let

$$G = M\left(\mathsf{N}_{\mathsf{Aut}(M)}\left(K_1\right) \cap \mathsf{N}_{\mathsf{Aut}(M)}\left(K_2\right) \cap \mathsf{N}_{\mathsf{Aut}(M)}\left(K_3\right)\right).$$

Since the cyclic subgroup of $\mathsf{Aut}(M)$ generated by the automorphism

$$\tau\colon (x_1, \ldots, x_k) \mapsto (x_k, x_1, \ldots, x_{k-1})$$

is transitive on the simple direct factors of M and normalises K_1, K_2, and K_3, the group M is a minimal normal subgroup of G. Moreover, since $\mathsf{C}_{\mathsf{Aut}(M)}(M) = 1$, M is the unique minimal normal subgroup of G.

If $G_0 = \mathsf{N}_{\mathsf{Aut}(M)}\left(K_1\right) \cap \mathsf{N}_{\mathsf{Aut}(M)}\left(K_2\right) \cap \mathsf{N}_{\mathsf{Aut}(M)}\left(K_3\right)$ then $MG_0 = G$. Also, since K_1, K_2, and K_3 are self-normalising in M (by Lemma 4.13 and the fact that A, B, C are maximal in T), $M \cap G_0 = K_1 \cap K_2 \cap K_3$. Consider the G-action on the coset space $\Omega = [G\colon G_0]$. Since the unique minimal normal subgroup M of G is not contained in G_0, this is a faithful action, and so G can be viewed as a permutation group on Ω. Further, since $MG_0 = G$, M is a transitive minimal normal subgroup of G. Letting ω be the trivial coset G_0, we find that

$M_\omega = M \cap G_\omega = M \cap G_0 = K_1 \cap K_2 \cap K_3$. Therefore, $\{K_1, K_2, K_3\}$ is a cartesian factorisation for M acted upon trivially by G_0. Consequently this action of G preserves an intransitive G-invariant cartesian decomposition given by the cartesian factorisation $\{K_1, K_2, K_3\}$.

The defining properties of $\mathcal{K}$ give us some useful constraints on T. For instance if the K_i involve no non-trivial, full strips, then $T_i = (K_j\sigma_i)(K_m\sigma_i)$ for all i, j, m such that $j \neq m$. In particular T has a proper factorisation, and so, for example, $T \not\cong \mathsf{PSU}_{2d+1}(q)$ unless $(d, q) \in \{(1, 3),\ (1, 5),\ (4, 2)\}$. Many sporadic simple groups can also be excluded; see the tables in (Liebeck *et al.* 1990).

In general it is difficult to give a complete description of cartesian decompositions that involve no strips. However we can give such a description in Section 10.4 when the initial intransitive cartesian decomposition $\mathcal{E}$ is homogeneous. Describing the inhomogeneous case would be more difficult than finding all factorisations of finite simple groups, as demonstrated by the following generic example.

Example 10.9 Let T be a non-abelian simple group (not necessarily finite), $k \geqslant 1$, and set $M = T^k$. Let $\{A, B\}$ be a non-trivial factorisation of the group T and set $K_1 = A^k$, $K_2 = B^k$. Then clearly $K_1 K_2 = T^k$, and the base group T^k is the unique minimal normal subgroup of $G = T \wr \mathsf{S}_k$. Consider the coset action of G on $\Omega = [G\colon G_0]$ where $G_0 = (A \cap B) \wr \mathsf{S}_k$. Then $K_1 \cap K_2 = (A \cap B)^k = M \cap G_0$, and K_1, K_2 are normalised by G_0. Thus $\{K_1, K_2\}$ is a G-invariant cartesian factorisation of M corresponding to a G-invariant intransitive cartesian decomposition of Ω with index 2.

The example above shows that a detailed description of all cartesian decompositions preserved by a finite innately transitive group would first require determining all factorisations of finite simple groups. But even assuming that such a classification is available, determining the relevant factorisations of finite characteristically simple groups would still be a difficult task. In the cases that we investigate in the remainder of this chapter the required factorisations of the T_i were readily available. The subgroups of these factorisations were almost simple or perfect which made possible an explicit description of the corresponding factorisations of M.

10.4 Intransitive homogeneous cartesian decompositions

The aim of this section is to describe homogeneous, intransitive cartesian decompositions preserved by a finite innately transitive group. Such cartesian decompositions need to be studied if we want to investigate embeddings of finite innately transitive groups into wreath products in product action.

The following is a classic divisibility result that proved to be very useful in the investigation of classical groups. The result was originally published by (Zsigmondy 1892); see (Huppert & Blackburn 1982, Chapter IX, 8.3 Theorem) for a proof.

Lemma 10.10 (Zsigmondy's Theorem) *For all integers $q \geqslant 2$ and $n \geqslant 3$, there exists a prime p which divides $q^n - 1$ but does not divide $q^m - 1$ for $0 < m < n$, except when $(q, n) = (2, 6)$.*

Let us use the notation of Section 10.1 and assume in addition that the cartesian decomposition $\mathcal{E}$ is homogeneous and that G is finite. Then, for each $i \in \underline{\ell}$, $m = |\Gamma_i|$ (independent of i), and there is an integer ℓ_i such that $|\Omega_i| = |M\colon K_i| = |\Gamma_1|^{\ell_i} = m^{\ell_i}$ for all $i \in \underline{s}$. The following theorem uses the concept of full factorisations of characteristically simple groups defined in Section A.2. If $M = T_1 \times \cdots \times T_k$ is a non-abelian characteristically simple group, then a set $\{K_1, K_2\}$ of proper subgroups of M is said to be a *full factorisation* of M, if $M = K_1K_2$, $K_i\sigma_j$ are proper subgroups for T_j for all i and j, and each prime divisor of $|T_i|$ divides $|K_i\sigma_j|$ for all i and j. Such factorisations are known for simple groups, see Theorem A.1, and are characterised for characteristically simple groups, see Theorem A.2.

Theorem 10.11 *If G is finite and $\mathcal{E}$ is homogeneous, then $K_j\sigma_i < T_i$ for all i and j. Further, in this case, $s = 2$ and $(M, \{K_1, K_2\})$ is a full factorisation.*

Proof Let us first prove that $K_j\sigma_i < T_i$ for all i and j. Suppose without loss of generality that $K_1\sigma_1 = T_1$. Then Theorem 10.5 implies that $s = 2$, K_1 is the direct product of strips of length 2, and $K_2' = (K_2\sigma_1)' \times \cdots \times (K_2\sigma_k)' \leqslant K_2$. Further, the isomorphism types of T and $K_2\sigma_i$ are as in Table 8.2. In particular, the isomorphism type of $(K_2\sigma_i)'$ is independent of i. Recall that there exist non-negative integers m, ℓ_1, ℓ_2 such that $[M\colon K_1] = m^{\ell_1}$ and $[M\colon K_2] = m^{\ell_2}$. Since $|K_1| \cong |T|^{k/2}$ we have $[M\colon K_1] = |T|^{k/2}$, and so all primes that divide

$|T|$ will also divide m. Since $K_2' \leqslant K_2$ and K_2' is the direct product of its projections $(K_2\sigma_i)'$, it follows that $|M\colon K_2'| = |T_1\colon (K_2\sigma_1)'|^k$ and this number is divisible by m. Therefore all prime divisors p of $|T|$ divide $|T_1\colon (K_2\sigma_1)'|$. This is not the case if $T \cong \mathsf{A}_6$ or $T \cong \mathsf{M}_{12}$ (take $p = 5$ in both cases). If $T \cong \mathsf{P}\Omega_8^+(q)$ and $K_2\sigma_1 \cong \Omega_7(q)$ then

$$|T| = \frac{1}{d^2}q^{12}(q^6-1)(q^4-1)^2(q^2-1) \quad \text{and} \quad |T_1\colon (K_2\sigma_1)'| = \frac{1}{d}q^3(q^4-1)$$

where $d = (2, q-1)$. By Zsigmondy's Theorem (Lemma 10.10), there exists a prime r, even if $q = 2$, dividing $q^6 - 1$ and not dividing $q^4 - 1$, whence r divides $|T|$ but not $|T_1\colon (K_2\sigma_1)'|$. When $T \cong \mathsf{Sp}_4(q)$ with q even, $q \geqslant 4$, then $(K_2\sigma_1)' \cong \mathsf{Sp}_2(q^2)$, so

$$|T| = q^4(q^4-1)(q^2-1) \quad \text{and} \quad |T_1\colon (K_2\sigma_1)'| = q^2(q^2-1).$$

Using Zsigmondy's theorem we find that $q^4 - 1$ has a prime divisor r that does not divide $q^2 - 1$. Thus r divides $|T|$ but not $|T_1\colon (K_2\sigma_1)'|$. Therefore $K_j\sigma_i < T_i$ for all i and j.

For all distinct $i,\ j \in \underline{s}$ we have $M = K_iK_j$, and hence $m^{\ell_i} = |M\colon K_i| = |K_j\colon K_i \cap K_j|$ divides $|K_j|$. Let p be a prime dividing $|M| = |M\colon K_j||K_j| = m^{\ell_j}|K_j|$ Then either p divides $|K_j|$ or p divides m, and in the latter case p also divides $m^{\ell_j} = |K_j\colon K_i \cap K_j|$, which divides $|K_j|$. This holds also for $|K_i|$ and hence (M, K_i, K_j) is a full factorisation.

By Proposition 10.1, G_ω normalises K_j, and since $G = MG_\omega$, G_ω acts transitively by conjugation on $\{T_1, \ldots, T_k\}$. It follows that, for $1 \leqslant i \leqslant k$, the projections $K_j\sigma_i$ are pairwise isomorphic, proper subgroups of T_i. Hence $|K_j|$ divides $|K_j\sigma_1|^k$. Now each prime p dividing $|T|$ divides $|K_j|$ and hence divides $|K_j\sigma_1|$. Set $Q_j = K_j\sigma_1$ for $j \in \underline{s}$.

If $s \geqslant 3$ then, since $\mathcal{K}$ is a cartesian factorisation, $(T_1, \{Q_1, \ldots, Q_s\})$ is a strong multiple factorisation (see the paragraph before Theorem A.4). Moreover, since $|T|$, $|Q_i|$, $|Q_j|$ are divisible by the same primes, $(T_1, \{Q_i, Q_j\})$ is a full factorisation for all $i \neq j$. Comparing Tables A.1 and 8.1, we find that no strong multiple factorisation of a finite simple group exists in which any two of the subgroups form a full factorisation. Hence we obtain that $s = 2$. □

Theorem 10.12 *Let G, M, $T_1, \ldots, T_k$, $\mathcal{E}$, $\mathcal{K}$ be as in Section 10.1, suppose that G is finite and that $\mathcal{E}$ is homogeneous. Then, for all $i \in \underline{k}$,*

	T	A	B
1	A_6	A_5	A_5
2	M_{12}	M_{11}	M_{11}, $\mathsf{PSL}_2(11)$
3	$\mathsf{P}\Omega_8^+(q)$	$\Omega_7(q)$	$\Omega_7(q)$
4	$\mathsf{Sp}_4(q)$, $q \geqslant 4$ and q even	$\mathsf{Sp}_4(q^2).2$	$\mathsf{Sp}_2(q^2).2$

Table 10.1. *The table for Theorem 10.12*

$(T_i, \{K_1\sigma_i, K_2\sigma_i\})$ is a factorisation $(T, \{A, B\})$ as in one of the lines of Table 10.1. If T is as in rows 1–3, then $K_i = K_i\sigma_1 \times \cdots \times K_i\sigma_k$ for $i = 1, 2$. Moreover, for these lines, $\mathsf{C}_{\mathsf{Sym}\,\Omega}(M) = 1$, and in particular G is quasiprimitive of type PA. *If T is as in row 4 then*

$$(K_i\sigma_1)' \times \cdots \times (K_i\sigma_k)' \leqslant K_i \leqslant K_i\sigma_1 \times \cdots \times K_i\sigma_k = \mathsf{N}_M(K_i) \quad (10.1)$$

for $i = 1, 2$. Further,

$$\mathsf{C}_{\mathsf{Sym}\,\Omega}(M) \cong (\mathsf{N}_M(K_1) \cap \mathsf{N}_M(K_2))/(K_1 \cap K_2).$$

In particular $\mathsf{C}_{\mathsf{Sym}\,\Omega}(M)$ is an elementary abelian 2-group of rank at most k, and all minimal normal subgroups of G different from M are elementary abelian 2-groups.

Proof Set $A = K_1\sigma_1$ and $B = K_2\sigma_1$, so that, by Theorem 10.11, $T_1 = AB$ is a full factorisation. We have to eliminate all full factorisations of T which are not contained in Table 10.1. These involve the group $T = \mathsf{Sp}_4(q)$ or $\mathsf{P}\Omega_8^+(2)$, and we consider these groups separately.

Suppose first that $T \cong \mathsf{Sp}_4(q)$ with q even, $q \geqslant 4$. If A and B are isomorphic to $\mathsf{Sp}_2(q^2).2$ then line 4 of Table 10.1 is valid. Suppose that A, say, is isomorphic to $\mathsf{Sp}_2(q^2)$. Then K_1 is isomorphic to $A^k \cong (\mathsf{Sp}_2(q^2))^k$ by Theorem A.2. As the factorisation $K_1K_2 = M$ holds, we must have, for all i, that $K_2\sigma_i \cong \mathsf{Sp}_2(q^2) \cdot 2$, and hence $|K_1| < |K_2|$. For a positive integer n and a prime p let n_p denote the exponent of the largest p-power dividing n. Recall that there is an integer m such that $|M \colon K_i| = m^{\ell_i}$ for $i = 1, 2$. For any odd prime p we have $|M \colon K_1|_p = |M \colon K_2|_p$, which implies that $\ell_1 = \ell_2$ and so $|K_1| = |K_2|$, which is a contradiction.

Suppose now that $T \cong \mathsf{P}\Omega_8^+(2)$. By Theorem A.2

$$(K_i\sigma_1)' \times \cdots \times (K_i\sigma_k)' = K_i'.$$

We read off from Table A.1 that in every case $|K_i \colon K_i'|$ is a 2-power, and $|T_i\colon K_j\sigma_i|_5 = 5$. Therefore $|M\colon K_i|_5 = 5^k$ for $i = 1, 2$, and so $\ell_1 = \ell_2$. This forces $|A| = |B|$ and inspection of Table A.1 yields that $A \cong B \cong \mathsf{Sp}_6(2) \cong \Omega_7(2)$. And so line 3 of Table 10.1 holds with $q = 2$.

Suppose that one of rows 1–3 of Table 10.1 is valid. The groups A and B in these rows are perfect, and so we only have to show that $\mathsf{C}_{\mathsf{Sym}\,\Omega}(M) = 1$. By Theorem 3.2,

$$\mathsf{C}_{\mathsf{Sym}\,\Omega}(M) \cong \mathsf{N}_M(M_\omega)/M_\omega = \mathsf{N}_M(K_1 \cap K_2)/(K_1 \cap K_2). \qquad (10.2)$$

It follows, however, from Proposition A.3 that in this case $K_1 \cap K_2$ is self-normalising in M, and so M is the unique minimal normal subgroup of G. Thus G is quasiprimitive. Suppose now that row 4 of Table 10.1 is valid. Then (10.1) follows from Theorem A.2 and Proposition A.3. By Proposition A.3,

$$\mathsf{N}_M(K_1 \cap K_2) = \mathsf{N}_M(K_1) \cap \mathsf{N}_M(K_2).$$

As $(\mathsf{N}_M(K_1) \cap \mathsf{N}_M(K_2))/(K_1' \cap K_2')$ is an elementary abelian group of order 2^k, by (10.2), so is $\mathsf{C}_{\mathsf{Sym}\,\Omega}(M)$, and so all minimal normal subgroups of G different from M are also elementary abelian groups of order at most 2^k. □

Examples of this kind of cartesian decompositions can be constructed as in Example 10.9 taking A and B as in one of the rows of Table 10.1.

10.5 The component transitive cartesian decompositions for finite groups

Recall the definitions of the sets $\mathcal{F}_i(\mathcal{E}, M, \omega)$ in (9.4). The set of transitive G-invariant cartesian decompositions is denoted by $\mathsf{CD}_{\mathrm{tr}}(G)$ (see the discussion after Definition 8.1) and, for G finite, this set is partitioned into the union of pairwise disjoint subsets $\mathsf{CD}_1(G)$, $\mathsf{CD}_{1\mathrm{S}}(G)$, $\mathsf{CD}_{\mathrm{S}}(G)$, $\mathsf{CD}_{2\sim}(G)$, $\mathsf{CD}_{2\not\sim}(G)$, $\mathsf{CD}_3(G)$; see Definition 9.6.

Theorem 10.13 *Using the notation of Section 10.1, suppose that G is finite. Then the following hold.*

(i) *If $\bar{\Xi}_i \in \mathsf{CD}_{\mathrm{S}}(G^{\Omega_i})$, for some $i \in \underline{s}$, then $s = 2$. Further, if, say, $\bar{\Xi}_1 \in \mathsf{CD}_{\mathrm{S}}(G^{\Omega_1})$, then (M, K_1, K_2) is a full strip factorisation, and $\bar{\Xi}_2 \in \mathsf{CD}_1(G^{\Omega_2})$.*
(ii) *If $\bar{\Xi}_i \in \mathsf{CD}_{2\not\sim}(G^{\Omega_i})$ for some $i \in \underline{s}$, then $s = 2$, and, for all $j \in \underline{k}$, the group T_j and the subgroups of $\mathcal{F}_j(\mathcal{E}, M, \omega)$ are as in Table 8.1. If $\bar{\Xi}_1 \in \mathsf{CD}_{2\not\sim}(G^{\Omega_1})$ then $\bar{\Xi}_2 \in \mathsf{CD}_1(G^{\Omega_1})$.*

(iii) *We have, for all* $i \in \underline{s}$, *that*

$$\bar{\Xi}_i \notin \mathsf{CD}_{1S}(G^{\Omega_i}) \cup \mathsf{CD}_{2\sim}(G^{\Omega_i}) \cup \mathsf{CD}_3(G^{\Omega_i}).$$

(iv) *If* $\mathcal{E}$ *is homogeneous then* $s = 2$, $\bar{\Xi}_i \in \mathsf{CD}_1(G^{\Omega_i})$ *for* $i = 1, 2$, *and* $(M, \{K_1, K_2\})$ *is a full factorisation.*

(v) *If* $s = 3$ *then* $\bar{\Xi}_i \in \mathsf{CD}_1(G^{\Omega_i})$ *for* $i = 1, 2, 3$, *and the pair* $(M, \{K_1, K_2, K_3\})$ *is a strong multiple factorisation.*

In this section we prove Theorem 10.13 working with the notation introduced in Section 10.1. We also assume in this section that G is finite.

Lemma 10.14 *Let* $T_1, \ldots, T_k$, $L_1, \ldots, L_\ell$, $K_1, \ldots, K_s$, $\bar{\Xi}_1, \ldots, \bar{\Xi}_s$, *and* $\Omega_1, \ldots, \Omega_s$ *be as in Section 10.1. If, for some* $i \in \underline{k}$ *and* $j \in \underline{s}$, $K_j\sigma_i$ *is a proper maximal subgroup of* T_i, *then* $\bar{\Xi}_j \in \mathsf{CD}_1(G^{\Omega_j})$.

Proof Since $G = MG_\omega$, the group G_ω is transitive by conjugation on the set $\{T_1, \ldots, T_k\}$, and, by Proposition 10.1, each of the K_j is normalised by G_ω. Thus it suffices to prove that if $K_1\sigma_1$ is a proper maximal subgroup of T_1, then $\bar{\Xi}_1 \in \mathsf{CD}_1(G^{\Omega_1})$. Assume without loss of generality that $\Xi_1 = \{\Gamma_1, \ldots, \Gamma_m\}$. By Proposition 10.1, the G-action on Ω_1 is equivalent to the G-action on $[M\colon K_1]$, and $\bar{\Xi}_1 \in \mathsf{CD}_{\mathrm{tr}}(G^{\Omega_1})$. Thus if $\bar{\Xi}_1 \in \mathsf{CD}_{\mathrm{S}}(G^{\Omega_1})$ then $K_1\sigma_1 = T_1$. If $\bar{\Xi}_1 \in \mathsf{CD}_{2\sim}(G^{\Omega_1}) \cup \mathsf{CD}_{2\not\sim}(G^{\Omega_1}) \cup \mathsf{CD}_3(G^{\Omega_1})$ then there are distinct $j_1, j_2 \in \underline{m}$ such that $L_{j_1}\sigma_1, L_{j_2}\sigma_1 < T_1$, $(L_{j_1}\sigma_1)(L_{j_2}\sigma_1) = T_1$, and $K_1\sigma_1 \leqslant L_{j_1}\sigma_1 \cap L_{j_2}\sigma_1$. Hence $K_1\sigma_1$ is not a maximal subgroup of T_1.

Suppose finally that $\bar{\Xi}_1 \in \mathsf{CD}_{1S}(G^{\Omega_1})$. Then, by Theorem 9.2, we may assume without loss of generality that there is a full strip X of length 2 involved in L_1 covering T_1 and T_2, and there are indices $j_1, j_2 \in \{2, \ldots, m\}$ such that $L_{j_1}\sigma_1 < T_1$, $L_{j_2}\sigma_2 < T_2$. Let $\alpha\colon T_1 \to T_2$ be the isomorphism such that $X = \{(t, t\alpha) \mid t \in T_1\}$. Since $K_1 \leqslant L_{j_1} \cap L_{j_2}$, K_1 projects to proper subgroups of both T_1 and T_2. As the L_i form a cartesian factorisation, and K_1 is an intersection of some of the L_i, we obtain from the factorisation property of Definition 8.1 that $T_1 \times T_2 = \{(t, t\alpha) \mid t \in T_1\}(K_1\sigma_{\{1,2\}})$. It follows from Lemma 8.16 that $(L_{j_1}\sigma_1)(L_{j_2}\sigma_2\alpha^{-1}) = T_1$. In particular $L_{j_1}\sigma_1$ and $L_{j_2}\sigma_2\alpha^{-1}$ are distinct subgroups of T_1. On the other hand, $K_1\sigma_1 \leqslant (L_1 \cap L_{j_1} \cap L_{j_2})\sigma_1 \leqslant (L_{j_1}\sigma_1) \cap (L_{j_2}\sigma_2\alpha^{-1})$. Hence $K_1\sigma_1$ cannot be a maximal subgroup of T_1. Thus the only remaining possibility is that $\bar{\Xi}_1 \in \mathsf{CD}_1(G^{\Omega_1})$. □

Recall that $\{L_1,\ldots,L_\ell\}$ is the original cartesian factorisation corresponding to the intransitive cartesian decomposition $\mathcal{E}$. The following lemma is an easy consequence of Lemma 8.9, and so its proof is left to the reader.

Lemma 10.15 *Let $L_1,\ldots,L_\ell$ be as in Section 10.1, and suppose that $I_1,\ldots,I_m$ are pairwise disjoint subsets of $\underline{\ell}$, and, for $i\in\underline{m}$, set $Q_i=\bigcap_{j\in I_i}L_j$. Then*

$$Q_i\left(\bigcap_{j\neq i}Q_j\right)=M\quad\textit{for all}\quad i\in\underline{m}.$$

Now we give the proof of Theorem 10.13.

Proof of Theorem 10.13 (i) Suppose first, without loss of generality, that $\bar{\Xi}_1\in\mathsf{CD}_{\mathrm{S}}(G^{\Omega_1})$. Then by Theorem 9.5, K_1 is a subdirect subgroup of M and it follows from Theorem 10.5 that $s=2$, and that (M,K_1,K_2) is a full strip factorisation. In particular, $K_2\sigma_i$ is a maximal subgroup of T_i for all i, and hence Lemma 10.14 implies that $\bar{\Xi}_2\in\mathsf{CD}_1(G^{\Omega_2})$, as required.

(ii) Next assume without loss of generality that $\bar{\Xi}_1\in\mathsf{CD}_{2\not\sim}(G^{\Omega_1})$, and that $\Xi_1=\{\Gamma_1,\ldots,\Gamma_m\}$. Note that there are $j_1,\ j_2\in\underline{m}$ such that $L_{j_1}\sigma_1,\ L_{j_2}\sigma_1<T_1$. If $K_2\sigma_1=T_1$, then, by Theorem 10.5, $s=2$ and $\bar{\Xi}_2\in\mathsf{CD}_{\mathrm{S}}(G^{\Omega_2})$, and so part (i) implies that $\bar{\Xi}_1\in\mathsf{CD}_1(G^{\Omega_1})$, which is a contradiction. Hence $K_2\sigma_1<T_1$. If $s\geqslant 3$ then the same argument shows that $K_3\sigma_1<T_1$. Further, since $K_i=\bigcap_{\Gamma_j\in\Xi_i}L_j$ holds for $i=1,\ 2$, we obtain from Lemma 10.15 that $\{L_{j_1},L_{j_2},K_2,K_3\}$ is an abstract cartesian factorisation of M. Thus $L_{j_1}\sigma_1,\ L_{j_2}\sigma_1,\ K_2\sigma_1,\ K_3\sigma_1$ form an abstract cartesian factorisation of the finite simple group T_1. As, by Lemma 8.14(i), such a factorisation has at most 3 subgroups, this yields a contradiction. Hence $s=2$. Similarly, if there are two indices $j_3,\ j_4\in\{m+1,\ldots,\ell\}$ such that $L_{j_3}\sigma_1,\ L_{j_4}\sigma_1<T_1$ then the subgroups $L_{j_1}\sigma_1,\ L_{j_2}\sigma_1,\ L_{j_3}\sigma_1,\ L_{j_4}\sigma_1$ form a strong multiple factorisation of T_1. This again is a contradiction and so $\bar{\Xi}_2\in\mathsf{CD}_1(G^{\Omega_2})\cup\mathsf{CD}_{1\mathrm{S}}(G^{\Omega_2})$. Thus there is a unique index $j_3\in\{m+1,\ldots,\ell\}$ such that $L_{j_3}\sigma_1<T_1$. The subgroups $L_{j_1}\sigma_1,\ L_{j_2}\sigma_1,\ L_{j_3}\sigma_1$ form a strong multiple factorisation of T_1 and so T_1 and these subgroups are as in Table 8.1. If $\bar{\Xi}_2\in\mathsf{CD}_{1\mathrm{S}}(G^{\Omega_2})$ then, by Theorem 9.11, T_1 must also be as in Table 8.2 and so $T_1\cong\mathsf{P}\Omega_8^+(3)$. Further, $L_{j_3}\sigma_1$, and hence $K_2\sigma_1$, must be maximal

subgroups of T_1. This, however, cannot be the case if $\bar{\Xi}_2 \in \mathsf{CD}_{1\mathrm{S}}(G^{\Omega_2})$, by Lemma 10.14. Thus the assertions in part (ii) all hold.

(iii) Suppose without loss of generality that

$$\bar{\Xi}_1 \in \mathsf{CD}_{1\mathrm{S}}(G^{\Omega_1}) \cup \mathsf{CD}_{2\sim}(G^{\Omega_1}) \cup \mathsf{CD}_3(G^{\Omega_1})$$

and that $\Xi_1 = \{\Gamma_1, \ldots, \Gamma_m\}$. It follows from part (i) that $\bar{\Xi}_i \notin \mathsf{CD}_{\mathrm{S}}(G^{\Omega_i})$ for all $i \in \{2, \ldots, s\}$. Thus for $i \in \underline{k}$ and $j \in \underline{s}$ the projection $K_j\sigma_i$ is proper in T_i. If $\bar{\Xi}_1 \in \mathsf{CD}_3(G^{\Omega_1})$ then there are pairwise distinct indices j_1, j_2, $j_3 \in \underline{m}$ such that $L_{j_1}\sigma_1$, $L_{j_2}\sigma_1, \mathcal{L}_{j_3}\sigma_1 < T_1$. By Lemma 10.15, the subgroups $L_{j_1}\sigma_1$, $L_{j_2}\sigma_1$, $L_{j_3}\sigma_1$, $K_2\sigma_1$ form a cartesian factorisation of T_1, which is a contradiction, by Lemma 8.14(i). Thus $\bar{\Xi}_1 \notin \mathsf{CD}_3(G^{\Omega_1})$.

Suppose next that $\bar{\Xi}_1 \in \mathsf{CD}_{2\sim}(G^{\Omega_1})$. Then there are distinct indices j_1, $j_2 \in \underline{m}$ such that $L_{j_1}\sigma_1$ and $L_{j_2}\sigma_1$ are proper isomorphic subgroups of T_1. On the other hand, as $K_2\sigma_1 < T_1$, the subgroups $L_{j_1}\sigma_1$, $L_{j_2}\sigma_1$, $K_2\sigma_1$ form a strong multiple factorisation of T_1. By Table 8.1 such a factorisation cannot contain two isomorphic subgroups, and so this is a contradiction. Thus $\bar{\Xi}_1$ cannot be an element of $\mathsf{CD}_{2\sim}(G^{\Omega_1})$.

Suppose finally that $\bar{\Xi}_1 \in \mathsf{CD}_{1\mathrm{S}}(G^{\Omega_1})$. Then, by Corollary 9.3, we may assume without loss of generality that there is a full strip X of length 2 involved in L_1 covering T_1 and T_2, and there are indices j_1, $j_2 \in \{2, \ldots, m\}$ such that $L_{j_1}\sigma_1 < T_1$, $L_{j_2}\sigma_2 < T_2$. Let $\alpha \colon T_1 \to T_2$ be the isomorphism such that $X = \{(t, t\alpha) \mid t \in T_1\}$. It follows from Lemma 8.16 that $(L_{j_1}\sigma_1)(L_{j_2}\sigma_2\alpha^{-1}) = T_1$. Theorem 10.5 and part (i) implies that $K_2\sigma_1 < T_1$. As $(L_1 \cap L_{j_1} \cap L_{j_2})K_2 = M$ and

$$(L_1 \cap L_{j_1} \cap L_{j_2})\sigma_1 \leqslant L_{j_1}\sigma_1 \cap L_{j_2}\sigma_2\alpha^{-1},$$

we obtain that $(L_{j_1}\sigma_1 \cap L_{j_2}\sigma_2\alpha^{-1})(K_2\sigma_1) = T_1$. Then Lemma 8.13 implies that $(T_1, \{L_{j_1}\sigma_1, L_{j_2}\sigma_2\alpha^{-1}, K_2\sigma_1\})$ is a strong multiple factorisation. By Table 8.1 distinct subgroups in such a factorisation cannot be isomorphic. This is a contradiction, and so $\bar{\Xi}_1 \notin \mathsf{CD}_{1\mathrm{S}}(G^{\Omega_1})$.

(iv) Suppose that $\mathcal{E}$ is homogeneous. Then it follows from Theorem 10.11 that G has exactly 2 orbits on $\mathcal{E}$ and so $s = 2$. The same result implies that K_1, K_2 form a full factorisation of M, and that $K_j\sigma_i$ is a maximal subgroup of T_i, for each i and j. Thus Lemma 10.14 gives $\bar{\Xi}_i \in \mathsf{CD}_1(G^{\Omega_i})$ for $i = 1, 2$.

(v) Finally suppose that $s = 3$. Then for each $i = 1, 2, 3$, by part (i), $\bar{\Xi}_i \notin \mathsf{CD}_{\mathrm{S}}(G^{\Omega_i})$ and, by part (iii), $\bar{\Xi}_i \notin \mathsf{CD}_{1\mathrm{S}}(G^{\Omega_i})$. If $\bar{\Xi}_i \notin \mathsf{CD}_1(G^{\Omega_i})$

for some i, then there must be 4 pairwise distinct indices j_1, j_2, j_3, $j_4 \in \underline{\ell}$ such that $L_{j_1}\sigma_1$, $L_{j_2}\sigma_1$, $L_{j_3}\sigma_1$, $L_{j_4}\sigma_1$ are proper subgroups of T_1. By (9.3), these subgroups form a cartesian factorisation of T_1, which is a contradiction, by Lemma 8.14(i). Thus $\bar{\Xi}_i \in \mathsf{CD}_1(G^{\Omega_i})$ for $i = 1, 2, 3$. It also follows from Theorem 10.7 that $(M, \{K_1, K_2, K_3\})$ is a strong multiple factorisation. □

Part III

Cartesian decompositions: Applications

11

Applications in permutation group theory

An important problem in the study of permutation groups is to describe, in as much detail as possible, inclusions among different classes of groups, in the sense of specifying the possible O'Nan–Scott types of primitive, quasiprimitive, or innately transitive permutation groups G, H with $G \leqslant H \leqslant \mathsf{Sym}\,\Omega$. For finite primitive groups this problem was solved by the first author in (Praeger 1990). The special case of describing the possible inclusions of finite primitive groups in wreath products in product action relied on the concept of a blow-up defined for finite primitive permutation groups by L. G. Kovács in his seminal paper (Kovács 1989a).

In the study of quasiprimitive permutation groups and their actions on combinatorial objects it is necessary to extend the results of (Praeger 1990) and to consider the class of inclusions of quasiprimitive groups in wreath products in product action. It turns out that this class of inclusions is much richer than that of the primitive groups. To describe this situation, we extend Kovács's blow-up concept to not necessarily finite quasiprimitive groups. In Sections 11.1–11.3, we study the connection between a quasiprimitive group and its components in terms of the blow-up concept. In particular, we investigate the extent to which a quasiprimitive permutation group can be recovered from knowledge of its components under a blow-up decomposition. In Section 11.4 we concentrate on finite quasiprimitive groups. Using the 6-Class Theorem 9.7, we deduce Theorems 11.16 and 11.17, giving quite precise information concerning the structure of finite quasiprimitive groups that are subgroups of wreath products in product action.

11.1 The blow-up construction

In this section we define the blow-up of a permutation group. Suppose that Ω is a set and $G_0 \leqslant \mathsf{Sym}\,\Omega$. Let $k \geqslant 2$, set $W = G_0 \wr \mathsf{S}_k$, and consider W as a permutation group acting in product action on $\Pi = \Omega^k$ (see equation (5.7)). Note that W preserves the natural cartesian decomposition $\mathcal{E}$ of Π defined in equation (5.8). We recall that $\mathcal{E} = \{\Omega_1, \ldots, \Omega_k\}$ where

$$\Omega_i = \{\{(\omega_1, \ldots, \omega_k) \in \Omega^k \mid \omega_i = \omega\} \mid \omega \in \Omega\}.$$

That is, two k-tuples are in the same block of Ω_i if and only if they have the same entry in the i-th position.

For a subgroup $G \leqslant W$ and $i \leqslant k$ we defined in Section 5.3 the i-th component of G as the permutation group induced on the partition Ω_i by the stabiliser G_{Ω_i}. In our rather special situation, we may view the components as permutation groups on Ω rather than on Ω_i as follows. The map $\omega \mapsto \{(\omega_1, \ldots, \omega_k) \in \Omega^k \mid \omega_i = \omega\}$ is a bijection from Ω onto Ω_i for each i. Thus we can naturally identify Ω_i with Ω. Let π denote the natural projection $W \to \mathsf{S}_k$ and consider π as a permutation representation of W on $\underline{k} = \{1, \ldots, k\}$. The representation π is permutationally equivalent to the representation of W on the cartesian decomposition $\mathcal{E}$ under the bijection $i \mapsto \Omega_i$. Thus the stabiliser W_0 of Ω_i in W can be written as

$$W_0 = G_0 \times (G_0 \wr \mathsf{S}_{k-1}) \tag{11.1}$$

where the first factor of this direct product acts on the i-th coordinate of $\Pi = \Omega^k$, while the second factor acts on the cartesian product Ω^{k-1} forgetting the i-th entry of the elements of Π. In particular 'S_{k-1}' stands for the symmetric group on $\underline{k} \setminus \{i\}$. Let π_0 denote the natural projection of W_0 onto its first factor G_0. Then the i-th component $G^{(i)}$ of G is permutationally isomorphic to $(G \cap W_0)\pi_0$ viewed as a permutation group on Ω. In this section we will identify the component $G^{(i)}$ with this permutation group acting on Ω.

Definition 11.1 A subgroup G of W is said to be a *blow-up* of G_0 if the following three conditions hold:

(i) the projection $G\pi$ is a transitive subgroup of S_k;

(ii) the component $G^{(i)}$, viewed as a permutation group on Ω as above, is equal to G_0 for all $i \in \underline{k}$;

(iii) $(\mathsf{Soc}\,G_0)^k \leqslant G$.

The number k is called the *blow-up index*.

If a subgroup G of W projects onto a transitive subgroup of S_k, then $G^{(i)} = G_0$ *for some* i if and only if the same holds *for all* i. Hence, in Definition 11.1, assuming (i), condition (ii) is equivalent to

(ii)' $G^{(1)} = G_0$.

First, for a blow-up G of a quasiprimitive group G_0, we establish a connection between the minimal normal subgroups of G_0 and those of G.

Lemma 11.2 *Suppose that G_0 is a quasiprimitive permutation group with* FCR*-socle acting on Ω and G is a blow-up of G_0 of index k. Suppose further that either* $\mathsf{Soc}\, G_0$ *is non-abelian or G is quasiprimitive. Then the rule $K \mapsto K^k$ defines a bijection from the set of minimal normal subgroups of G_0 to the set of minimal normal subgroups of G.*

Proof To prove this we adapt the argument preceding (Kovács 1989a, (2.1)). Let $W = G_0 \wr \mathsf{S}_k$ and set $M_0 = \mathsf{Soc}\, G_0$. Then, by Definition 11.1, G contains $(M_0)^k = (\mathsf{Soc}\, G_0)^k$. Since G_0 is quasiprimitive, $\mathsf{C}_{G_0}(M_0) \leqslant M_0$ (Theorem 3.6), and so $\mathsf{C}_W((M_0)^k) \leqslant (M_0)^k$ (see Lemma 5.2(i)). Thus, by Lemma 3.4, each minimal normal subgroup of G must lie in $(M_0)^k$, and hence must lie in $(G_0)^k$.

Suppose first that M_0 is non-abelian. Then, being an FCR-socle, M_0, and also $(M_0)^k$, is a direct product of finitely many non-abelian simple groups. Given that $(M_0)^k \leqslant G$, we deduce from Theorem 4.16(iv) that each minimal normal subgroup of G is a direct product of the G-conjugates of some simple direct factor of $(M_0)^k$. Since G projects onto a transitive subgroup of S_k, each minimal normal subgroup K of G is of the form $(K_0)^k$, where K_0 is a characteristically simple normal subgroup of G_0. Moreover, the minimality of K implies that G_0 is transitive on the simple direct factors of K_0, and so K_0 is a non-abelian minimal normal subgroup of G_0. Conversely, let K_0 be a non-abelian minimal normal subgroup of G_0. Then, since G_0 has an FCR-socle, G_0 is transitive by conjugation on the simple direct factors of K_0 and G is transitive on $\underline{k}$, and hence G is transitive on the simple direct factors of $(K_0)^k$. Thus $(K_0)^k$ is a non-abelian minimal normal subgroup of G.

Suppose now that M_0 is abelian. Then, being an FCR-socle, M_0 is a finite elementary abelian group, and is regular on Ω (Theorem 3.6). By assumption, in this case, G is quasiprimitive on $\Pi = \Omega^k$, and hence primitive of type HA (see Section 7.1). Since M_0 is regular on Ω, $(M_0)^k$ is regular on Π, and hence $(M_0)^k$ must be the unique minimal normal subgroup of G (Theorem 3.6). Therefore $\mathsf{Soc}\, G = (M_0)^k$. As both G and G_0 have a unique minimal normal subgroup, namely $(M_0)^k$ and M_0 respectively, this gives the required result. □

We will show in Theorem 11.4 that when $\mathsf{Soc}\, G_0$ is non-abelian, then the blow-up construction preserves quasiprimitivity and primitivity, but the following example shows that this is not true when $\mathsf{Soc}\, G_0$ is abelian. The example also demonstrates that the extra condition in the case when $\mathsf{Soc}\, G_0$ is abelian is necessary in Lemma 11.2.

Example 11.3 Suppose that $G_0 = \langle(1,2,3)\rangle$ acting on $\Omega = \{1,2,3\}$ and let $G = G_0 \wr D_8$ where D_8 is the dihedral group acting on the set $\{1,2,3,4\}$ preserving the partition $\{\{1,2\},\{3,4\}\}$. Then G is a blow-up of G_0 with blow-up index 4 and G is a permutation group acting on $\Pi = \Omega^4$. Further, G_0 is primitive with an abelian socle (namely itself), but G has three minimal normal subgroups

$$\begin{aligned} M_1 &= \{(x,x,x,x) \mid x \in G_0\}; \\ M_2 &= \{(x,x,x^2,x^2) \mid x \in G_0\}; \\ M_3 &= \{(x,x^2,y,y^2) \mid x,\ y \in G_0\}, \end{aligned}$$

and in particular G is not quasiprimitive. Hence the correspondence $K_0 \mapsto (K_0)^k$ in Lemma 11.2 is not always one-to-one if $\mathsf{Soc}(G_0)$ is abelian, but G is not quasiprimitive. This example also shows that the condition in Theorem 11.4(iii) that $\mathsf{Soc}(G_0)$ is non-abelian is necessary.

Theorem 11.4 *Suppose that G_0 is a permutation group on Ω and let G be a blow-up of G_0 with index k.*

(i) *Suppose that every non-trivial normal subgroup of G_0 contains a minimal normal subgroup of G_0. If G is quasiprimitive, then so is G_0. Further, if in addition G_0 has* FCR-*socle, then $\mathsf{Soc}\, G = (\mathsf{Soc}\, G_0)^k$.*
(ii) *If G is primitive, then G_0 is primitive and not cyclic of prime order.*
(iii) *If G_0 is quasiprimitive with* FCR-*socle and $\mathsf{Soc}(G_0)$ is non-abelian, then G is quasiprimitive with* FCR-*socle.*
(iv) *If G_0 is primitive and $\mathsf{Soc}\, G_0$ is non-trivial and non-regular, then G is primitive. Moreover, if $\omega \in \Omega$ and $\alpha = (\omega,\ldots,\omega) \in \Omega^k$, then $(G_\alpha)^{(i)} = (G_0)_\omega$ for each i.*

Proof (i) Since G is a blow-up of G_0, G contains $(\mathsf{Soc}\,G_0)^k$. Suppose that G_0 is not quasiprimitive. Then there exists a non-trivial normal subgroup K_0 of G_0 with K_0 intransitive on Ω. By assumption, K_0 contains a minimal normal subgroup M_0 of G_0, and M_0 is intransitive since K_0 is. Then $M_0 \leqslant \mathsf{Soc}\,G_0$, and so $(M_0)^k \leqslant (G_0)^k$ and $(M_0)^k$ is a non-trivial normal subgroup of $G_0 \wr \mathsf{S}_k$ contained in G and is intransitive on Ω^k. This contradicts the quasiprimitivity of G. Hence G_0 is quasiprimitive. If G_0 has FCR-socle, then, by Lemma 11.2, for a minimal normal subgroup M of G, $M = (M_0)^k$ where M_0 is a minimal normal subgroup of G_0 which implies the assertion.

(ii) If G is primitive, then so is the larger group $W = G_0 \wr \mathsf{S}_k$. This implies, by Theorem 5.18, that G_0 is primitive and not cyclic of prime order.

(iii) Now assume that G_0 is quasiprimitive with non-abelian FCR-socle M_0. Let N be a non-trivial normal subgroup of G. We first claim that N contains a minimal normal subgroup of G. Since G_0 is quasiprimitive, either M_0 is the unique minimal normal subgroup of G_0 or $M_0 = M_{01} \times M_{02}$ where M_{01} and M_{02} are minimal normal subgroups of G_0 (Theorem 3.6). In the latter case M_{01} and M_{02} are both regular, $M_{01} = \mathsf{C}_{\mathsf{Sym}\,\Omega}(M_{02})$ and $M_{02} = \mathsf{C}_{\mathsf{Sym}\,\Omega}(M_{01})$. In the first case, when M_0 is a minimal normal subgroup of G_0, $(M_0)^k$ is a minimal normal subgroup of G (Lemma 11.2) and $\mathsf{C}_{G_0}(M_0) = 1$, and so $\mathsf{C}_G((M_0)^k) = 1$. Since $(M_0)^k$ is a minimal normal subgroup and N is a non-trivial normal subgroup of G, we have, by Lemma 3.4, that either $(M_0)^k \leqslant N$ or $N \leqslant \mathsf{C}_G((M_0)^k)$. On the other hand, we have already shown that $\mathsf{C}_G((M_0)^k) = 1$, and hence $(M_0)^k \leqslant N$ must hold. Therefore, in the first case, N contains the minimal normal subgroup $(M_0)^k$. In the second case, $(M_{01})^k$ and $(M_{02})^k$ are minimal normal subgroups of G, by Lemma 11.2. If $(M_{01})^k \leqslant N$, then we are done, and so we may assume that $(M_{01})^k \nleqslant N$. Then $N \leqslant \mathsf{C}_G((M_{01})^k) = (M_{02})^k$. Since $(M_{02})^k$ is a minimal normal subgroup of G, we must have that $N = (M_{02})^k$. Therefore, in all cases, N contains a minimal normal subgroup of G.

By the previous paragraph, to show that G is quasiprimitive, we only need to show that every minimal normal subgroup of G is transitive. By Lemma 11.2, a minimal normal subgroup of G is of the form $(K_0)^k$, where K_0 is a minimal normal subgroup of G_0. Since G_0 is quasiprimitive, K_0 is transitive on Ω, and hence $(K_0)^k$ is transitive on Ω^k. Therefore G is quasiprimitive on Ω^k. Further, $\mathsf{Soc}\,G$ is the direct product of such minimal normal subgroups $(K_0)^k$, and each such K_0 is FCR. Since G_0 has at most two minimal normal subgroups, we obtain that $\mathrm{Soc}\, G$ is an FCR-group.

(iv) The argument presented here follows the proof of (Kovács 1989a, Theorem 1). Suppose that G_0 is primitive and $\mathrm{Soc}\, G_0$ is not regular. Let M_0 denote the socle of G_0. Since G is a blow-up of G_0, $(M_0)^k \leqslant G$. First note that M_0 is transitive on Ω, as G_0 is primitive, and so $(M_0)^k$ is transitive on Ω^k. Since $(M_0)^k \leqslant G$, the group G is transitive. Let $W = G_0 \wr \mathsf{S}_k$ and suppose that $\omega \in \Omega$ and let $\alpha \in \Omega^k$ be the point $\alpha = (\omega, \ldots, \omega)$. Note that $W_\alpha = (G_0)_\omega \wr \mathsf{S}_k$. Let W_0, π, and π_0 be defined as at the beginning of this section before the definition of blow-ups.

Claim. $(G_\alpha)^{(i)} = (G_0)_\omega$ for all i.

Proof of the claim. We prove this claim without loss of generality for $i = 1$. Since $G_\alpha \leqslant W_\alpha$, $(G_\alpha)^{(1)} \leqslant (W_\alpha)^{(1)} = (G_0)_\omega$. Thus we need to prove that $(G_0)_\omega \leqslant (G_\alpha)^{(1)}$.

Let $g_0 \in (G_0)_\omega$. Since G is a blow-up of G_0, $G^{(1)} = G_0$, and so there is some $g \in G \cap W_0$ such that $g\pi_0 = g_0$ and g fixes ω in its action on the first coordinate. Let $M = (M_0)^k$; since M is transitive, $g = mh$ where $m \in M$ and $h \in G_\alpha$. Since $M \leqslant W_0$, $m \in W_0$, and hence $h \in W_0$. Further, using the decomposition (11.1), m can be written as $m = m'm''$ where $m' \in M^{(1)} = M_0$ and $m'' \in \prod_{j \geqslant 2} M^{(j)}$. Then

$$g_0 = g\pi_0 = (mh)\pi_0 = (m\pi_0)(h\pi_0) = m'(h\pi_0).$$

Thus $m' = g_0(h^{-1}\pi_0)$. As this element lies, on the one hand, in M_0 and, on the other hand, in $(G_0)_\omega$, we obtain that $m' \in (M_0)_\omega$. Now, by assumption, $M_0 \leqslant G$, and hence m' can be viewed as an element of G that preserves ω in the first coordinate and acts trivially on the other coordinates. Thus $m'h \in G_\alpha$. Since $g_0 = m'(h\pi_0) = (m'h)\pi_0$, we find that $g_0 \in (G_\alpha \cap W_0)\pi_0 = G_\alpha^{(1)}$. Thus $(G_0)_\omega \leqslant G_\alpha^{(1)}$ and hence the claim is proved. $\square$

Let us now show that G_α is a maximal subgroup of G. Assume that H is a subgroup of G such that $G_\alpha \leqslant H \leqslant G$. Then $(G_0)_\omega = G_\alpha^{(i)} \leqslant H^{(i)}$. Since G_0 is primitive, $(G_0)_\omega$ is a maximal subgroup, and so $H^{(i)} = (G_0)_\omega$ or $H^{(i)} = G_0$. If $H^{(i)} = (G_0)_\omega$ for all i, then H is contained in $(G_0)_\omega \wr \mathsf{S}_k = W_\alpha$, and hence H is contained in $G \cap W_\alpha = G_\alpha$, which shows that $H = G_\alpha$.

Suppose now that $H^{(i)} = G_0$ for some i and assume without loss of generality that $i = 1$. Consider the decomposition (11.1) taking

$i = 1$. If X is a subgroup of G_0, then we write $X \times 1$ to denote the corresponding subgroup of W_0 via the decomposition (11.1). Similarly, if X is a subgroup of the second factor of (11.1), then $1 \times X$ denotes the corresponding subgroup of W_0. We claim that $M_0 \times 1$ is contained in H. Let $m \in (M_0)_\omega \times 1$. Since G is a blow-up of G_0, $(M_0)^k \leqslant G$, and so $M_0 \times 1 \leqslant G$. Hence $m \in G$, and, since m stabilises ω in the first coordinate and acts trivially on the other coordinates, we obtain that $m \in G_\alpha$, and, in turn, that $m \in H$. Thus

$$(M_0)_\omega \times 1 \leqslant H.$$

We claim that the normal closure of $(M_0)_\omega$ in G_0 is M_0. Indeed, M_0 is a normal subgroup of G_0 that contains $(M_0)_\omega$. On the other hand, M_0 is either a minimal normal subgroup of G_0 or the direct product of two minimal normal subgroups. In the latter case $(M_0)_\omega$, being non-trivial, is not contained in a minimal normal subgroup of G_0 (since the minimal normal subgroups of G_0 are regular), and so the claim is valid in both cases. Since H contains $(M_0)_\omega \times 1$ and projects onto G_0, H must contain the normal closure of $(M_0)_\omega \times 1$ in $G_0 \times 1$, which is equal to $M_0 \times 1$. Thus $M_0 \times 1 \leqslant H$, as claimed. Since G, being a blow-up, projects onto a transitive subgroup of S_k, so do G_α and H. Thus H contains $(M_0)^k = (\mathsf{Soc}\, G_0)^k$. Thus H contains G_α and $(\mathsf{Soc}\, G_0)^k$. Since G_0 is primitive, $\mathsf{Soc}\, G_0$ is transitive on Ω, and so $(\mathsf{Soc}\, G_0)^k$ is transitive on Ω^k. Thus, by Lemma 2.11, $G = (\mathsf{Soc}\, G_0)^k G_\alpha$, which implies that $G \leqslant H$, and hence $H = G$.

Therefore either $H = G_\alpha$ or $H = G$, which shows that G_α is a maximal subgroup of G. Thus G is primitive, as claimed. □

We interpret the previous results as saying that, given a group G that is a blow-up of a group G_0, the structure of the socle of G is strongly related to the structure of the socle of G_0. Also there is a strong link between possible primitivity or quasiprimitivity of G and the primitivity or quasiprimitivity of G_0. Notice that the conditions of Theorem 11.4(iv) for primitive G_0 require that $\mathsf{Soc}\, G_0$ should not be regular, whereas the conditions for the corresponding claim for quasiprimitive G_0 in Theorem 11.4(iii) only require the weaker condition that $\mathsf{Soc}\, G_0$ should not be abelian. Thus the concept of a blow-up in fact behaves better with respect to quasiprimitivity than it does to primitivity.

The following example, originally given in (Kovács 1989a, page 310), shows that the condition that $\mathsf{Soc}\, G_0$ is non-regular is necessary in Theorem 11.4(iv).

Example 11.5 Suppose that G_0 is a primitive group on Ω with a regular socle M_0. We will construct a blow-up G of G_0 such that G is not primitive. Choose a point $\omega \in \Omega$. Since M_0 is a regular normal subgroup of G_0, we have $G_0 = M_0 \rtimes (G_0)_\omega$. Define $W = G_0 \wr \mathsf{S}_2$ and let $H = \{(h,h) \mid h \in (G_0)_\omega\}$ be the diagonal subgroup in W isomorphic to $(G_0)_\omega$. Note that H is a subgroup of $\mathsf{Sym}\,\Omega^2$ centralised by the top group S_2 of W, and hence we may consider $H \times \mathsf{S}_2$ as a subgroup of W. Further, the subgroup of W generated by $(M_0)^2$ and $H \times \mathsf{S}_2$ is equal to their semidirect product $(M_0)^2 \rtimes (H \times \mathsf{S}_2)$; set $G = (M_0)^2 \rtimes (H \times \mathsf{S}_2)$. Then G is a subgroup of W acting in product action on Ω^2. Thus G is a blow-up of G_0 with index 2. Further, if $\alpha = (\omega, \omega)$, then

$$G_\alpha = G \cap ((G_0)_\omega \wr \mathsf{S}_2) = \{(h,h) \mid h \in (G_0)_\omega\} \times \mathsf{S}_2 = H \times \mathsf{S}_2.$$

On the other hand, $G_\alpha = H \times \mathsf{S}_2$ is not a maximal subgroup of G, as it is properly contained in the proper subgroup $\{(g,g) \mid g \in G_0\} \times \mathsf{S}_2$. Thus G is not primitive.

Since G in the previous example is arbitrary subject to the condition that its socle is regular, we obtain the following corollary, which shows that Theorem 11.4(iv) cannot be made sharper.

Corollary 11.6 *If G_0 is a primitive permutation group with a regular socle, then there exists a blow-up G of G_0 that is not primitive.*

The following concept will help us recognise when a permutation group is permutationally isomorphic to a blow-up. Suppose that $\mathcal{E}$ is a G-invariant cartesian decomposition for some permutation group G. We say that $\mathcal{E}$ is a *blow-up decomposition* for G if $\mathcal{E}$ is transitive and $\mathcal{E}$ is M-normal for some transitive normal subgroup M of G such that, for all $\Gamma \in \mathcal{E}$, we have $M^\Gamma = \mathsf{Soc}(G^\Gamma)$.

Proposition 11.7 *Suppose that $\mathcal{E}$ is a blow-up decomposition for a group G and let G_0 denote a component of G. Then G is permutationally isomorphic to a blow-up of G_0 with blow-up index $|\mathcal{E}|$.*

Proof Let $\Gamma \in \mathcal{E}$ and consider G_0 as a permutation group on Γ and set $k = |\mathcal{E}|$. By Theorem 5.14, G can be embedded into $G_0 \wr \mathsf{S}_k$ acting on Γ^k. Denote the image of G under this embedding by $\overline{G}$. Since G is transitive on $\mathcal{E}$, the projection of $\overline{G}$ into S_k is transitive. Further, $\overline{G}^{(i)} \cong G^\Gamma = G_0$. By the normality of the decomposition, $\prod_{\Gamma' \in \mathcal{E}} \mathsf{Soc}(G^{\Gamma'}) \leqslant G$, and hence the factor $\mathsf{Soc}(G^\Gamma)$, considered as a subgroup of the direct product $\prod_{\Gamma'\in\mathcal{E}} \mathsf{Soc}(G^{\Gamma'})$, is contained in G. Therefore $(\mathsf{Soc}\, G_0)^k \leqslant \overline{G}$, and so $\overline{G}$ is a blow-up of G_0 with blow-up index k, as claimed. □

11.2 Normal decompositions and blow-up decompositions

The observant reader may ask whether, for a permutation group G, it is possible that a transitive, G-invariant, normal cartesian decomposition with quasiprimitive components is *not* a blow-up decomposition. This question is answered in the affirmative by the results of this section.

Suppose that $G \leqslant \mathsf{Sym}\,\Omega$ is a permutation group with a non-abelian, non-simple, regular, minimal normal FCR-subgroup M_1. By Theorem 3.2(i), the centraliser $\mathsf{C}_{\mathsf{Sym}\,\Omega}(M_1)$ is isomorphic to M_1, and so it is isomorphic to T^k where T is a non-abelian simple group, and $k \geqslant 2$. If $\mathsf{C}_G(M_1)$ is a proper subdirect subgroup of $\mathsf{C}_{\mathsf{Sym}\,\Omega}(M_1)$, then, using the terminology of (Bamberg & Praeger 2004), G is said to be an *innately transitive group of diagonal quotient type*. In this case G has two minimal normal subgroups M_1 and N_1 where $M_1 \cong T^k$ and N_1 is a subdirect subgroup of $\mathsf{C}_{\mathsf{Sym}\,\Omega}(M_1) \cong T^k$. Thus N_1 is the direct product of non-trivial full strips in $\mathsf{C}_{\mathsf{Sym}\,\Omega}(M_1)$. Each of these full strips is isomorphic to T and they are permuted transitively by the conjugation action of G. Thus these strips cover the same number of factors in T^k. If m denotes this number of factors, then $m \mid k$ and $N_1 \cong T^{k/m}$; see (Bamberg & Praeger 2004, Section 3) for more details, and note that T need not be finite. Further, N_1 is semiregular and intransitive, and, in particular, G is not quasiprimitive.

Theorem 11.8 *Let G be a permutation group on a set Ω and suppose that $\mathcal{E}$ is a transitive, normal, G-invariant cartesian decomposition of Ω, and that, for $\Gamma \in \mathcal{E}$, the component G^Γ is quasiprimitive with* FCR-*socle. Then exactly one of the following possibilities holds.*

(i) *G is permutationally isomorphic to a blow-up of $G_0 = G^\Gamma$;*
(ii) *G is quasiprimitive of type* Tw *and G^Γ is primitive of type* HS *or* HC.
(iii) *G is innately transitive with diagonal quotient type and G^Γ is primitive of type* HS *or* HC.

Proof Set $G_0 = G^\Gamma$ and $\ell = |\mathcal{E}|$. By assumption, G_0 is quasiprimitive. By Theorem 5.14, G is permutationally isomorphic to a subgroup of

$G_0 \wr \mathsf{S}_\ell$ in its product action on Γ^ℓ, and hence we assume that $G \leqslant G_0 \wr \mathsf{S}_\ell$. As $\mathcal{E}$ is a normal cartesian decomposition, there exists a transitive normal subgroup M of G such that $M \leqslant G_{(\mathcal{E})} \leqslant (G_0)^\ell$ and $M = (M_0)^\ell$ where M_0 is a normal subgroup of G_0. Note that M_0 is transitive on Γ, since G_0 is quasiprimitive. If $\mathsf{Soc}\, G_0 \leqslant M_0$, then by the transitivity of G on $\mathcal{E}$, we have $\mathsf{Soc}(G_0)^\ell \leqslant G$. Hence $\mathcal{E}$ is a blow-up decomposition and part (i) holds.

Assume now that $\mathsf{Soc}\, G_0 \not\leqslant M_0$. Let M_1 be a minimal normal subgroup of G_0. Since $\mathsf{Soc}\, G_0$ is an FCR-group, so is M_1. Further, by Lemma 7.1, either $\mathsf{C}_{G_0}(M_1) = 1$ or $\mathsf{C}_{G_0}(M_1)$ is transitive. If $\mathsf{C}_{G_0}(M_1) = 1$ or M_1 is abelian, then M_1 is the unique minimal normal subgroup of G_0 and Lemma 3.4(i) implies that $\mathsf{Soc}\, G_0 = M_1 \leqslant M_0$. Hence $\mathsf{C}_{G_0}(M_1) \neq 1$, and so, $\mathsf{C}_{G_0}(M_1)$ is transitive and G_0 is primitive by Lemma 7.1. As discussed in Section 7.3, the O'Nan–Scott type of G_0 is HS or HC, and G_0 has exactly two minimal normal subgroups M_1, $N_1 = \mathsf{C}_{G_0}(M_1)$, both regular, non-abelian and transitive on Γ. Since $\mathsf{C}_{G_0}(M_1) \cap \mathsf{C}_{G_0}(N_1) = N_1 \cap M_1 = 1$, we have that either $M_1 \leqslant M_0$ or $N_1 \leqslant M_0$. Without loss of generality we may assume that $M_1 \leqslant M_0$. Then $(M_1)^\ell \leqslant G$ and therefore, by definition, $\mathcal{E}$ is $(M_1)^\ell$-normal, so we may assume that $M = (M_1)^\ell$. Thus M is a regular non-abelian minimal normal subgroup of G. Let $C = \mathsf{C}_{\mathsf{Sym}\,\Omega}(M)$ and note that $\mathsf{Soc}\, G \leqslant M \times C$. By Theorem 3.6(iii), $C \cong M$. Since $N_1 = \mathsf{C}_{\mathsf{Sym}\,\Gamma}(M_1) \leqslant G_0$, we have $C = (N_1)^\ell \leqslant G_0 \wr \mathsf{S}_\ell$; therefore $\mathsf{Soc}(G_0 \wr \mathsf{S}_\ell) = M \times C$. We may write $C = \prod_{s \in S} T_s$, where S is a set of size $k\ell$ with some $k \geqslant 1$, and each T_s is isomorphic to a non-abelian simple group T. As G_0 is transitive on the set of minimal normal subgroups of N_1 and G is transitive on $\mathcal{E}$, we have that G induces a transitive permutation group on S of degree $k\ell$. If $C \leqslant G$ then $\mathsf{Soc}\, G = M \times C$, and so G is quasiprimitive with two minimal normal subgroups. In this case $\mathsf{Soc}(G_0) = M_1 \times N_1$ and $\mathsf{Soc}(G_0)^\ell = M \times C \leqslant G$. Therefore $\mathcal{E}$ is a blow-up decomposition and part (i) holds. If $C \cap G = 1$ then G has a unique minimal normal subgroup, which is regular. Therefore G is quasiprimitive of type Tw and part (ii) holds.

Thus we may assume that $1 < C \cap G < C$. In this case $1 \neq (C \cap G)^\Gamma \trianglelefteq G_0$ and $(C \cap G)^\Gamma \leqslant C^\Gamma$. Since G^Γ is quasiprimitive, $(C \cap G)^\Gamma$ must be transitive. Recall that C^Γ is regular. Thus $(C \cap G)^\Gamma = C^\Gamma \cong T^k$, and also $N = C \cap G = \mathsf{C}_G(M) \neq 1$. Therefore, N is a proper subdirect subgroup of C where C is viewed as a direct product of its minimal normal subgroups, and hence N is a direct product of pairwise disjoint non-trivial full strips of C (Theorem 4.16(iii)). Thus G is innately transitive with diagonal quotient type.

These possibilities are mutually exclusive. Indeed $\mathcal{E}$ is a blow-up decomposition if and only if (i) holds; $\mathcal{E}$ is not a blow-up decomposition and G has a unique minimal normal subgroup if and only if (ii) holds; $\mathcal{E}$ is not a blow-up decomposition and G has two minimal normal subgroups if and only if (iii) holds. □

Next we construct an example to show that the situation described by Theorem 11.8(iii) does occur.

Example 11.9 Let T be a non-abelian finite simple group, let H be any subgroup of the holomorph

$$\mathrm{Hol}(T^k) = T^k \rtimes \mathsf{Aut}(T^k) = T^k \rtimes (\mathrm{Aut}(T) \wr \mathsf{S}_k)$$

such that H has two minimal normal subgroups M_1 and N_1 where $M_1 \cong N_1 \cong T^k$. Then H, considered as a permutation group on $\Gamma = T^k$, is a primitive group of type HS if $k = 1$ or type HC if $k \geqslant 2$. Let $\ell \geqslant 2$, and let G be a subgroup of $H \wr \mathsf{S}_\ell$ in its product action on $\Omega = \Gamma^\ell$, such that G contains $M_1^\ell \cong (T^k)^\ell$, as a regular normal subgroup, and $G = M_1^\ell(H\delta \times \mathsf{S}_\ell)$ where δ is the diagonal embedding $\delta\colon H \to H^\ell$ defined by $h\delta = (h, \ldots, h)$. Then $\mathsf{Soc}\, G = M_1^\ell \times N_1\delta$, and, in addition, $N_1\delta$ is a semiregular and intransitive minimal normal subgroup of G. Thus G is innately transitive with diagonal quotient type in its action on Γ^ℓ. Moreover, the component of G induced on Γ is H, and, as noted above, H is primitive.

Theorem 11.8 demonstrates that the quasiprimitivity of the components of a transitive normal cartesian decomposition for G often implies that G also is quasiprimitive. More precisely, we have the following result.

Corollary 11.10 *Suppose that $G \leqslant \mathsf{Sym}\,\Omega$ is a permutation group and $\mathcal{E}$ is a transitive, normal G-invariant cartesian decomposition of Ω such that, for $\Gamma \in \mathcal{E}$, the component G^Γ is quasiprimitive with* FCR*-socle but not of type* HA, HS *or* HC. *Then $\mathcal{E}$ is a blow-up decomposition and G is a quasiprimitive group.*

Proof As the type of a component is not HS or HC, it follows from Theorem 11.8 that $\mathcal{E}$ is a blow-up decomposition. Then Theorem 11.4(iii) implies that G is quasiprimitive. □

The next example shows that a quasiprimitive group may have non-quasiprimitive components with respect to a normal cartesian decomposition.

Example 11.11 Let T be a non-abelian finite simple group and let P be a finite group with a core-free subgroup Q and a homomorphism $\varphi\colon Q \to \mathsf{Aut}(T)$ such that φ induces a non-trivial, proper subgroup R of $\mathsf{Inn}(T)$. Then the twisted wreath product $W = T\,\mathrm{twr}_\varphi\, P$ is a quasiprimitive permutation group acting on $\Omega = T^{|P:Q|}$; see Section 6.1.3. Further, the natural cartesian decomposition of Ω is clearly $(\mathsf{Soc}\,W)$-normal. However, a component of W has a regular minimal normal subgroup isomorphic to T, and an intransitive normal subgroup isomorphic to R. Thus a component of W is not quasiprimitive.

The last theorem of this section shows that in the case of finite primitive groups most G-invariant cartesian decompositions are blow-up decompositions.

Theorem 11.12 *Suppose that G is a finite primitive permutation group of type* CD *or of type* PA *acting on a set Ω and let $\mathcal{E}$ be a G-invariant cartesian decomposition of Ω. If G has type* PA, *then also assume that the simple direct factor of its socle is not isomorphic to A_6, M_{12}, $\mathsf{Sp}_4(2^a)$, with $a \geqslant 2$, or $\mathsf{P}\Omega_8^+(q)$. Then $\mathcal{E}$ is a blow-up decomposition.*

Proof By Theorem 9.19, $\mathcal{E} \in \mathsf{CD}_1(G)$ if the type of G is PA, and $\mathcal{E} \in \mathsf{CD}_{\mathrm{S}}(G)$ if the type of G is CD. Thus $\mathcal{E}$ is a normal cartesian decomposition, and in particular G is transitive on $\mathcal{E}$. Suppose that $\Gamma \in \mathcal{E}$ and $k = |\mathcal{E}|$. Then, by Theorem 5.14, G is permutationally isomorphic to a subgroup of $(G^\Gamma) \wr \mathsf{S}_k$. Since G is primitive, so is $(G^\Gamma) \wr \mathsf{S}_k$, and so Theorem 5.18 implies that G^Γ is primitive also. Further, M^Γ is a non-regular, minimal normal subgroup of G^Γ, and hence Corollary 11.10 implies that $\mathcal{E}$ is a blow-up decomposition. □

11.3 The blow-up construction and the O'Nan–Scott Theorem

In the discussion of the O'Nan–Scott Theorem in Chapter 7, we left open the problem of characterising primitive groups of type PA and quasiprimitive groups of type CD (including the primitive groups of type

CD). Now we fill this gap and characterise, up to permutational isomorphism, these and other groups as blow-ups.

Theorem 11.13 *Suppose that G is a quasiprimitive group acting on Ω with an* FCR*-plinth.*

(i) *If the type of G is* HC *then G is permutationally isomorphic to a blow-up of a primitive group G_0 of type* HS.
(ii) *If G is of type* CD*, then G is permutationally isomorphic to a blow-up of a quasiprimitive group G_0 of type* SD*. Further, G is primitive if and only if G_0 is primitive.*
(iii) *If G is primitive of type* PA*, then G is permutationally isomorphic to a blow-up of a primitive group G_0 of type* AS.

Proof In each of the cases of Theorem 11.13, the group G has a non-abelian, non-simple minimal normal FCR-subgroup M. Hence $M = T_1 \times \cdots \times T_k$ where the T_i are isomorphic non-abelian simple groups and $k \geqslant 2$. Let $\omega \in \Omega$ be fixed.

(i) Suppose that G has type HC. Then M is regular, and, by Theorem 3.10, G can be embedded into $\mathsf{Hol}\, M$. Thus we assume without loss of generality that G is a subgroup of $\mathsf{Hol}\, M$ acting on the set M. The two minimal normal subgroups of G are $M\varrho$ and $M\lambda$ where ϱ and λ are the right and left regular representations of M, respectively. For $i \leqslant k$, let M_i denote the product $\prod_{j \neq i} T_j$. Since the minimal normal subgroup $M\varrho$ of G is regular, G preserves a normal cartesian decomposition $\mathcal{E} = \{\Gamma_1, \ldots, \Gamma_k\}$ where Γ_i is the set of $M_i\varrho$-orbits for each i (Corollary 4.25). We claim that a component G^{Γ_i} is primitive of type HS and that $\mathcal{E}$ is a blow-up decomposition. The groups $(M\varrho)^{\Gamma_i}$ and $(M\lambda)^{\Gamma_i}$ are normal subgroups of G^{Γ_i}. Further, if $j \neq i$, then $T_j \leqslant M_i$, and so $T_j\varrho$ acts trivially on Γ_i. If $T_i\varrho$ acted trivially on Γ_i, then $M\varrho$ would act trivially on Γ_i, which is impossible, as $M\varrho$ is transitive on M. Thus $(M\varrho)^{\Gamma_i} = T_i$. Since the orbits of $1 \in M$ under $M_i\varrho$ and $M_i\lambda$ are both equal to M_i, we obtain that the partition Γ_i coincides with the set of $M_i\lambda$-orbits on M. Thus, similarly to the argument concerning $(M\varrho)^{\Gamma_i}$, we obtain for $j \neq i$, that $T_j\lambda$ acts trivially on Γ_i and $T_i\lambda$ acts faithfully on Γ_i. Therefore $(M\lambda)^{\Gamma_i} = T_i\lambda$. Since $T_i\varrho$ and $T_i\lambda$ are simple and non-abelian, they must be distinct minimal normal subgroups of G^{Γ_i}. Thus G^{Γ_i} contains two distinct minimal normal subgroups. By Corollary 3.13, G is primitive, and the O'Nan–Scott type of G is HS (see Section 7.3). Therefore the conditions of Theorem 11.8 hold and, since the type of G is HC, case (i) of Theorem 11.8 is valid. Thus $\mathcal{E}$ is a blow-up decomposition for G. Therefore, by Proposition 11.7, G is permutationally isomorphic to a blow-up of $G_0 = G^{\Gamma_i}$ which is a primitive permutation group of type HS.

(ii) Suppose that G has type CD. Then M_ω is the direct product of pairwise disjoint non-trivial full strips $X_1, \ldots, X_\ell$ (see Section 7.4). For $i \leqslant \ell$, set $M_i = \prod_{T_j \in \mathsf{Supp}\, X_i} T_j$. Then $\{M_1, \ldots, M_\ell\}$ is a direct decomposition of M such that $M_\omega = \prod_i (M_\omega \cap M_i)$. Set $\overline{M}_i = \prod_{j \neq i} M_j$, and let Γ_i be the partition of Ω into the orbits of $\overline{M}_i$. By Theorem 4.24, $\mathcal{E} = \{\Gamma_i \mid i \in \underline{\ell}\}$ is a G-invariant M-normal cartesian decomposition of Ω. Further, the G_ω-action on $\mathcal{E}$ is equivalent to its conjugation action on the set $\{X_1, \ldots, X_\ell\}$ of strips involved in M_ω via the bijection $X_i \mapsto \Gamma_i$. Consider the components G^{Γ_i} and M^{Γ_i}. We have by the definition of the component that $M^{\Gamma_i} = M_i$. Suppose that $T_j, T_m \leqslant M_i$. The group G_ω permutes transitively the set $\{T_1, \ldots, T_k\}$ preserving the system of imprimitivity that corresponds to the partition $\{\mathsf{Supp}\, X_s \mid s \leqslant \ell\}$. Thus there exists $g \in G_\omega$ that normalises X_i such that $(T_j)^g = T_m$. Since g normalises X_i, it stabilises Γ_i, and so $g \in G_{\Gamma_i}$. Hence G_{Γ_i} permutes transitively the simple direct factors of M_i, which shows that $M_i = M^{\Gamma_i}$ is a minimal normal subgroup of G^{Γ_i}. For each $i \leqslant \ell$, let γ_i be the part in Γ_i that contains ω. By the definition of M_i, the strip X_i lies in $M_\omega \cap M_i$, and hence X_i is contained in $(M_i)_{\gamma_i}$. Conversely, if $m \in (M_i)_{\gamma_i}$, then, since M_i fixes each γ_r with $r \neq i$ setwise, the element m also fixes γ_r, and so m stabilises the intersection $\bigcap_{s \geqslant 1} \gamma_s$. Since this intersection is the singleton $\{\omega\}$, we find that $m \in M_i \cap M_\omega = X_i$. Thus $(M_i)_{\gamma_i} = X_i$, and so G^{Γ_i} is a quasiprimitive group of type SD. Now Corollary 11.10 implies that $\mathcal{E}$ is a blow-up decomposition for G. Thus G is permutationally isomorphic to a blow-up of $G_0 = G^{\Gamma_1}$ which is a quasiprimitive group of SD type. Theorem 11.4 implies that G is primitive if and only if G_0 is primitive.

(iii) Let $\sigma_i \colon M \to T_i$ denote the i-th coordinate projection. By Theorem 7.7, $M_\omega = \prod_i (M_\omega \sigma_i) = \prod_i (M_\omega \cap T_i)$ and hence G preserves an M-normal cartesian decomposition $\mathcal{E} = \{\Gamma_1, \ldots, \Gamma_k\}$ where Γ_i is the set of $\mathsf{C}_M(T_i)$-orbits. Consider the components G^{Γ_i} and M^{Γ_i}. Let $\overline{T_i} = \mathsf{C}_M(T_i) = \prod_{j \neq i} T_j$. The subgroup $\overline{T_i}$ acts trivially on Γ_i, and so $M^{\Gamma_i} = T_i$. An argument similar to the one used in part (ii) shows that if $\gamma_i \in \Gamma_i$ such that $\omega \in \gamma_i$, then $(T_i)_{\gamma_i} = M_\omega \cap T_i$. The group G is contained in $W = G^{\Gamma_i} \wr \mathsf{S}_k$. Since G is primitive and not cyclic, W is primitive , and so G^{Γ_i} is primitive and $M^{\Gamma_i} = T_i$ is transitive

(Theorem 5.18). Since T_i is a non-regular, non-abelian, simple minimal normal subgroup of G^{Γ_i}, we obtain that G^{Γ_i} is of type AS. Now Theorem 11.8 shows that $\mathcal{E}$ is a blow-up decomposition. Thus G is permutationally isomorphic to a blow-up of G^{Γ_i} which is a primitive group of type AS. □

The following example shows that a quasiprimitive group of type PA need not be a blow-up.

Example 11.14 Let A be an almost simple group with socle T such that $A/T \cong C_2$ and let $G_0 = A \times C_2$. Let $a \in A \setminus T$ and let b be a generator of the direct factor C_2 of G_0. Suppose that H is a core-free subgroup of G_0 such that $T \cap H \neq 1$. Since H is core-free, we can consider G_0 as a permutation group on the right coset space $\Omega = [G_0 \colon H]$. Then T is a non-regular transitive minimal normal subgroup of G_0, and so G_0 is innately transitive on Ω with a unique minimal normal subgroup T which is non-abelian and simple. Let $W = G_0 \wr C_2$ acting in product action on Ω^2. Then T^2 is a non-regular, transitive minimal normal subgroup of W. Let G be the group generated by the normal subgroup $(T \times T) \rtimes C_2$, and by the elements (a, b), $(b, a) \in (G_0)^2$. Then T^2 is a minimal normal subgroup of G and T^2 is transitive. Further, $\mathsf{C}_G(T^2) = \mathsf{C}_W(T^2) \cap G = \langle (b, 1), (1, b) \rangle \cap G = 1$. Thus T^2 is the unique minimal normal subgroup of G, and so G is quasiprimitive of type PA. Further G preserves the natural cartesian decomposition of Ω^2 with component G_0. On the other hand, G_0 is not quasiprimitive, as its direct factor C_2 is an intransitive normal subgroup.

11.4 The inclusion problem for finite permutation groups

As discussed in the introduction of this chapter, the inclusion problem for innately transitive, quasiprimitive, or primitive permutation groups is the problem of describing, in as much detail as possible, the permutation groups G, X such that $G \leqslant X$ and both G and X are innately transitive, or they are both quasiprimitive, or both are primitive. The inclusion problem has several deep applications in combinatorics and graph theory. For a given permutation group $G \leqslant \mathsf{Sym}\,\Omega$, the problem of describing the subgroups of $\mathsf{Sym}\,\Omega$ that contain G and that are permutationally isomorphic to $\mathsf{Sym}\,\Gamma \wr \mathsf{S}_\ell$ acting in product action on Γ^ℓ is equivalent to the problem of identifying all cartesian decompositions of Ω preserved by G.

As mentioned in the introduction of this chapter, in this final section we focus on finite groups. Thus the main theorems of this section will be stated under the following hypothesis. Recall that we defined the O'Nan–Scott types of primitive and quasiprimitive groups in Sections 7.1–7.5.

Hypothesis 11.15 Let H be a finite quasiprimitive permutation group of type As acting on a set Γ. That is, H has a unique minimal normal subgroup U, and U is a non-abelian simple group and is transitive on Γ. Set $W = H \wr \mathsf{S}_\ell$ for some $\ell \geqslant 2$, and consider W as a permutation group in product action acting on $\Omega = \Gamma^\ell$. Let $N = U_1 \times \cdots \times U_\ell$ be the unique minimal normal subgroup of W; note that $N \cong U^\ell$. Assume that G is an innately transitive subgroup of W with a non-abelian plinth $M = T_1 \times \cdots \times T_k$ where $T_1, \ldots, T_k$ are finite, non-abelian simple groups all isomorphic to a group T. For $i \in \underline{k}$, let σ_i denote the natural projection $M \to T_i$, and, for $i \in \underline{\ell}$, let μ_i denote the natural projection $N \to U_i$.

The following two theorems are proved in this section.

Theorem 11.16 *Assume that Hypothesis 11.15 holds. Then $M \leqslant N$. Further, if G projects onto a transitive subgroup of S_ℓ, then exactly one of the following holds.*

(i) *$k = \ell$; the T_i and the U_i can be indexed so that $T_i \leqslant U_i$ for all i.*
(ii) *$\ell = 2k$; T and U are as in Table 11.1; the T_i and the U_i can be indexed so that $T_i \leqslant U_{2i-1} \times U_{2i}$ for all i.*
(iii) *Neither case (i) nor case (ii) holds and $U = \mathsf{Alt}\,\Gamma$.*

Theorem 9.13(i) implies that the O'Nan–Scott type of the group G in Hypothesis 11.15 cannot be SD. The following theorem gives additional information when G has type CD complementing the information given by Theorem 9.13(ii).

Theorem 11.17 *Assume Hypothesis 11.15 and suppose that G has type* CD*. Then the following all hold.*

(i) *G projects onto a transitive subgroup of S_ℓ;*
(ii) *$U = \mathsf{Alt}\,\Gamma$;*

	T	U
1	A_6	A_6
2	M_{12}	M_{12}
3	M_{12}	A_{12}
4	$\mathsf{P}\Omega_8^+(q)$	$\mathsf{P}\Omega_8^+(q)$
5	$\mathsf{P}\Omega_8^+(q)$	A_n where $n = \lvert\mathsf{P}\Omega_8^+(q) : \Omega_7(q)\rvert$
6	$\mathsf{P}\Omega_8^+(2)$	$\mathsf{Sp}_8(2)$
7	$\mathsf{Sp}_4(2^a)$, $a \geqslant 2$	$\mathsf{Sp}_{4b}(2^{a/b})$ where $b \mid a$
8	$\mathsf{Sp}_4(2^a)$, $a \geqslant 2$	A_n where $n = \lvert\mathsf{Sp}_4(2^a) : \mathsf{Sp}_2(2^{2a}) \cdot 2\rvert$

Table 11.1. *Table for Theorem 11.16*

(iii) $M \leqslant N$; $k = m\ell$ *for some* $m \geqslant 2$ *and the* U_i *and the* T_i *can be indexed so that* $T_{(i-1)m+1} \times \cdots \times T_{im} \leqslant U_i$ *for all* $i \in \underline{\ell}$.

In Section 11.4 we prove Theorems 11.16 and 11.17. Throughout the section we assume that Hypothesis 11.15 holds. By Lemma 5.10, W is contained in the stabiliser in $\mathsf{Sym}\,\Gamma^\ell$ of the natural cartesian decomposition $\mathcal{E} = \{\Gamma_1, \ldots, \Gamma_\ell\}$ of Γ^ℓ where

$$\Gamma_i = \{\{(\gamma_1, \ldots, \gamma_\ell) \mid \gamma_i = \gamma\} \mid \gamma \in \Gamma\} \quad \text{for} \quad i \in \underline{\ell}.$$

Fix $\gamma \in \Gamma$ and let $\omega = (\gamma, \ldots, \gamma)$. Let $\mathcal{K} = \{K_1, \ldots, K_\ell\}$ be the G_ω-invariant cartesian factorisation of M with respect to ω corresponding to $\mathcal{E}$, as defined in Lemma 8.6. It follows from the definition of the product action that the W-actions on $\mathcal{E}$ and on the set of simple direct factors of N are equivalent.

The proof of the next lemma uses the result, usually referred to as Schreier's Conjecture, that the outer automorphism group of a finite simple group is soluble (see (Dixon & Mortimer 1996, page 133)). The validity of Schreier's Conjecture follows from the finite simple group classification.

Lemma 11.18 *The first assertion of Theorem 11.16 holds, namely, the plinth* M *of* G *is contained in* N.

Proof Let B denote the base group H^ℓ of W. It follows from the definition of the product action that the pointwise stabiliser in W of $\mathcal{E}$ coincides with B, and so Theorem 8.3 implies that $M \leqslant B$. Now, as M is a minimal normal subgroup of G, either $M \leqslant N$ or $M \cap N = 1$. Suppose that $M \cap N = 1$. Then $(MN)/N \cong M/(M \cap N) \cong M$. On the other hand, $B/N \cong (H/U)^\ell$. By Schreier's Conjecture, H/U is a soluble group, and therefore so is B/N. This is a contradiction, since B/N has a subgroup isomorphic to the insoluble group M. Therefore $M \leqslant N$. □

The following theorem will be used frequently in this section. Its proof, which depends on the classification of finite simple groups, can be found in (Baddeley & Praeger 2003, Theorem 1.4).

Theorem 11.19 *Let X be a finite almost simple group and suppose that $X = AB$ where A is a proper subgroup of X not containing $\mathsf{Soc}\,X$ and B is a non-abelian characteristically simple group that is not simple. Then $\mathsf{Soc}\,X = \mathsf{A}_n$ and $A \cap \mathsf{Soc}\,X = \mathsf{A}_{n-1}$ where $n = |X : A|$ and $n \geqslant 10$.*

Note that M is a characteristically simple group and, by Theorem 4.16(iv), each of its normal subgroups is a product of some of the T_i. If L is a normal subgroup of M then the quotient M/L can naturally be identified with the subgroup $\prod_{T_i \not\leqslant L} T_i$, and in future this identification will be used without further comment. Recall, for a subgroup $K < M$, that $\mathsf{Core}_M(K) = \bigcap_{m \in M} K^m$ is the largest normal subgroup of M contained in K. Since each Γ_i is an M-invariant partition, the group M induces a permutation group, denoted by M^{Γ_i}, on Γ_i.

Under Hypothesis 11.15, there is a natural isomorphism $U \to U_i$ for each $i \in \underline{\ell}$, and we consider the U_i as subgroups of $\mathsf{Sym}\,\Gamma$. Note that at this stage we do not assume that G projects onto a transitive subgroup of S_ℓ. Recall that $\mathcal{K} = \{K_1, \ldots, K_\ell\}$ is the G-invariant cartesian factorisation $\mathcal{K}_\omega(\mathcal{E})$ defined before Lemma 11.18 and that $\sigma_i \colon M \to T_i$ and $\mu_i \colon N \to U_i$ are the natural projections.

Lemma 11.20 *Let $i \in \underline{\ell}$. Then*

$$M^{\Gamma_i} = \prod_{T_j \not\leqslant K_i} T_j \cong T^{s_i} \qquad (11.2)$$

for some $s_i \geqslant 1$, and $M\mu_i$ is permutationally isomorphic to M^{Γ_i}. Moreover, $M\mu_i$ is a transitive subgroup of U_i, and if $s_i \geqslant 2$ then $U = \mathsf{Alt}\,\Gamma$.

Proof Note that K_i is the stabiliser of a point for the M-action on Γ_i, and the kernel of this action is $\mathsf{Core}_M(K_i)$. Now $T_j \leqslant \mathsf{Core}_M(K_i)$ if and only if $T_j \leqslant K_i$, and so $M^{\Gamma_i} = \prod_{T_j \not\leqslant K_i} T_j \cong T^{s_i}$ for some s_i. Since

$K_i < M$, there exists at least one index j such that $T_j \not\leqslant K_i$, and so $s_i \geqslant 1$. Therefore (11.2) holds.

Let α denote the bijection $\Gamma \to \Gamma_i$ given by

$$\alpha\colon \gamma \mapsto \{(\gamma_1, \dots, \gamma_\ell) \mid \gamma_i = \gamma\} \quad \text{for all} \quad \gamma \in \Gamma.$$

The map μ_i can be considered as a permutation representation of M in $\mathsf{Sym}\,\Gamma$. We claim that the M-actions on Γ (by μ_i) and on Γ_i are equivalent via the bijection α. Let $\gamma \in \Gamma$ and $m \in M$. Then

$$(\gamma m)\alpha = \{(\gamma_1, \dots, \gamma_\ell) \mid \gamma_i = \gamma m\} = \{(\gamma_1, \dots, \gamma_\ell) \mid \gamma_i = \gamma\}m = (\gamma\alpha)m.$$

This proves the claim. Hence $M\mu_i \cong M^{\Gamma_i} \cong T^{s_i}$ where s_i is as in the previous paragraph. As Γ_i is an M-invariant partition of Ω and as M is transitive on Ω, the group M is transitive on Γ_i, and so $M\mu_i$ is transitive on Γ. Thus the factorisation $U_i = (M\mu_i)(U_i)_\gamma$ holds. If $s_i \geqslant 2$, then $M\mu_i$ is a non-abelian characteristically simple group that is not simple, and so Theorem 11.19 implies that $U = \mathsf{Alt}\,\Gamma$. □

It is part of the conditions of Theorem 11.16 that G should project onto a transitive subgroup of S_ℓ. This is equivalent to stating that $\mathcal{E}$ is a transitive G-invariant cartesian decomposition of Ω. Using the notation introduced in Section 8.1, this can be expressed as $\mathcal{E} \in \mathsf{CD}_{\mathrm{tr}}(G)$. By the 6-Class Theorem 9.7, $\mathcal{E}$ belongs to precisely one of the six classes $\mathsf{CD}_1(G)$, $\mathsf{CD}_{\mathrm{S}}(G)$, $\mathsf{CD}_{1\mathrm{S}}(G)$, $\mathsf{CD}_{2\sim}(G)$, $\mathsf{CD}_{2\not\sim}(G)$, and $\mathsf{CD}_3(G)$ of transitive G-invariant cartesian decompositions of Ω; see Definition 9.6. In what follows, we analyse the embedding $G \leqslant W$ considering separately each of the possible classes for $\mathcal{E}$.

Lemma 11.21 *Suppose that $\mathcal{E} \in \mathsf{CD}_{\mathrm{tr}}(G)$. If $M\mu_i \cong T^{s_i}$ with $s_i \geqslant 2$ for some $i \in \underline{\ell}$, then Theorem 11.16 (iii) is valid.*

Proof Suppose that $M\mu_i \cong T^{s_i}$ and $s_i \geqslant 2$ for some $i \geqslant 1$. It follows from Lemma 11.20 that in this case $U = \mathsf{Alt}\,\Gamma$. Also, note that there are exactly s_i indices j such that $T_j\mu_i \cong T$. On the other hand, in cases (i) and (ii) of Theorem 11.16, for each $i \in \underline{\ell}$ there is a unique $j \in \underline{k}$ such that $T_j\mu_i \cong T$. Therefore neither case (i) nor case (ii) of Theorem 11.16 holds, and so all assertions of Theorem 11.16(iii) are valid. □

Lemma 11.22 *The following are valid.*

(i) *If $\mathcal{E} \in \mathsf{CD}_{\mathrm{S}}(G) \cup \mathsf{CD}_{1\mathrm{S}}(G) \cup \mathsf{CD}_3(G) \cup \mathsf{CD}_{2\not\sim}(G)$, then $M\mu_i \cong T^s$*

with s independent of i and $s \geqslant 2$, and Theorem 11.16(iii) is valid.

(ii) *If $\mathcal{E} \in \mathsf{CD}_1(G)$ then either case (i) or case (iii) of Theorem 11.16 is valid.*

(iii) *If $\mathcal{E} \in \mathsf{CD}_{2\sim}(G)$ then either case (ii) or case (iii) of Theorem 11.16 is valid.*

Proof In each case, G is transitive on $\mathcal{E}$, and M is a normal subgroup of G, and so the permutation groups M^{Γ_i} are pairwise permutationally isomorphic. Thus $M^{\Gamma_i} \cong T^s$ for some $s \geqslant 1$ independent of i, and, by Lemma 11.20, $M\mu_i \cong T^s$. If $s \geqslant 2$ then Lemma 11.21 implies that case (iii) of Theorem 11.16 holds.

(i) Let $\mathcal{E} \in \mathsf{CD_S}(G) \cup \mathsf{CD_{1S}}(G) \cup \mathsf{CD}_3(G) \cup \mathsf{CD}_{2\not\sim}(G)$. By the discussion in the previous paragraph, it suffices to verify that $M\mu_1 \cong T^s$ with $s \geqslant 2$. Note that, by Lemma 11.20, $M\mu_1 = \prod_{T_j \not\leqslant K_1} T_j$. If $\mathcal{E} \in \mathsf{CD_S}(G) \cup \mathsf{CD_{1S}}(G)$, then K_1 involves a non-trivial, full strip, and so $s \geqslant 2$ follows immediately.

Suppose next that $\mathcal{E} \in \mathsf{CD}_3(G)$. Then $\mathcal{F}_1(\mathcal{E}, M, \omega) = \{A, B, C\}$ (see the definition in (9.4)) for some subgroups A, B, and C of T_1, such that A, B, and C form a cartesian factorisation of T_1. Hence Lemma 8.14(ii) yields that the subgroups A, B, and C represent three distinct isomorphism types. Then there are pairwise distinct indices $j_1, j_2, j_3 \in \underline{\ell}$ such that $K_{j_1}\sigma_1 = A$, $K_{j_2}\sigma_1 = B$, $K_{j_3}\sigma_1 = C$. Let $g_1, g_2, g_3 \in G_\omega$ such that $(K_{j_1})^{g_1} = (K_{j_2})^{g_2} = (K_{j_3})^{g_3} = K_1$. Let $i_1, i_2, i_3 \in \underline{k}$ such that $(T_1)^{g_1} = T_{i_1}$, $(T_1)^{g_2} = T_{i_2}$, $(T_1)^{g_3} = T_{i_3}$. Then $K_1\sigma_{i_1} = (K_{j_1}\sigma_1)^{g_1} = A^{g_1}$, $K_1\sigma_{i_2} = (K_{j_2}\sigma_1)^{g_2} = B^{g_2}$, and $K_1\sigma_{i_3} = (K_{j_3}\sigma_1)^{g_3} = C^{g_3}$. As A, B, and C are pairwise non-isomorphic, it follows that i_1, i_2, i_3 are also pairwise distinct. Thus (11.2) implies that $M\mu_1 \cong T^s$ for some $s \geqslant 3$.

Suppose finally that $\mathcal{E} \in \mathsf{CD}_{2\not\sim}(G)$ such that $\mathcal{F}_1(\mathcal{E}, M, \omega) = \{A, B\}$ where $A, B \leqslant T_1$ are not isomorphic. Then it follows that there are indices $j_1, j_2 \in \underline{\ell}$ such that $K_{j_1}\sigma_1 = A$ and $K_{j_2}\sigma_1 = B$. Let $g_1, g_2 \in G_\omega$ be such that $(K_{j_1})^{g_1} = (K_{j_2})^{g_2} = K_1$. Then $(T_1)^{g_1} = T_{i_1}$ and $(T_1)^{g_2} = T_{i_2}$ for some $i_1, i_2 \in \underline{k}$. Hence $K_1\sigma_{i_1} = (K_{j_1}\sigma_1)^{g_1} = A^{g_1}$ and $K_1\sigma_{i_2} = (K_{j_2}\sigma_1)^{g_2} = B^{g_2}$. As A and B are not isomorphic, we have $i_1 \neq i_2$. Thus (11.2) implies that $M\mu_i \cong T^s$ where $s \geqslant 2$, and the result follows.

(ii) As noted above, if $M\mu_i \cong T^s$ with $s \geqslant 2$ then Theorem 11.16(iii) holds. Assume now that $M\mu_i \cong T$. As $\mathcal{E} \in \mathsf{CD}_1(G)$, for all $i \in \underline{k}$, there is a unique $j \in \underline{k}$ such that $T_i \not\leqslant K_j$. This means that for all $i \in \underline{k}$

	T	$(T_1)_{\gamma_1}$, $(T_1)_{\gamma_2}$	$(T_1)_\omega$
1	A_6	A_5, $\tau(\mathsf{A}_5)$ $(\tau \notin \mathsf{S}_6)$	D_{10}
2	M_{12}	M_{11}, $\tau(\mathsf{M}_{11})$ $(\tau \notin \mathsf{M}_{12})$	$\mathsf{PSL}_2(11)$
3	$\mathsf{P}\Omega_8^+(q)$	$\Omega_7(q)$, $\tau(\Omega_7(q))$ (τ a triality)	$\mathsf{G}_2(q)$
4	$\mathsf{Sp}_4(2^a)$, $a \geqslant 2$	$\mathsf{Sp}_2(2^{2a}) \cdot 2$, $\mathsf{O}_4^-(q)$	$D_{q^2+1} \cdot 2$

Table 11.2. *The stabilisers in* T_1

there is a unique $j \in \underline{\ell}$ such that $T_i\mu_j \neq 1$, and so $T_i \leqslant U_j$. On the other hand, as $M\mu_i \cong T$, for all $j \in \underline{\ell}$ there is a unique $i \in \underline{k}$ such that $T_i \leqslant U_j$. Therefore $\ell = k$ and the T_i and the U_j can be indexed so that $T_i \leqslant U_i$ for all i, and so Theorem 11.16(i) holds.

(iii) As in part (ii), either Theorem 11.16(iii) holds or $M\mu_i \cong T$ for all i; assume the latter. As $\mathcal{E} \in \mathsf{CD}_{2\sim}(G)$, for all $i \in \underline{k}$, there are exactly two indices $j \in \underline{\ell}$ such that $T_i\mu_j \neq 1$. If $j_1, j_2 \in \underline{\ell}$ are these indices then we obtain that $T_i \leqslant U_{j_1} \times U_{j_2}$. On the other hand, as $M\mu_j \cong T$, for all $j \in \underline{\ell}$ there is a unique $i \in \underline{k}$ such that $T_i\mu_j \neq 1$. Counting the pairs in the set

$$\{(i,j) \mid i \in \underline{k},\ j \in \underline{\ell},\ T_i \not\leqslant K_j\}$$

we obtain that $\ell = 2k$ and the T_i and the U_i can be indexed so that $T_i \leqslant U_{2i-1} \times U_{2i}$ for all i.

Note that $U_1 \times U_2$ can be viewed as a permutation group acting on $\Gamma_1 \times \Gamma_2$ preserving the natural cartesian decomposition formed by the $(U_1 \times U_2)$-invariant partitions Γ_1 and Γ_2. Choose $\gamma_1 \in \Gamma_1$, $\gamma_2 \in \Gamma_2$ and set $\omega = (\gamma_1, \gamma_2)$. Then Theorem 8.17 implies that the isomorphism type T of T_1, and the stabilisers $(T_1)_{\gamma_1}$, $(T_1)_{\gamma_2}$ and $(T_1)_\omega$ are as in Table 11.2. Hence T_1 acts primitively on both Γ_1 and Γ_2. Thus $(U_1)^{\Gamma_1}$ and $(U_2)^{\Gamma_2}$ are primitive permutation groups. Since $(T_1)^{\Gamma_1} \leqslant U_1^{\Gamma_1}$ and $(T_2)^{\Gamma_2} \leqslant U_2^{\Gamma_2}$, the results of (Liebeck *et al.* 1987) yield that U and T are as in the corresponding columns of Table 11.1 at the beginning of the section. Thus Theorem 11.16(ii) holds. □

The proof of Theorem 11.16 is now very easy.

Proof of Theorem 11.16 Noting that the condition that G projects onto a transitive subgroup of S_ℓ is the same as $\mathcal{E} \in \mathsf{CD}_{\mathrm{tr}}(G)$, the theorem follows from the 6-Class Theorem (Theorem 9.7) and Lemma 11.22. □

Finally we prove Theorem 11.17.

Proof of Theorem 11.17 Let G and W be as in Theorem 11.17. By Theorem 9.13(ii), the cartesian decomposition $\mathcal{E}$ corresponding to the product action of W belongs to $\mathsf{CD}_S(G)$. In particular, $\mathcal{E} \in \mathsf{CD}_{\mathrm{tr}}(G)$, and so G projects onto a transitive subgroup of S_ℓ. Thus part (i) is proved. It follows from Lemma 11.22 that the inclusion $G \leqslant W$ is as in Theorem 11.16(iii), and so $U = \mathsf{Alt}\,\Gamma$, proving part (ii).

By Lemma 11.18, $M \leqslant N$. Let $\omega \in \Omega$ and consider the cartesian factorisation $\mathcal{K}_\omega(\mathcal{E})$ of M. If $K \in \mathcal{K}_\omega(\mathcal{E})$ then $M_\omega \leqslant K$, and so K is a subdirect subgroup of M. Thus Theorem 4.16 implies that K is the direct product of pairwise disjoint, full strips. By Theorem 9.1, if X_1 and X_2 are non-trivial, full strips involved in $\mathcal{K}_\omega(\mathcal{E})$ then X_1 and X_2 are disjoint. Thus if $T_i \not\leqslant K_j$ for some $i \in \underline{k}$ and $j \in \underline{\ell}$ then $T_i \leqslant K_{j'}$ for all $j' \in \underline{\ell} \setminus \{j\}$. Hence the $T_1, \ldots, T_k$ and the $U_1, \ldots, U_\ell$ can be indexed such that

$$\begin{aligned} M^{\Gamma_1} &= T_1 \times \cdots \times T_m \\ &\vdots \\ M^{\Gamma_\ell} &= T_{(\ell-1)m+1} \times \cdots \times T_{\ell m}. \end{aligned}$$

Thus $k = \ell m$ and by Lemmas 11.20 and 11.22, $m \geqslant 2$. Also

$$\begin{aligned} T_1 \times \cdots \times T_m &\leqslant U_1 \\ &\vdots \\ T_{(\ell-1)m+1} \times \ldots \times T_{\ell m} &\leqslant U_\ell, \end{aligned}$$

and statement (iii) is also valid. □

12
Applications to graph theory

Graphs that satisfy certain symmetry conditions play an important role in permutation group theory. Several interesting classes of graphs arise from group theoretic constructions, for example, orbital graphs (studied in Section 2.7) or Cayley graphs. The study of such graphs often leads to a deeper insight into the structure of permutation groups and even sometimes of abstract groups. More background concerning the interaction between graph theory and permutation group theory can be found in the textbooks (Biggs 1993, Godsil & Royle 2001).

In this final chapter, we present several graph theoretical applications of the theory of cartesian decompositions. In most of this chapter, the groups and graphs need not be finite. We will only assume finiteness in Theorem 12.10 and in Section 12.7. Some important classes of graphs, such as Hamming graphs and graph products, are defined on vertex sets which are set theoretical cartesian products. Thus the vertex sets of such graphs admit natural cartesian decompositions and the full automorphism groups of the graphs often preserve these cartesian decompositions. This is the case for instance for not necessarily finite Hamming graphs, as shown in Theorem 12.3. In the case of finite Hamming graphs we find, in addition, that our blow-up concept introduced in Section 11.1 can be used to describe the arc-transitive subgroups of the automorphism groups (Theorem 12.10). We also obtain in Section 12.6 several results for 2-distance-transitive graphs (Theorem 12.21) whose vertex sets admit a cartesian decomposition that is invariant under the group action.

Another important class of symmetric graphs is formed by the s-arc-transitive graphs for $s \geqslant 2$ (see the definition on page 39). In fact, the values of s for which s-arc-transitivity can hold are very restricted: by the celebrated results of Weiss and Tutte, if $\mathfrak{G}$ is a finite s-arc-transitive graph with valency at least 3, then $s \leqslant 7$ (Weiss 1981), while if $\mathfrak{G}$ is a finite graph of valency 3, then $s \leqslant 5$ (Tutte 1947). These results were generalised by Seifter for infinite graphs who proved that if $\mathfrak{G}$ is a connected locally finite graph of valency at least 3 with polynomial growth then $\mathfrak{G}$ can be at most 7-arc-transitive; see (Seifter 1991, Theorem 4.3). Tutte's result relies on elementary methods, while Weiss's result, and hence also Seifter's generalisation, use the finite simple group classification. The problem of classifying finite s-arc-transitive graphs has a long history; see for instance (Seress 2007) for an excellent survey. The final Section 12.7 of this chapter is devoted to the study of finite 2-arc-transitive graphs (Theorem 12.22) whose vertex sets admit cartesian decompositions that are invariant under the group action. Using the 6-Class Theorem (Theorem 9.7) we investigate what type of cartesian decompositions can occur on such vertex sets. Theorem 12.22 excludes four of the six types and presents some interesting examples.

12.1 Graph theoretic definitions

For the basic definitions and notation concerning graphs, refer to Section 2.7. In this chapter all graphs are undirected. Recall that a path of length k in a graph $\mathfrak{G}$ is a sequence $(\alpha_0, \ldots, \alpha_k)$ of vertices such that either $k = 0$ or α_i and α_{i+1} are adjacent for all $i \geqslant 0$.

The *distance* $d(u, v)$ between two vertices u and v in a graph is the length of the shortest path that connects u and v. If α and β lie in distinct connected components and no such path exists, then the distance is undefined. The distance function on a connected graph is a metric and, in particular, the *triangle inequality*

$$d(u, w) \leqslant d(u, v) + d(v, w) \tag{12.1}$$

holds for all vertices u, v, w.

The complete graph K_X on a vertex set Ω is the graph $(\Omega, \Omega^{\{2\}})$. A *clique* in a graph is a subset C of vertices such that each pair of vertices of C is adjacent, whereas an *independent set* is a subset D such that no pair of vertices of D is adjacent. A vertex is *isolated* if it lies on no edge. A graph $\mathfrak{G}$ is said to be *bipartite* if its vertex set can be written as a disjoint union $\mathsf{V}(\mathfrak{G}) = V_1 \cup V_2$ such that whenever $u, v \in V_1$ or $u, v \in V_2$ then u and v are not adjacent. That is, V_1 and V_2 are independent sets. The *complete bipartite graph* $K_{n,m}$ is the graph with vertex set $\{(1, j) \mid 1 \leqslant j \leqslant n\} \cup \{(2, j) \mid 1 \leqslant j \leqslant m\}$ such that two vertices (i_1, j_1) and (i_2, j_2) are adjacent if and only if $i_1 \neq i_2$. A graph is *regular* if there is a positive integer k such that each of its vertices lies on exactly k edges. The constant k is called the *valency* of the graph.

12.2 Cartesian graph products of complete graphs

In this section we consider a graph that is constructed from a cartesian decomposition of a set Ω.

Suppose that Ω is a set and $\mathcal{E} = \{\Gamma_1, \ldots, \Gamma_\ell\}$ is a cartesian decomposition of Ω. We will call the elements of the Γ_i *blocks*. Let us consider the set of blocks $\mathcal{L} = \bigcup_{i \leqslant \ell} \Gamma_\ell$. The pair $(\Omega, \mathcal{L})$, with the incidence given by inclusion, can be viewed as a point-line incidence geometry, but apart from the short Section 12.3 on its connection with chamber systems, we will not make a detailed study of this geometry. We only note that if Ω is finite, $\ell = 2$ and $|\Gamma_1| = |\Gamma_2|$, then $(\Omega, \mathcal{L})$ is a thin generalised quadrangle, often called a grid of type $(\sqrt{|\Omega|} - 1, 1)$ (see (Payne & Thas 2009)). Since two blocks of $\mathcal{L}$ belong to the same partition Γ_i if and only if they are disjoint, the set $\mathcal{L}$ of blocks alone determines the cartesian decomposition $\mathcal{E}$. Let us define an undirected graph $\mathfrak{Gr}(\mathcal{E})$ using the cartesian decomposition $\mathcal{E}$.

Definition 12.1 The undirected graph $\mathfrak{Gr}(\mathcal{E})$ is the graph whose vertex set is Ω such that two distinct vertices α, $\beta \in \Omega$ are adjacent if and only if there are $\ell - 1$ distinct blocks of $\mathcal{L}$ that contain both α and β.

If α and β are adjacent in $\mathfrak{Gr}(\mathcal{E})$, then the $\ell - 1$ blocks of $\mathcal{L}$ that contain both α and β must come from $\ell - 1$ different partitions Γ_i. By the definition of a cartesian decomposition, if α, $\beta \in \Omega$ are contained in ℓ distinct blocks of $\mathcal{L}$, then $\alpha = \beta$. Thus the number $\ell - 1$ in the definition of adjacency for $\mathfrak{Gr}(\mathcal{E})$ is the maximum number of blocks of $\mathcal{L}$ that may simultaneously contain two distinct points of Ω.

Suppose that $\mathfrak{G}_1$ and $\mathfrak{G}_2$ are undirected graphs. The *cartesian product* $\mathfrak{G}_1 \Box \mathfrak{G}_2$ of $\mathfrak{G}_1$ and $\mathfrak{G}_2$ is defined as the graph whose vertex set is the cartesian product $\mathsf{V}(\mathfrak{G}_1) \times \mathsf{V}(\mathfrak{G}_2)$ and distinct vertices (α_1, α_2), (β_1, β_2) of $\mathfrak{G}_1 \Box \mathfrak{G}_2$ are adjacent if and only if either $\alpha_1 = \beta_1$ and α_2 is adjacent to β_2 in $\mathfrak{G}_2$ or $\alpha_2 = \beta_2$ and α_1 is adjacent to β_1 in $\mathfrak{G}_1$. We can extend this notion of graph product to an arbitrary finite number of graphs defining recursively $\mathfrak{G}_1 \Box \cdots \Box \mathfrak{G}_\ell = (\mathfrak{G}_1 \Box \cdots \Box \mathfrak{G}_{\ell-1}) \Box \mathfrak{G}_\ell$. Two vertices $(\alpha_1, \ldots, \alpha_\ell)$ and $(\beta_1, \ldots, \beta_\ell)$ in $\mathfrak{G}_1 \Box \cdots \Box \mathfrak{G}_\ell$ are adjacent if and only if they agree in $\ell - 1$ coordinates and, if $\alpha_i \neq \beta_i$ for some i, then α_i is adjacent to β_i in $\mathfrak{G}_i$. Since $(\mathfrak{G}_1 \Box \mathfrak{G}_2) \Box \mathfrak{G}_3 = \mathfrak{G}_1 \Box (\mathfrak{G}_2 \Box \mathfrak{G}_3)$, the cartesian product is an associative operation on the set of graphs, and hence using parenthesis is not required when considering the cartesian product of more than two graphs. See (Hammack, Imrich & Klavžar 2011, Section 5.1), (Imrich & Klavžar 2000, Section 1.4) for more background on the cartesian product construction for graphs.

If X is a set, then, as above, K_X denotes the complete graph $(X, X^{\{2\}})$ on X. The graph $K_X \square \cdots \square K_X$ (ℓ copies) is known as the *Hamming graph* on the vertex set X^ℓ and is denoted by $H(\ell, X)$. When X is finite and $X = \underline{q}$, then we write K_q for K_X and $H(\ell, q)$ for $H(\ell, X)$.

Lemma 12.2 *Suppose that $\mathcal{E} = \{\Gamma_1, \ldots, \Gamma_\ell\}$ is a cartesian decomposition of a set Ω and set $\Pi = \prod_{i \leqslant \ell} \Gamma_i$. Define the map $\psi \colon \Omega \to \Pi$ by $\omega \mapsto (\gamma_1, \ldots, \gamma_\ell)$, where, for each i, γ_i is the unique block of Γ_i that contains ω. Then $\psi \colon \mathfrak{Gr}(\mathcal{E}) \to K_{\Gamma_1} \square \cdots \square K_{\Gamma_\ell}$ is a graph isomorphism.*

Proof It follows from the definition of the cartesian decomposition $\mathcal{E}$ (Definition 1.1) that ψ is a bijection (see equation (1.1)). Note that Π is the vertex set of the graph $\mathfrak{G} = K_{\Gamma_1} \square \cdots \square K_{\Gamma_\ell}$. Let α, $\beta \in \Omega$. The definition of $\mathfrak{Gr}(\mathcal{E})$ implies that α and β are adjacent in $\mathfrak{Gr}(\mathcal{E})$ if and only if they are contained in $\ell - 1$ blocks of $\mathcal{L}$. This amounts to saying that $\alpha\psi$ and $\beta\psi$ have $\ell - 1$ coordinates in common, which is equivalent to the statement that $\alpha\psi$ and $\beta\psi$ are adjacent in $\mathfrak{G}$. Thus ψ is a graph isomorphism and $\mathfrak{Gr}(\mathcal{E}) \cong \mathfrak{G}$. □

Conversely, for sets $X_1, \ldots, X_\ell$, the vertex set $\prod_{i \geqslant 1} X_i$ of the cartesian product $K_{X_1} \square \cdots \square K_{X_\ell}$ admits the natural cartesian decomposition $\mathcal{E}_0 = \{\Gamma_1, \ldots, \Gamma_\ell\}$ defined by, for each i,

$$\Gamma_i = \{\gamma_{i,x} \mid x \in X_i\} \quad \text{where} \quad \gamma_{i,x} = \{(x_1, \ldots, x_\ell) \mid x_i = x\}. \qquad (12.2)$$

By Lemma 12.2, $\mathfrak{Gr}(\mathcal{E}_0) \cong K_{\Gamma_1} \square \cdots \square K_{\Gamma_\ell} \cong K_{X_1} \square \cdots \square K_{X_\ell}$.

The following theorem characterises the automorphism group of the cartesian product of complete graphs. A more general theorem for connected graphs was proved by Sabidussi (Sabidussi 1959/1960); see also Vizing (Vizing 1963).

Theorem 12.3 *Let $X_1, \ldots, X_\ell$ be sets such that $|X_i| \geqslant 2$ for all $i \in \underline{\ell}$, and let $\mathfrak{G}$ denote the cartesian product $K_{X_1} \square \cdots \square K_{X_\ell}$. If $\mathcal{E}_0$ is the natural cartesian decomposition of the vertex set Ω of $\mathfrak{G}$ as defined in (12.2), then $\mathsf{Aut}(\mathfrak{G}) = (\mathsf{Sym}\,\Omega)_{\mathcal{E}_0}$.*

Proof We introduce some notation that will be used in this proof. Let Ω denote the vertex set of $\mathfrak{G}$. If $v \in \Omega$ and $i \in \underline{\ell}$, let $v[i]$ denote the i-th entry of v, so $v = (v[1], \ldots, v[\ell])$. For $i \in \underline{\ell}$ and $\alpha \in X_i$, define

$$\gamma_{i,\alpha} = \{v \in \Omega \mid v[i] = \alpha\}.$$

Let $\Gamma_i = \{\gamma_{i,\alpha} \mid \alpha \in X_i\}$ and $\mathcal{E}_0 = \{\Gamma_1, \ldots, \Gamma_\ell\}$. Then $\mathcal{E}_0$ is the natural cartesian decomposition of Ω as defined in (12.2).

Set $\mathcal{L} = \{\gamma_{i,\alpha} \mid 1 \leqslant i \leqslant \ell,\ \alpha \in X_i\}$. By the definition of $\mathfrak{G}$, two vertices u and v are adjacent in $\mathfrak{G}$ if and only if there are $\ell - 1$ blocks in $\mathcal{L}$ that contain both u and v. The group $(\mathsf{Sym}\,\Omega)_{\mathcal{E}_0}$ leaves the set $\mathcal{L}$ invariant as a set of subsets of vertices. Hence if $g \in (\mathsf{Sym}\,\Omega)_{\mathcal{E}_0}$ and u and v are distinct vertices of $\mathfrak{G}$, then u, v are contained in $\ell - 1$ members of $\mathcal{L}$ if and only if ug, vg are contained in $\ell - 1$ distinct members of $\mathcal{L}$. Therefore, $(\mathsf{Sym}\,\Omega)_{\mathcal{E}_0} \leqslant \mathsf{Aut}(\mathfrak{G})$.

The proof of the converse inclusion that $\mathsf{Aut}(\mathfrak{G}) \leqslant (\mathsf{Sym}\,\Omega)_{\mathcal{E}_0}$ will be carried out by induction on ℓ. If $\ell = 1$, then $\mathcal{E}_0$ contains a unique partition whose blocks are the singletons $\{\omega\}$ for $\omega \in \Omega$, and so $\mathcal{E}_0$ is the trivial cartesian decomposition of Ω that is preserved by $\mathsf{Sym}\,\Omega$, and is obviously preserved by $\mathsf{Aut}(\mathfrak{G})$. Assume now that $\ell \geqslant 2$ and $\mathsf{Aut}(\mathfrak{G}) \leqslant (\mathsf{Sym}\,\Omega)_{\mathcal{E}_0}$ holds whenever $\mathfrak{G}$ is a cartesian product of at most $\ell - 1$ complete graphs each of size at least two. The proof that $\mathsf{Aut}(\mathfrak{G}) \leqslant (\mathsf{Sym}\,\Omega)_{\mathcal{E}_0}$ holds for ℓ factors is rather more complex and, to help the readers, we break the proof into digestible claims. The induction hypothesis will not be used until much later in the proof. Recall that the distance in $\mathfrak{G}$ between two vertices u and v is denoted by $d(u, v)$.

Claim 1. The following assertions are equivalent for distinct vertices u, v of $\mathfrak{G}$ and for $i \leqslant \ell$:

(i) $d(u, v) \leqslant i$;
(ii) $u[j] = v[j]$ for at least $\ell - i$ distinct indices $j \in \underline{\ell}$.

Proof of Claim 1. We prove this claim by induction on i. The claim is true for $i = 1$ by the definition of $\mathfrak{G}$. Assume that $i \geqslant 2$ and the claim is true for $i - 1$. First suppose that $u[j] = v[j]$ for $\ell - i$ distinct indices j. Suppose without loss of generality that $u[1] = v[1], \ldots, u[\ell - i] = v[\ell - i]$. If $u[\ell - i + 1] = v[\ell - i + 1]$, then u and v agree in $\ell - i + 1$ positions, and hence $d(u, v) \leqslant i - 1$, by the induction hypothesis. Thus we may assume that $u[\ell - i + 1] \neq v[\ell - i + 1]$. Set $u' = (u[1], \ldots, u[\ell - i], v[\ell - i + 1], u[\ell - i + 2], \ldots, u[\ell])$. Then, by the definition of $\mathfrak{G}$, $d(u, u') = 1$, and u', v agree in the first $\ell - i + 1$

positions. Hence, by the induction hypothesis, $d(u',v) \leqslant i-1$. Thus, by (12.1), $d(u,v) \leqslant d(u,u') + d(u',v) \leqslant i$, as required.

Conversely assume that $d(u,v) \leqslant i$. Then there exists a vertex u' such that u and u' are adjacent and $d(u',v) \leqslant i-1$. By the induction hypothesis, $u'[j] = v[j]$ for at least $\ell-(i-1)$ distinct indices j, and, by the definition of $\mathfrak{G}$, $u[j] = u'[j]$ for $\ell-1$ indices j. Thus there must be at least $\ell-i$ indices j such that $u[j] = v[j]$. Therefore the equivalence of the statements (i) and (ii) is established. □

Claim 2. Suppose that C is a clique in $\mathfrak{G}$. Then there exists a subset $J \subset \underline{\ell}$ with $|J| = \ell-1$, and for each $i \in J$ there exists $\alpha_i \in X_i$ such that $u[i] = \alpha_i$ for all $u \in C$.

Proof of Claim 2. If $|C| = 2$, the claim is true by the definition of adjacency in $\mathfrak{G}$. Assume now that $|C| \geqslant 3$ and consider three distinct vertices u, v and w of C. Then each pair in $\{u,v,w\}$ is adjacent. Thus there exists $J \subset \underline{\ell}$ with $|J| = \ell-1$ such that $u[i] = v[i]$ for all $i \in J$. Similarly, there exists some index set J' with $|J'| = \ell-1$, such that $u[i] = w[i]$ for all $i \in J'$. Suppose that $J \neq J'$ and so $\underline{\ell} \setminus J = \{i_1\}$ and $\underline{\ell} \setminus J' = \{i_2\}$ with $i_1 \neq i_2$. This implies that $i_1 \in J'$ and $i_2 \in J$, and hence that $w[i_1] = u[i_1] \neq v[i_1]$ and $v[i_2] = u[i_2] \neq w[i_2]$, and so $d(v,w) \geqslant 2$, which is a contradiction. Hence $J = J'$ and so the claim holds. □

By Claim 2, if C is a maximal clique in $\mathfrak{G}$, then there exists a subset $J \subseteq \underline{\ell}$ with $|J| = \ell-1$ and elements $\alpha_i \in X_i$ such that

$$C = \{v \mid v[i] = \alpha_i \text{ for all } i \in J\}.$$

Recall that $\ell \geqslant 2$ and so $J \neq \emptyset$. If $\underline{\ell} \setminus J = \{j\}$, then we say that the clique C *omits the index* j. Also $|C| = |X_j|$. For $i \in \underline{\ell}$, let $\mathcal{C}_i$ denote the set of maximal cliques in $\mathfrak{G}$ that omit the index i. If $C \in \mathcal{C}_i$ and $c \in C$, then for all $j \neq i$, the entry $c[j]$ is independent of the vertex c. We write $C[j]$ to denote this common entry.

We construct graphs $\mathfrak{G}_0, \mathfrak{G}_1, \ldots, \mathfrak{G}_{\ell-1}$ as follows. The vertex sets of these graphs are the same as the vertex set of $\mathfrak{G}$. In $\mathfrak{G}_i$, two distinct vertices u and v are adjacent if and only if $|\{j \mid u[j] = v[j]\}| = i$; that is, $d(u,v) = \ell - i$ in $\mathfrak{G}$. This implies that each automorphism of $\mathfrak{G}$ is an automorphism of $\mathfrak{G}_i$ for all i. We also have that $\mathfrak{G} = \mathfrak{G}_{\ell-1}$.

Suppose that C_1 and C_2 are two maximal cliques in $\mathfrak{G}$. For $i = 0,\ldots,\ell-1$, let $\mathfrak{C}_i$ denote the subgraph of $\mathfrak{G}_i$ induced by $C_1 \cup C_2$. Thus we obtain a sequence of graphs $\mathfrak{C}_0,\ldots,\mathfrak{C}_{\ell-1}$ and each graph in this sequence has the same vertex set as the vertex set of the union

$C_1 \cup C_2$. Suppose that k_0 is the smallest index such that the graph $\mathfrak{C}_{k_0}$ has at least one edge. Note that k_0 is well defined and $k_0 \leqslant \ell - 1$, since each C_i is a clique of $\mathfrak{C}_{\ell-1}$.

Claim 3. Suppose that C_1 and C_2 are distinct cliques which omit the same index i. Then $k_0 \leqslant \ell - 2$ and $\mathfrak{C}_{k_0}$ is a regular bipartite graph of valency $|X_i| - 1$. In particular, $\mathfrak{C}_{k_0}$ has no isolated vertices.

Proof of Claim 3. Assume without loss of generality that C_1 and C_2 omit the index 1 and, for $j = 2, \ldots, \ell$, let $\alpha_j = C_1[j]$ and $\beta_j = C_2[j]$. Let $k = |\{j \in \{2, \ldots, \ell\} \mid \alpha_j = \beta_j\}|$ and note that $k \leqslant \ell - 2$, since $C_1 \neq C_2$. Consider a graph $\mathfrak{C}_s$ for $0 \leqslant s \leqslant \ell - 2$. Since C_1 and C_2 are cliques in $\mathfrak{G}_{\ell-1}$, C_1 and C_2 are independent sets in $\mathfrak{C}_s$; that is, if u, $v \in C_1$ or if u, $v \in C_2$, then u and v are not adjacent in $\mathfrak{C}_s$. As $C_1 \cap C_2 = \emptyset$ and $C_1 \cup C_2 = \mathsf{V}(\mathfrak{C}_s)$, the graph $\mathfrak{C}_s$ is bipartite with bipartition $C_1 \cup C_2$. Let us assume that $\mathfrak{C}_s$ has an edge $\{u, v\}$ and suppose without loss of generality that $u \in C_1$ and $v \in C_2$. Then $u = (\alpha_1, \alpha_2, \ldots, \alpha_\ell)$ and $v = (\beta_1, \beta_2, \ldots, \beta_\ell)$ for some α_1, $\beta_1 \in X_1$. Being adjacent in $\mathfrak{C}_s$, u and v have s coordinates in common and $s = \delta_{\alpha_1,\beta_1} + k$, where $\delta_{\alpha_1,\beta_1} = 1$ if $\alpha_1 = \beta_1$ and $\delta_{\alpha_1,\beta_1} = 0$ otherwise. Thus either $\delta_{\alpha_1,\beta_1} = 0$ and $s = k$, or $\delta_{\alpha_1,\beta_1} = 1$ and $s = k + 1$. Since $|X_1| \geqslant 2$, both of the cases $\delta_{\alpha_1,\beta_1} = 0$ and $\delta_{\alpha_1,\beta_1} = 1$ are possible. Thus the minimal value for s such that $\mathfrak{C}_s$ has at least one edge is obtained when $\delta_{\alpha_1,\beta_1} = 0$ and $s = k$. Therefore $k_0 = k$. As noted above, the graph $\mathfrak{C}_k$ is bipartite with bipartition $C_1 \cup C_2$. A vertex $u \in C_1$ is adjacent to $v \in C_2$ in the graph $\mathfrak{C}_k$ if and only if $u[1] \neq v[1]$. Therefore each vertex of $\mathfrak{C}_k$ is adjacent to precisely $|X_1| - 1$ vertices, which completes the proof of this claim. □

Claim 4. Suppose that C_1 and C_2 omit different indices. Then $k_0 \leqslant \ell - 2$, and $\mathfrak{C}_{k_0}$ has at least one isolated vertex. In particular, $\mathfrak{C}_{k_0}$ is not a regular graph.

Proof of Claim 4. Suppose without loss of generality that C_1 omits index 1 and C_2 omits index 2. If $\ell = 2$, then $C_1 = \{(\beta', \beta) \mid \beta' \in \Gamma_1\}$ and $C_2 = \{(\alpha, \alpha') \mid \alpha' \in \Gamma_2\}$ for some $\beta \in \Gamma_2$ and $\alpha \in \Gamma_1$. As in Claim 3, C_1 and C_2 are independent sets of $\mathfrak{C}_0$, while the vertices (β', β) and (α, α') are adjacent if and only if $\beta' \neq \alpha$ and $\beta \neq \alpha'$. This shows that $k_0 = 0$ and (α, β) is an isolated vertex, and so the claim holds for $\ell = 2$.

Assume now that $\ell \geqslant 3$. Let, for $i \in \{2, \ldots, \ell\}$, $\alpha_i = C_1[i]$, and, for all $i \in \{1, 3, \ldots, \ell\}$, $\beta_i = C_2[i]$. Set $k = |\{i \in \{3, \ldots, \ell\} \mid \alpha_i = \beta_i\}|$. Consider, for $0 \leqslant s \leqslant \ell - 2$, the graph $\mathfrak{C}_s$. As in Claim 3, the sets C_1 and C_2 are independent sets of $\mathfrak{C}_s$. Suppose that $\{u, v\}$ is an edge of $\mathfrak{C}_s$. Then we may assume without loss of generality that $u \in C_1$ and $v \in C_2$. Writing $u = (\alpha, \alpha_2, \alpha_3, \ldots, \alpha_\ell)$ and $v = (\beta_1, \beta, \beta_3, \ldots, \beta_\ell)$ and considering that u and v are adjacent in $\mathfrak{C}_s$, we have that $s = \delta_{\alpha,\beta_1} + \delta_{\alpha_2,\beta} + k$. Thus the least value for s such that $\mathfrak{C}_s$ has at least one edge is k and this occurs as $|X_1|$, $|X_2| \geqslant 2$. Thus $k_0 = k \leqslant \ell - 2$. Further, $(\beta_1, \alpha_2, \alpha_3, \ldots, \alpha_\ell) \in C_1$ and $(\beta_1, \alpha_2, \beta_3, \ldots, \beta_\ell) \in C_2$ are both isolated (and distinct if $k \neq \ell - 2$) vertices of $\mathfrak{C}_s$. Therefore the claim holds for $\ell \geqslant 3$. □

As discussed after Claim 2, the sets $\mathcal{C}_1, \ldots, \mathcal{C}_\ell$ form a partition of the set of maximal cliques of the graph $\mathfrak{G}$.

Claim 5. The partition of the set of maximal cliques of $\mathfrak{G}$ formed by $\mathcal{C}_1, \ldots, \mathcal{C}_\ell$ is preserved by $\mathsf{Aut}(\mathfrak{G})$.

Proof of Claim 5. Suppose that C_1, $C_2 \in \mathcal{C}_i$, let $g \in \mathsf{Aut}(\mathfrak{G})$, and, for $i \in \{1, 2\}$, set $D_i = C_i g$. Suppose that k_0 is the smallest index k such that the subgraph $\mathfrak{C}_k$ induced by $C_1 \cup C_2$ in $\mathfrak{G}_k$ has an edge. Similarly, let m_0 be the smallest index m such that the subgraph $\mathfrak{Y}_m$ induced by $D_1 \cup D_2$ in $\mathfrak{G}_m$ has an edge. As we noted before Claim 3, $g \in \mathsf{Aut}(\mathfrak{G}_j)$ for all $j \in \{0, \ldots, \ell - 1\}$. Thus $k_0 = m_0$ and $\mathfrak{C}_{k_0}$ is isomorphic to $\mathfrak{Y}_{k_0}$. On the other hand, by Claim 3, $\mathfrak{C}_{k_0}$ has no isolated vertices, since C_1, $C_2 \in \mathcal{C}_i$, and hence, by Claim 4, the cliques D_1 and D_2 that induce the graph $\mathfrak{Y}_{k_0}$ must belong to the same clique class, say $\mathcal{C}_j$. Thus g maps $\mathcal{C}_i$ to $\mathcal{C}_j$. □

Claim 6. Let $g \in (\mathsf{Sym}\,\Omega)_{\mathcal{E}_0}$. Then, for all i, $j \in \underline{\ell}$, $\Gamma_i g = \Gamma_j$ if and only if $\mathcal{C}_i g = \mathcal{C}_j$.

Proof of Claim 6. Suppose that $i \in \underline{\ell}$ and let $C \in \mathcal{C}_i$. Assume that $\Gamma_i g = \Gamma_{j_1}$ and that $Cg \in \mathcal{C}_{j_2}$. We are required to prove that $j_1 = j_2$. The clique C has precisely one vertex in each block of Γ_i (by the definition of $\mathcal{E}_0$), while, for $m \neq i$, the clique C is contained in the single block $\gamma_{m,C[m]}$ of Γ_m. Similarly, Cg contains precisely one vertex in each block of Γ_{j_2}, and, for $m \neq j_2$, the image Cg is contained in a block of Γ_m. Since $\Gamma_i g = \Gamma_{j_1}$, Cg must also contain a unique vertex in each block of Γ_{j_1}, which forces $j_1 = j_2$. Therefore we obtain that $\Gamma_i g = \Gamma_j$ if and only if $\mathcal{C}_i g = \mathcal{C}_j$. □

Suppose that $i \in \underline{\ell}$, and for $j \neq i$, let $\alpha_j \in X_j$. Then there is a unique clique $C \in \mathcal{C}_i$ such that $C[j] = \alpha_j$ for all $j \neq i$. Let us denote this clique by C_α where $\alpha = (\alpha_1, \ldots, \alpha_{i-1}, \alpha_{i+1}, \ldots, \alpha_\ell)$. If $\mathfrak{G}_0$ denotes the graph $K_{X_i} \Box \cdots \Box K_{X_{i-1}} \Box K_{X_{i+1}} \Box \cdots \Box K_{X_\ell}$, then there is a natural bijection ϑ from $\mathcal{C}_i$ to the vertex set Ω_0 of $\mathfrak{G}_0$ given by $\vartheta \colon C_\alpha \mapsto \alpha$.

The set $\mathcal{C}_i$ can be considered as a graph by making C_1, $C_2 \in \mathcal{C}_i$ adjacent if and only if the vectors $(C_1[1], \ldots, C_1[i-1], C_1[i+1], \ldots, C_1[\ell])$ and $(C_2[1], \ldots, C_2[i-1], C_2[i+1], \ldots, C_2[\ell])$ agree in $\ell-2$ coordinates. Thus ϑ becomes a graph isomorphism. Further, the adjacency between distinct C_1 and C_2 is equivalent to the following: C_1 and C_2 are adjacent in this new graph if and only if each vertex of C_1 is adjacent in $\mathfrak{G}$ to a unique vertex of C_2. Hence, if $g \in \mathsf{Aut}(\mathfrak{G})$ such that $\mathcal{C}_i g = \mathcal{C}_i$, then g induces an automorphism of the graph $\mathcal{C}_i$. Note that $\mathcal{C}_i$, being in bijective correspondence with Ω_0, admits a natural cartesian decomposition $\mathcal{E}$ defined as follows. For $j \in \underline{\ell} \setminus \{i\}$ and $\alpha \in X_j$, set

$$\sigma_{j,\alpha} = \{C \in \mathcal{C}_i \mid C[j] = \alpha\}. \tag{12.3}$$

Letting $\Delta_j = \{\sigma_{j,\alpha} \mid \alpha \in X_j\}$, we have

$$\mathcal{E} = \{\Delta_1, \ldots, \Delta_{i-1}, \Delta_{i+1}, \ldots, \Delta_\ell\}.$$

Note that $\mathcal{E}$, being a cartesian decomposition of $\mathcal{C}_i$, depends on i, but we would like to keep our notation simple, and so we choose not to incorporate this fact. Now we invoke the inductive hypothesis introduced before Claim 1: if $g \in \mathsf{Aut}(\mathfrak{G})$ such that $\mathcal{C}_i g = \mathcal{C}_i$, then g preserves the cartesian decomposition $\mathcal{E}$ of $\mathcal{C}_i$.

Claim 7. If $g \in \mathsf{Aut}(\mathfrak{G})$ such that $\mathcal{C}_i g = \mathcal{C}_i$, and if $\Gamma_j \in \mathcal{E}_0 \setminus \{\Gamma_i\}$, then there exists $k \in \underline{\ell} \setminus \{i\}$ such that $\Gamma_j g = \Gamma_k$.

Proof of Claim 7. Suppose that $\gamma_{j,\alpha} \in \Gamma_j$ is a block in the cartesian decomposition $\mathcal{E}_0$. Then (12.3) gives a corresponding block $\sigma_{j,\alpha} \in \Delta_j$ in the cartesian decomposition $\mathcal{E}$ of $\mathcal{C}_i$. We showed above that $\mathcal{C}_i g = \mathcal{C}_i$ implies that g preserves $\mathcal{E}$. Thus there must exist $k \in \underline{\ell} \setminus \{i\}$ and $\beta \in X_k$ such that $(\sigma_{j,\alpha})g = \sigma_{k,\beta}$. Further, the index k depends only on j and not on α. As

$$\gamma_{j,\alpha} = \bigcup_{C \in \sigma_{j,\alpha}} C,$$

we have

$$(\gamma_{j,\alpha})g = \bigcup_{C \in \sigma_{j,\alpha}} (Cg) = \bigcup_{C \in \sigma_{k,\beta}} C = \gamma_{k,\beta}.$$

Thus $\Gamma_j g = \Gamma_k$. □

The proof of the theorem is finally concluded by proving the following claim.

Claim 8. $\mathsf{Aut}(\mathfrak{G}) \leqslant (\mathsf{Sym}\,\Omega)_{\mathcal{E}_0}$.

Proof of Claim 8. Suppose that $g \in \mathsf{Aut}(\mathfrak{G})$ and let $\Gamma_j \in \mathcal{E}_0$. Choose an index $i \in \underline{\ell} \setminus \{j\}$. By Claim 5, there exists $k \in \underline{\ell}$ such that $C_i g = C_k$. Then C_i and C_k contain cliques of the same size, which implies that $|X_i| = |X_k|$. Thus there exists $h \in (\mathsf{Sym}\,\Omega)_{\mathcal{E}_0}$ such that $\Gamma_i h = \Gamma_k$, and so, by Claim 6, $C_i h = C_k$. Since $(\mathsf{Sym}\,\Omega)_{\mathcal{E}_0} \leqslant \mathsf{Aut}(\mathfrak{G})$, the element $g_0 = gh^{-1}$ lies in $\mathsf{Aut}(\mathfrak{G})$ and stabilises C_i. By Claim 7, there exists $m \in \underline{\ell} \setminus \{i\}$ such that $\Gamma_j gh^{-1} = \Gamma_j g_0 = \Gamma_m$, and so $\Gamma_j g = \Gamma_m h$. Since h preserves $\mathcal{E}_0$, we have $\Gamma_m h \in \mathcal{E}_0$, and so $\Gamma_j g \in \mathcal{E}_0$. □

The last claim implies that $\mathsf{Aut}(\mathfrak{G}) \leqslant (\mathsf{Sym}\,\Omega)_{\mathcal{E}_0}$. Combining this with the other inclusion verified earlier in this proof, we obtain that $\mathsf{Aut}(\mathfrak{G}) = (\mathsf{Sym}\,\Omega)_{\mathcal{E}_0}$. □

Corollary 12.4 *If $\mathcal{E}$ is a cartesian decomposition of a set Ω, then $\mathsf{Aut}(\mathfrak{Gr}(\mathcal{E})) = (\mathsf{Sym}\,\Omega)_{\mathcal{E}}$. If $\mathcal{E}$ is homogeneous, then $\mathsf{Aut}(\mathfrak{Gr}(\mathcal{E})) \cong \mathsf{Sym}\,\Gamma_1 \wr \mathsf{S}_\ell$ for $\Gamma_1 \in \mathcal{E}$. In particular, $\mathsf{Aut}(H(\ell, \Gamma)) = \mathsf{Sym}\,\Gamma \wr \mathsf{S}_\ell$.*

Proof If $\psi\colon \mathfrak{Gr}(\mathcal{E}) \to K_{\Gamma_1} \square \cdots \square K_{\Gamma_\ell}$ is the map defined in Lemma 12.2, then ψ is a graph isomorphism. Further, ψ takes $\mathcal{E}$ into the natural cartesian decomposition of $\Gamma_1 \times \cdots \times \Gamma_\ell$ (see the discussion after Lemma 12.2). Thus Theorem 12.3 implies the first assertion of the corollary, while the second and the third assertions about homogeneous cartesian decompositions follow from Lemma 5.10. □

Remark 12.5 Let $\mathcal{E} = \{\Gamma_1, \ldots, \Gamma_\ell\}$ be a cartesian decomposition of a set Ω, and let $e = \{\omega_1, \omega_2\}$ be an edge of the graph $\mathfrak{Gr}(\mathcal{E})$. Then, by definition, there are $\ell - 1$ blocks in $\bigcup_{i \leqslant \ell} \Gamma_i$ that contain both ω_1 and ω_2. Since two blocks in the same Γ_i must be disjoint, the $\ell - 1$ blocks that contain ω_1 and ω_2 must come from pairwise distinct partitions of $\mathcal{E}$. Hence there is a unique index $i_0 \in \underline{\ell}$ such that no block of Γ_{i_0} contains both ω_1 and ω_2. In this way, the edge e can be labelled (or coloured) with the index i_0 and $\mathfrak{Gr}(\mathcal{E})$ can be viewed as an edge-coloured graph. For $i \in \underline{\ell}$, let E_i denote the set of edges of $\mathfrak{Gr}(\mathcal{E})$ that are coloured with colour i. Set $W = (\mathsf{Sym}\,\Omega)_{\mathcal{E}}$. Then the W-actions on $\mathcal{E} = \{\Gamma_1, \ldots, \Gamma_\ell\}$ and on the set $\{E_1, \ldots, E_\ell\}$ are equivalent under the bijection $\Gamma_i \mapsto E_i$ for all $i \in \underline{\ell}$. In other words, W permutes the colours of the edge set $\mathsf{E}(\mathfrak{Gr}(\mathcal{E}))$. Therefore, by Theorem 12.3, every automorphism of $\mathfrak{Gr}(\mathcal{E})$ induces a permutation on the set $\{E_1, \ldots, E_\ell\}$ of edge-colour classes of $\mathsf{E}(\mathfrak{Gr}(\mathcal{E}))$.

12.3 Cartesian decompositions and chamber systems

Chamber systems were introduced by Tits (Tits 1981) (see also (Scharlau 1995)) with the objective of developing a combinatorial approach to understanding the geometry of the groups of Lie type, including the finite simple groups of Lie type. By definition, a *chamber system* is a pair $(\Omega, \{\sim_i \mid i \in I\})$ where Ω is a set, I is an index set, and $\sim_i$ is an equivalence relation on Ω, for all $i \in I$.

Let Ω be a set and let $\mathcal{E} = \{\Gamma_1, \ldots, \Gamma_\ell\}$ be a cartesian decomposition of Ω. Each partition Γ_i defines an equivalence relation of Ω whose equivalence classes are the blocks of Γ_i. Thus, as was also noted by Aschbacher in (Aschbacher 2009a, Aschbacher 2009b), the pair $(\Omega, \mathcal{E})$ can directly be interpreted as a chamber system. We present, in this section, a different chamber system associated with the cartesian decomposition $\mathcal{E}$ which is more closely linked with the geometrical origins of the concept.

Definition 12.6 Let Ω be a set and let $\mathcal{E} = \{\Gamma_1, \ldots, \Gamma_\ell\}$ be a cartesian decomposition of Ω. For $i \in \underline{\ell}$, let $\sim_i$ be the equivalence relation defined as follows: for $\omega_1,\ \omega_2 \in \Omega$, we let $\omega_1 \sim_i \omega_2$ if and only if

$$\text{for all } j \in \underline{\ell} \setminus \{i\} \text{ there exists } \gamma_j \in \Gamma_j \text{ such that } \{\omega_1, \omega_2\} \subseteq \bigcap_{j \neq i} \gamma_j.$$

The pair $\mathcal{C}(\mathcal{E}) = (\Omega, \{\sim_i \mid i \in \underline{\ell}\})$ is called the *chamber system associated with* $\mathcal{E}$.

Let $\mathcal{E}$ be a cartesian decomposition of a set Ω and consider the associated chamber system $\mathcal{C}(\mathcal{E})$. For $\omega_1,\ \omega_2 \in \Omega$ we write $\omega_1 \sim \omega_2$ if and only if $\omega_1 \sim_i \omega_2$ for at least one of the equivalence relations $\sim_i$ of $\mathcal{C}(\mathcal{E})$. Note that $\omega_1 \sim \omega_2$ if and only if ω_1 and ω_2 are adjacent in the graph $\mathfrak{Gr}(\mathcal{E})$ defined in Section 12.2. Several concepts defined for chamber systems can be interpreted in the context of the graph $\mathfrak{Gr}(\mathcal{E})$. For instance, a *gallery* of length n in the chamber system $\mathcal{C}(\mathcal{E})$ is defined as a tuple $(\omega_0, \omega_1, \ldots, \omega_n)$ such that $\omega_{i-1} \sim \omega_i$ for all $i \in \underline{n}$. Hence a gallery of $\mathcal{C}(\mathcal{E})$ is the same thing as a path in $\mathfrak{Gr}(\mathcal{E})$.

Let $\mathcal{C}(\mathcal{E})$ be the chamber system associated with a cartesian decomposition $\mathcal{E}$ of a set Ω and let $g \in \mathsf{Sym}\,\Omega$. The permutation g is said to be a *weak automorphism* of $\mathcal{C}(\mathcal{E})$ if for all $\omega_1,\ \omega_2 \in \Omega$, the relation $\omega_1 \sim \omega_2$ implies $\omega_1 g \sim \omega_2 g$. Further, g is said to be an *automorphism* of $\mathcal{C}(\mathcal{E})$ if there exists a permutation g^* of the set $\underline{\ell}$ such that $\omega_1 \sim_i \omega_2$ implies $\omega_1 g \sim_{ig^*} \omega_2 g$ for all $\omega_1,\ \omega_2 \in \Omega$ and for all $i \in \underline{\ell}$. This is the usual notion of automorphism used in the theory of chamber systems as for example in (Scharlau 1995).

Since the relation $\sim$ coincides with the adjacency in $\mathfrak{Gr}(\mathcal{E})$ and considering the edge-colouring of $\mathfrak{Gr}(\mathcal{E})$ defined in Remark 12.5, we obtain the following easy lemma.

Lemma 12.7 *If $\mathcal{C}(\mathcal{E})$ is as above and $g \in \mathsf{Sym}\,\Omega$, then the following hold:*

(i) *g is a weak automorphism of $\mathcal{C}(\mathcal{E})$ if and only if g is an automorphism of $\mathfrak{Gr}(\mathcal{E})$.*
(ii) *g is an automorphism of $\mathcal{C}(\mathcal{E})$ if and only if g is an automorphism of $\mathfrak{Gr}(\mathcal{E})$ that permutes the edge-colour classes $E_1, \ldots, E_\ell$ defined in Remark 12.5.*

As was noted in Remark 12.5, every automorphism of the graph $\mathfrak{Gr}(\mathcal{E})$ permutes the edge-colour classes, and hence we obtain the following theorem.

Theorem 12.8 *Let Ω be a set, let $\mathcal{E}$ be a cartesian decomposition of Ω, and let $\mathcal{C}(\mathcal{E})$ be the corresponding chamber system, as in Definition 12.6. Then the group of automorphisms of $\mathcal{C}(\mathcal{E})$ coincides with the group of weak automorphisms of $\mathcal{C}(\mathcal{E})$. Further, this automorphism group coincides with $\mathsf{Aut}(\mathfrak{Gr}(\mathcal{E}))$.*

Thus the seemingly more restrictive notion of automorphism for $\mathcal{C}(\mathcal{E})$ is equivalent to the more natural notion of weak automorphism for $\mathcal{C}(\mathcal{E})$.

12.4 Arc-transitive automorphism groups of Hamming graphs

Suppose that $\mathcal{E}$ is a cartesian decomposition of a set Ω and let $\mathfrak{Gr}(\mathcal{E})$ denote the corresponding graph defined in Definition 12.1. In what follows, we will consider the case when a group G of automorphisms of $\mathfrak{Gr}(\mathcal{E})$ acts arc-transitively on $\mathfrak{Gr}(\mathcal{E})$. Recall that, for a group $G \leqslant \mathsf{Sym}\,\Gamma \wr \mathsf{S}_\ell$ and for $i \leqslant \ell$, the components $G^{(i)}$ are as defined in Section 11.1.

Theorem 12.9 *Suppose that $G \leqslant \mathsf{Aut}(\mathfrak{Gr}(\mathcal{E}))$ such that G is arc-transitive. Then $\mathcal{E}$ is a homogeneous cartesian decomposition and $\mathfrak{Gr}(\mathcal{E}) \cong H(\ell, \Gamma)$ with some set Γ such that $|\Gamma| = |\Gamma_i|$ for all i. Further, considering G as a subgroup of $\mathsf{Aut}(H(\ell, \Gamma)) = \mathsf{Sym}\,\Gamma \wr \mathsf{S}_\ell$, the following hold:*

(i) *G projects onto a transitive subgroup of S_ℓ;*
(ii) *a component $G^{(i)}$ is 2-transitive on Γ.*

Proof By Lemma 12.2, we may view $\mathfrak{Gr}(\mathcal{E})$ as the cartesian product $\mathfrak{G} = K_{\Gamma_1}\Box\cdots\Box K_{\Gamma_\ell}$ of complete graphs. Suppose that $\alpha = (\alpha_1,\ldots,\alpha_\ell)$ is a vertex of $\mathfrak{Gr}(\mathcal{E})$. Then the neighbourhood of α in $\mathfrak{G}$ is

$$\mathfrak{G}(\alpha) = \{(\beta_1,\ldots,\beta_\ell) \mid |\{i \mid \beta_i = \alpha_i\}| = \ell - 1\}.$$

Set

$$\Sigma_i = \{(\beta_1,\ldots,\beta_\ell) \mid \beta_i \neq \alpha_i \text{ and } \beta_j = \alpha_j \text{ for all } j \neq i\}.$$

Then $|\Sigma_i| = |\Gamma_i| - 1$ and $\mathfrak{G}(\alpha) = \bigcup_{i\leqslant\ell}\Sigma_i$. Assume that G is arc-transitive on $\mathfrak{G}$. Then G_α is transitive on $\mathfrak{G}(\alpha)$. Further, Theorem 12.3 implies that G preserves the cartesian decomposition $\mathcal{E}$, and so the Σ_i form a G_α-invariant partition of $\mathfrak{G}(\alpha)$. Therefore, by Lemma 2.12, G_α permutes transitively the sets Σ_i, which implies that $|\Sigma_i| = |\Sigma_j|$ for all i, j, and, in turn, that $|\Gamma_i| = |\Gamma_j|$ for all i, j. Thus $\mathcal{E}$ is a homogeneous cartesian decomposition. By Lemma 12.2, $\mathfrak{G} \cong H(\ell,\Gamma)$ for some Γ such that $|\Gamma| = |\Gamma_i|$ for all i. By Corollary 12.4, $\mathsf{Aut}(\mathfrak{G}) = \mathsf{Sym}\,\Gamma \wr \mathsf{S}_\ell$ in product action.

In the rest of the proof we assume that $\mathfrak{G} = H(\ell,\Gamma)$. Assume that $\gamma \in \Gamma$ and consider the vertex $\alpha = (\gamma,\ldots,\gamma)$. Define the neighbourhood $\mathfrak{G}(\alpha)$ and, for $1 \leqslant i \leqslant \ell$, the subsets Σ_i as above and set $W = \mathsf{Sym}\,\Gamma \wr \mathsf{S}_\ell$. Then, as noted above, $\mathfrak{G}(\alpha) = \Sigma_1 \cup \cdots \cup \Sigma_\ell$ is a W_α-invariant partition. The permutation representation of W_α on $\underline{\ell}$ given by the natural projection $W_\alpha \to \mathsf{S}_\ell$ is equivalent to the action of W_α on the set $\{\Sigma_1,\ldots,\Sigma_\ell\}$. Now, as G_α is transitive on $\mathfrak{G}(\alpha)$, G_α must be transitive on the set $\{\Sigma_1,\ldots,\Sigma_\ell\}$, and so G_α projects onto a transitive subgroup of S_ℓ. Thus G projects onto a transitive subgroup of S_ℓ.

We show that the components of G are 2-transitive. Since G projects to a transitive subgroup of S_ℓ, it suffices to prove that $G^{(1)}$ is 2-transitive. Let H be the stabiliser in G_α of the block Σ_1. Then H acts transitively on Σ_1 and H is a subgroup of the stabiliser of the point 1 under the permutation representation of W on $\underline{\ell}$. Thus $H^\Gamma \leqslant G^{(1)}$. Further, H stabilises γ in its action on Γ, and so $H^\Gamma \leqslant (G^{(1)})_\gamma$. Therefore $(G^{(1)})_\gamma$ is transitive on $\Gamma\backslash\{\gamma\}$. By Lemma 2.31, $G^{(1)}$ is 2-transitive on Γ. □

In the remainder of the section, we investigate arc-transitive actions of groups on the graph $\mathfrak{Gr}(\mathcal{E})$ for a cartesian decomposition $\mathcal{E}$ of a *finite* set Ω. Hence we may assume that $\mathfrak{G} = H(\ell, \Gamma)$ and $G \leqslant \mathsf{Sym}\,\Gamma \wr \mathsf{S}_\ell$ acting in product action on the vertex set Γ^ℓ of $\mathfrak{G}$. By Theorem 12.9, $G^{(1)}$ is a finite 2-transitive subgroup of $\mathsf{Sym}\,\Gamma$. By Theorem 3.21, a finite 2-transitive group is primitive of O'Nan–Scott type HA or AS. Here we only consider the case when the type of $G^{(1)}$ is AS. Replacing G by a conjugate in $\mathsf{Sym}\,\Omega$ and $\mathfrak{G}$ by an isomorphic copy if necessary, we may assume, by Theorem 5.14, that the components $G^{(i)}$ are all equal; that is $G \leqslant G^{(1)} \wr \mathsf{S}_\ell$.

Theorem 12.10 *Suppose as above that Γ is a finite set with $|\Gamma| \geqslant 2$ and that G is an arc-transitive subgroup of $\mathsf{Aut}(H(\ell, \Gamma))$ such that $G^{(1)}$ is a finite 2-transitive permutation group of O'Nan–Scott type AS. Assume, in addition, that $G \leqslant G^{(1)} \wr \mathsf{S}_\ell$. Setting $T = \mathsf{Soc}\, G^{(1)}$, one of the following holds:*

(i) *$T^\ell \leqslant G$ and G is a blow-up of $G^{(1)}$;*
(ii) *$T \cong \mathsf{A}_6$ or $T \cong \mathsf{M}_{12}$ and $G \cap T^\ell$ is a direct product of pairwise disjoint non-trivial full strips $X_1, \ldots, X_{\ell/2}$ each of length 2.*

The proof of the theorem relies on the following lemma which is a consequence of the classification of finite 2-transitive permutation groups. A permutation group is said to be *half-transitive* if its orbits have the same cardinality. By Lemma 2.20, a normal subgroup of a transitive permutation group is half-transitive.

Lemma 12.11 *Suppose that Γ is a finite set and S is an almost simple 2-transitive subgroup of $\mathsf{Sym}\,\Gamma$. Suppose that $r \geqslant 2$, and that X is a full strip in $(\mathsf{Soc}\, S)^r$ covering all the factors of S^r acting in product action on Γ^r. Then one of the following holds:*

(i) *X is not half-transitive, and hence $\mathsf{N}_{\mathsf{Sym}(\Gamma^r)}(X)$ is intransitive on Γ^r;*
(ii) *$r = 2$, and $\mathsf{Soc}\, S \in \{\mathsf{A}_6, \mathsf{M}_{12}\}$, and X is transitive.*

Proof Set $T = \mathsf{Soc}\, S$ and let us assume that the group X is half-transitive. Since X is a full strip, there exist, for $i = 2, \ldots, r$, $\alpha_i \in \mathsf{Aut}(T)$ such that

$$X = \{(x, x\alpha_2, \ldots, x\alpha_r) \mid x \in T\}. \tag{12.4}$$

By possibly reordering the factors of S^r, we may assume without loss of generality that there exists $r_0 \in \underline{r}$ such that the automorphisms

$\alpha_2, \ldots, \alpha_{r_0}$ are permutational isomorphisms (that is, they preserve the T-conjugacy class of point stabilisers T_γ for $\gamma \in \Gamma$), while $\alpha_{r_0+1}, \ldots, \alpha_r$ are not permutational isomorphisms. We allow the possibility that $r_0 = 1$ and in this case none of the α_i are permutational isomorphisms, whereas $r_0 = r$ occurs when each of the α_i are permutational isomorphisms.

Claim 1. $r_0 < r$.

Proof of Claim 1. Seeking a contradiction, assume that $r_0 = r$; that is, $\alpha_2, \ldots, \alpha_r$ are permutational isomorphisms, and so, for $\gamma \in \Gamma$, the image $T_\gamma \alpha_i$ of a point stabiliser is also a point stabiliser for all i. Let $\gamma_1 \in \Gamma$ be fixed, and choose $\gamma_2, \ldots, \gamma_r \in \Gamma$ such that $T_{\gamma_1}\alpha_i = T_{\gamma_i}$ for all $i \geqslant 2$. Suppose that $\delta \in \Gamma \setminus \{\gamma_1\}$. Since S is 2-transitive, we find inspecting (Cameron 1999, Table 7.4) that either T is 2-transitive, or $T \cong \mathsf{PSL}_2(8)$, $S = \mathsf{Ree}(3)$, and $|\Gamma| = 28$. In both cases, T is primitive and hence $T_\delta \neq T_{\gamma_1}$. Setting $\alpha = (\gamma_1, \gamma_2, \ldots, \gamma_r)$ and $\beta = (\delta, \gamma_2, \ldots, \gamma_r)$ we obtain that

$$\begin{aligned} X_\alpha &= \{(x, x\alpha_2 \ldots, x\alpha_r) \mid x \in T_{\gamma_1}\}; \\ X_\beta &= \{(x, x\alpha_2 \ldots, x\alpha_r) \mid t \in T_{\gamma_1} \cap T_\delta\}. \end{aligned}$$

Thus $|X\colon X_\alpha| \neq |X\colon X_\beta|$ and so $|\alpha X| \neq |\beta X|$ which implies that X is not half-transitive: a contradiction. Claim 1 follows. □

Now for some $\gamma \in \Gamma$, the image $T_\gamma \alpha_{r_0+1}$ is not a point stabiliser. On the other hand, considering the action of T on the coset space $\Delta = [T\colon T_\gamma \alpha_{r_0+1}]$, we obtain a primitive action of T such that the point stabilisers in T of a point of Γ and of a point of Δ are not conjugate. Further, this T-action on Δ is 2-transitive unless $T \cong \mathsf{PSL}_2(8)$. Since $\mathsf{PSL}_2(8)$ has a unique conjugacy class of maximal subgroups of index 28, $T \not\cong \mathsf{PSL}_2(8)$, and hence T must be 2-transitive on both Γ and Δ. Using (Cameron 1999, Table 7.4), T must be isomorphic to one of the following groups: $\mathsf{PSL}_d(q)$ with $d \geqslant 3$, $\mathsf{PSL}_2(11)$ (as a subgroup of S_{11}), A_7 (as a subgroup of S_{15}), HS, A_6, or M_{12}. In all cases, the group T has precisely two conjugacy classes of subgroups such that the permutation representations on the right coset spaces are 2-transitive. Thus the subgroups $T_\gamma \alpha_{r_0+1}, \ldots, T_\gamma \alpha_r$ are conjugate in T. Therefore we may assume without loss of generality that $T_\gamma \alpha_i = T_\gamma$ for $i = 2, \ldots, r_0$, and that $\delta \in \Delta$ such that $T_\gamma \alpha_i^{-1} = T_\delta$ for $i \geqslant r_0 + 1$.

Claim 2. T is isomorphic to either A_6 or M_{12}.

Proof of Claim 2. Suppose that this is not the case and so T is isomorphic to one of the groups $\mathsf{PSL}_d(q)$ with $d \geqslant 3$, $\mathsf{PSL}_2(11)$, A_7, or HS. Let $\delta \in \Delta$. We claim that T_δ has two orbits Γ_1, Γ_2 in Γ and $|\Gamma_1| \neq |\Gamma_2|$. The claim can be verified using Magma (Bosma, Cannon & Playoust 1997) or GAP (The GAP Group 2013) for $\mathsf{PSL}_2(11)$, A_7 and HS. Indeed, in these cases $\{|\Gamma_1|, |\Gamma_2|\} = \{5,6\}$, $\{7,8\}$, $\{50,126\}$, respectively.

Let us now verify this claim for $\mathsf{PSL}_d(q)$ with $d \geqslant 3$ and suppose that $T = \mathsf{PSL}_d(q)$. We may assume that Γ is the set of points, while Δ is the set of hyperplanes in the projective space $\mathsf{PG}(d-1,q)$. Fix $\delta \in \Delta$. Now T_δ is transitive on the set of points $\gamma_0 \in \delta$ and also on the set of points $\gamma_0 \notin \delta$. Further,

$$\begin{aligned} |\{\gamma_0 \mid \gamma_0 \text{ is a point in } \mathsf{PG}(d-1,q) \text{ in } \delta\}| &= \frac{q^{d-1}-1}{q-1}; \\ |\{\gamma_0 \mid \gamma_0 \text{ is a point in } \mathsf{PG}(d-1,q) \text{ not in } \delta\}| &= \frac{q^d - q^{d-1}}{q-1}. \end{aligned}$$

Thus the two sets have different sizes, as claimed.

Suppose that $\gamma_1 \in \Gamma$ such that γ_1 is not a member of the T_δ-orbit of γ. Since the T_δ-orbits have different cardinalities, $|T_\gamma \cap T_\delta| \neq |T_{\gamma_1} \cap T_\delta|$. For $i = 2, \ldots, r_0$, let $\gamma_i \in \Gamma$ such that $T_{\gamma_1}\alpha_i = T_{\gamma_i}$. Set $\alpha = (\gamma, \gamma, \ldots, \gamma)$ and $\beta = (\gamma_1, \gamma_2 \ldots, \gamma_{r_0}, \gamma, \ldots, \gamma)$. Then

$$\begin{aligned} X_\alpha &= \{(x, x\alpha_2, \ldots, x\alpha_r) \mid x \in T_\gamma \cap T_\delta\}; \\ X_\beta &= (x, x\alpha_2, \ldots, x\alpha_r) \mid x \in T_{\gamma_1} \cap T_\delta\}. \end{aligned}$$

Therefore $|X_\alpha| = |T_\gamma \cap T_\delta| \neq |T_{\gamma_1} \cap T_\delta| = |X_\beta|$, and we obtain that $|\alpha X| \neq |\beta X|$. Hence X is not half-transitive, which is a contradiction. Claim 2 follows. □

Claim 3. $r = 2$.

Proof of Claim 3. Assume that $r \geqslant 3$. Now γ and δ are the unique points in Γ and in Δ, respectively, stabilised by $T_\gamma \cap T_\delta$. By possibly interchanging Γ and Δ, we may assume that $r_0 \geqslant 2$. Let $\gamma' \in \Gamma \setminus \{\gamma\}$ and for $i = 2, \ldots, r_0$, let γ_i such that $T_{\gamma'}\alpha_i = T_{\gamma_i}$. Set $\alpha = (\gamma, \ldots, \gamma)$ and $\beta = (\gamma, \gamma_2, \ldots, \gamma_{r_0}, \gamma, \ldots, \gamma)$. Then

$$\begin{aligned} X_\alpha &= \{(x, x\alpha_2, \ldots, x\alpha_r) \mid x \in T_\gamma \cap T_\delta\}; \\ X_\beta &= \{(x, x\alpha_2, \ldots, x\alpha_r) \mid x \in T_\gamma \cap T_{\gamma'} \cap T_\delta\}. \end{aligned}$$

Thus $|X \colon X_\alpha| \neq |X \colon X_\beta|$, and hence $|\alpha X| \neq |\beta X|$. Hence X is not half-transitive in this case. This contradiction proves Claim 3. □

Since $r = 2$, our assumptions imply that α_2 is not a permutational isomorphism, and so $T_\gamma \alpha_2^{-1} = T_\delta$. Since $T \cong \mathsf{A}_6$ or $T \cong \mathsf{M}_{12}$, we have that T_δ is transitive on Γ. Thus $|X : X_{(\gamma,\delta)}| = |\Gamma|^2$, which implies that $X = \{(x, x\alpha_2) \mid x \in T\}$ is transitive on Γ^2. □

Proof of Theorem 12.10. Set $K = G^{(1)}$, $T = \mathsf{Soc}\, K$, let H be the projection of G in S_ℓ. By assumption, $G \leqslant K \wr H$. Let B denote the base group K^ℓ of the wreath product $K \wr H$. By Theorem 5.15, the component $(G \cap B)^{(1)}$ is transitive on Γ. Further, the component $(G \cap B)^{(1)}$ is normal in $K = G^{(1)}$, and hence $T \leqslant (G \cap B)^{(1)}$. Therefore $G \cap B$ is a subdirect subgroup of L^ℓ for some $L \leqslant K$ such that $T \leqslant L$. Suppose, for a group X, that $X^{[i]}$ denotes the i-th term of its derived series. Let $\sigma_i : K^\ell \to K$ denote the i-th coordinate projection. Since L is almost simple with socle T, L/T is a solvable group by Schreier's Conjecture (Dixon & Mortimer 1996, p. 133), and so there exists some a such that $L^{[a]} = T$. Applying the coordinate projection σ_i, we obtain that $(G \cap B)^{[a]}\sigma_i = ((G \cap B)\sigma_i)^{[a]} = L^{[a]} = T$. Hence $(G \cap B)^{[a]}$ is a subdirect subgroup of T^ℓ, and so $G \cap T^\ell$ is also a subdirect subgroup of T^ℓ. By Theorem 4.16, $G \cap T^\ell$ is a direct product of pairwise disjoint full strips $X_1, \ldots, X_m$ for some $m \geqslant 1$.

Suppose that $T_1, \ldots, T_\ell$ are the internal simple direct factors of T^ℓ and let $K_1, \ldots, K_\ell$ be the corresponding almost simple factors of $B = K^\ell$. Since $X_1 \times \cdots \times X_m = G \cap T^\ell \trianglelefteq G$, the group G permutes the X_i by conjugation and hence the sets $\{i \mid i \in \mathrm{Supp}\, X_i\}$ form a G-invariant partition of $\underline{\ell}$. Since G is transitive on $\underline{\ell}$, the strips $X_1, \ldots, X_m$ form a G-conjugacy class and in particular $|\mathrm{Supp}\, X_i|$ is independent of i. Set $s = \ell/m$. We may assume that $\mathrm{Supp}\, X_i = \{T_{(i-1)s+1}, \ldots, T_{is}\}$ for all i.

Set $X = X_1 \times \cdots \times X_m$. As G is transitive on Γ^ℓ, X is half-transitive on Γ^ℓ. If an X_i is not half-transitive, considered as a subgroup of Γ^s, and it has orbits in Γ^s of sizes a and b with $a \neq b$, then X will have orbits in $\Gamma^\ell = (\Gamma^s)^m$ of sizes a^m and b^m. This implies that each X_i must be half-transitive. By Lemma 12.11, $s \leqslant 2$, and if $s = 2$, then $T \cong \mathsf{A}_6$ or $T \cong \mathsf{M}_{12}$. If $s = 1$, then (i) must hold, while if $s = 2$, then (ii) must be valid. □

Let us next show that part (ii) of Theorem 12.10 does in fact occur.

Example 12.12 Suppose that T is isomorphic to either A_6 or to M_{12}. Assume that T is a permutation group acting on Γ, where $|\Gamma| = 6$, or $|\Gamma| = 12$, respectively. Let $\mathfrak{G}$ denote the Hamming graph $H(\ell, \Gamma)$.

Let A, $B \leqslant T$ be two proper subgroups of T such that $|A| = |B|$ and $AB = T$. Then, by Lemma 8.14, $A \cong B \cong \mathsf{A}_5$ when $T \cong \mathsf{A}_6$ and $A \cong B \cong \mathsf{M}_{11}$ when $T \cong \mathsf{M}_{12}$. We may also assume that A is a point stabiliser T_γ for some $\gamma \in \Gamma$. By Lemma 8.14, there exists $\vartheta \in \mathsf{Aut}(T)$ such that ϑ interchanges A and B. Set $X = \{(t, t\vartheta) \mid t \in T\}$. Then X is a non-trivial full strip in T^2 and is transitive on Γ^2. Further, for $\alpha = (\gamma, \gamma)$, a point stabiliser X_α is transitive on

$$\mathfrak{G}(\alpha) = \{(\gamma, \beta) \mid \beta \in \Gamma\} \cup \{(\beta, \gamma) \mid \beta \in \Gamma\}.$$

Let $\tau \in \mathsf{Aut}(T^2)$ given by $(t_1, t_2) \mapsto (t_2, t_1\vartheta^2)$. Then τ normalises X and hence we may define $G_0 = X \rtimes \langle \tau \rangle$. Since G_0 projects onto the top group S_2 of the wreath product $T \wr \mathsf{S}_2$, the stabiliser $(G_0)_\alpha$ is transitive on $\mathfrak{G}(\alpha)$, and so, by Lemma 2.37(i), G_0 is an arc-transitive subgroup of $\mathsf{Aut}(H(\ell, \Gamma))$.

To obtain more complex examples, we may consider the wreath product $G = G_0 \wr H$ for some transitive permutation group $H \leqslant \mathsf{S}_\ell$. Then G is an arc-transitive subgroup of $H(2\ell, \Gamma)$.

Remark 12.13 Regular embeddings of Hamming graphs into surfaces can be determined by studying arc-transitive subgroups of automorphisms with a prescribed stabiliser structure. Orientable embeddings are related to arc-transitive automorphism groups with cyclic stabilisers, while non-orientable embeddings can be studied through arc-transitive automorphism groups with dihedral stabilisers; see (Jones 2011, Jones & Kwon 2012). If ℓ, $q \geqslant 2$ and $G \leqslant \mathsf{Aut}(H(q, \ell))$ is an arc-transitive subgroup with abelian stabiliser G_α, then the component $G^{(1)}$ is itself a permutation group with abelian stabiliser. Since $G^{(1)}$ is 2-transitive, and hence primitive, by Theorem 12.9, Proposition 3.17 implies that $G^{(1)}$ has an elementary abelian minimal normal subgroup. Thus, by the discussion in Section 7.1, the O'Nan–Scott type of $G^{(1)}$ is HA and we also obtain that q is a prime-power (cf. (Jones 2011, Theorem 2.1)).

A similar but a bit more complicated argument shows that if G is arc-transitive, as above, but G_α is dihedral, then either $G^{(1)}$ is a 2-transitive group of type HA and q is a prime-power, or Theorem 12.10(ii) is valid with $\ell = 2$, $T \cong \mathsf{A}_6$ and $H(\ell, q) = H(2, 6)$. Examples can be constructed for both the cyclic and the dihedral cases with GAP (The GAP Group 2013) or Magma (Bosma *et al.* 1997).

12.5 Arc-transitive graphs and cartesian decompositions

In this section we investigate the class of connected G-arc-transitive graphs for innately transitive permutation groups G such that the vertex set of the graph can be expressed as a cartesian product which is G-invariant. We will work under the following hypothesis.

Hypothesis 12.14 Let Γ be a set with $|\Gamma| \geqslant 2$, let $\ell \geqslant 2$, and set $\Omega = \Gamma^\ell$. Set $W = \mathsf{Sym}\,\Gamma \wr \mathsf{S}_\ell$ and consider W as a permutation group on Ω acting in product action and let $\pi\colon W \to \mathsf{S}_\ell$ denote the natural projection. Suppose that G is an innately transitive subgroup of W with plinth $M = T_1 \times \cdots \times T_k \cong T^k$, where the T_i and T are pairwise isomorphic finite simple groups. Let $\mathcal{E} = \{\Gamma_1, \ldots, \Gamma_\ell\}$ denote the natural cartesian decomposition of Ω (see Section 5.2.2). Fix $\gamma \in \Gamma$, set $\omega = (\gamma, \ldots, \gamma)$, and let $\mathcal{K} = \mathcal{K}_\omega(\mathcal{E}) = \{K_1, \ldots, K_\ell\}$ be the corresponding cartesian factorisation of M as defined in Lemma 8.6. Assume in addition that $\mathfrak{G}$ is a connected graph with vertex set Ω and that G is a subgroup of $\mathsf{Aut}(\mathfrak{G})$.

Assume that Hypothesis 12.14 is valid. For $j \leqslant \ell$, let $G^{(j)}$ be the component as defined in Section 5.3. As in Section 11.1, we will view the component $G^{(j)}$ as a permutation group acting on Γ. If $G\pi$ is transitive, then we may assume, by Theorem 5.14, that $G \leqslant G^{(1)} \wr (G\pi)$. In this case, by possibly replacing G and $\mathfrak{G}$ by (permutationally) isomorphic copies, we may assume without loss of generality that $G \leqslant G^{(1)} \wr (G\pi)$. Set $L = G^{(1)}$.

Let $\alpha = (\alpha_1, \ldots, \alpha_\ell)$ and $\beta = (\beta_1, \ldots, \beta_\ell)$ be adjacent vertices of $\mathfrak{G}$. For each i such that $\alpha_i \neq \beta_i$, define $\mathfrak{G}_i$ as the graph with vertex set Γ and edge set $\{\alpha_i, \beta_i\}L$. If $\alpha_i = \beta_i$, then define $\mathfrak{G}_i$ as the empty graph with vertex set Γ. The group L acts vertex-transitively and edge-transitively on $\mathfrak{G}_i$ for each i. Consider

$$\mathcal{S}(\alpha, \beta) := \{\mathfrak{G}_i \mid 1 \leqslant i \leqslant \ell\} \tag{12.5}$$

as a multiset containing ℓ elements; that is, $\mathfrak{G}_i$ may occur with multiplicity greater than one. Let $I(\alpha, \beta) := \{i \mid \alpha_i \neq \beta_i\}$. Then $\mathfrak{G}_i$ is not the empty graph (that is, $\mathfrak{G}_i$ has an edge) if and only if $i \in I(\alpha, \beta)$. Since $\{\alpha, \beta\}$ is an edge in $\mathfrak{G}$, $I(\alpha, \beta)$ is non-empty, so at least one of the $\mathfrak{G}_i$ is not an empty graph.

Lemma 12.15 *Assume Hypothesis 12.14, and also that $G\pi$ is a transitive subgroup of S_ℓ, that $G \leqslant L \wr (G\pi)$ with $L = G^{(1)}$, and that*

$\mathfrak{G}$ is G-edge-transitive. Then the multiset $\mathcal{S}(\alpha,\beta)$, defined in (12.5), is independent of the choice of the edge $\{\alpha,\beta\} \in \mathsf{E}(\mathfrak{G})$. Moreover, if $h_1,\ldots,h_\ell \in G^{(1)}$ and $\sigma \in G\pi$ are such that $g = (h_1,\ldots,h_\ell)\sigma \in G$, then $I(\alpha g,\beta g) = I(\alpha,\beta)\sigma$.

Proof Let $\{\alpha',\beta'\}$ be an edge in $\mathsf{E}(\mathfrak{G})$ where $\alpha' = (\alpha_1',\ldots,\alpha_\ell')$ and $\beta' = (\beta_1',\ldots,\beta_\ell')$. Then there is an element $g = (h_1,\ldots,h_\ell)\sigma \in G$ such that each $h_i \in L$, $\sigma \in (G\pi)$, and $\{\alpha,\beta\}g = \{\alpha',\beta'\}$. Suppose that $\alpha g = \alpha'$ and $\beta g = \beta'$; the other case (where $\alpha g = \beta'$ and $\beta g = \alpha'$) can be handled similarly. Then $\alpha' = (\alpha_{1\sigma^{-1}}h_{1\sigma^{-1}},\ldots,\alpha_{\ell\sigma^{-1}}h_{\ell\sigma^{-1}})$ and $\beta' = (\beta_{1\sigma^{-1}}h_{1\sigma^{-1}},\ldots,\beta_{\ell\sigma^{-1}}h_{\ell\sigma^{-1}})$. Define the graphs $\mathfrak{G}_i$ for the edge $\{\alpha,\beta\}$. For each i, $\mathsf{E}(\mathfrak{G}_i)$ is the L-orbit $\{\alpha_i,\beta_i\}L = \{\alpha_i h_i,\beta_i h_i\}L$, and hence $\mathsf{E}(\mathfrak{G}_i)$ is the L-orbit containing $\{\alpha_{i\sigma}',\beta_{i\sigma}'\}$. It follows that $I(\alpha',\beta') = I(\alpha,\beta)\sigma$, and that $\mathcal{S}(\alpha,\beta)$ is also the multiset of L-orbits containing the pairs $\{\alpha_i',\beta_i'\}$ for $1 \leqslant i \leqslant \ell$; that is, $\mathcal{S}(\alpha,\beta)$ is independent of the edge $\{\alpha,\beta\}$. □

Thus, under the conditions of Lemma 12.15, the multiset $\mathcal{S}(\alpha,\beta)$ is independent of the edge $\{\alpha,\beta\}$ of $\mathfrak{G}$ and we denote it by $\mathcal{S}(\mathfrak{G})$. The graph $\mathfrak{G}$ determines uniquely the multiset $\mathcal{S}(\mathfrak{G})$ of L-edge-transitive graphs.

Lemma 12.16 *Assume Hypothesis 12.14, and also that $\mathfrak{G}$ is M-arc-transitive, that $G\pi$ is a transitive subgroup of S_ℓ and that $G \leqslant G^{(1)} \wr (G\pi)$. Let $\omega = (\gamma,\ldots,\gamma) \in \Omega$ as in Hypothesis 12.14, $\beta = (\gamma_1,\ldots,\gamma_\ell) \in \mathfrak{G}(\omega)$ and set $L = G^{(1)}$. Then there is some $h \in (L_\gamma)^\ell$ such that $\beta h = (\gamma_1,\ldots,\gamma_1)$. Thus, in the image graph $\mathfrak{G}h$, the tuples $(\gamma,\ldots,\gamma)$ and $(\gamma_1,\ldots,\gamma_1)$ are adjacent. Furthermore, $G^h \leqslant \mathsf{Aut}(\mathfrak{G}h)$ with $G^h \leqslant L \wr (G\pi)$.*

Proof Since $\mathfrak{G}$ is M-arc-transitive, M_ω is transitive on $\mathfrak{G}(\omega)$. As $M_\omega \leqslant (L_\gamma)^\ell$, the subset $\mathfrak{G}(\omega)$ satisfies

$$\mathfrak{G}(\omega) \subseteq \beta_1 L_\gamma \times \cdots \times \beta_\ell L_\gamma$$

where $\beta_i L_\gamma$ is the L_γ-orbit containing γ_i and the product is the cartesian product of sets. We claim that $\gamma_1 L_\gamma = \gamma_i L_\gamma$ for all i. Choose $i \in \underline{\ell}$. Since M is transitive on Ω, $G = MG_\omega$. Also, since $G\pi$ is transitive on $\underline{\ell}$, and since, by Theorem 8.3, M is in the kernel of π, $G_\omega\pi$ must be transitive on $\underline{\ell}$. Thus there is an element $g \in G_\omega$ such that

$g = (k_1, \ldots, k_\ell)\sigma$ with $k_1, \ldots, k_\ell \in L_\gamma$, $\sigma \in G\pi$ and $1\sigma = i$. Now

$$\beta g = (\gamma_1, \ldots, \gamma_\ell)(k_1, \ldots, k_\ell)\sigma = (\gamma_1 k_1, \ldots, \gamma_\ell k_\ell)\sigma.$$

Thus the i-th entry of βg is $\gamma_1 k_1$ and the point $\gamma_1 k_1$ is contained in $\gamma_1 L_\gamma$. However, the i-th entry of βg is also contained in $\gamma_i L_\gamma$, and hence $\gamma_i L_\gamma = \gamma_1 L_\gamma$. Thus there is an element $h = (h_1, \ldots, h_\ell) \in (L_\gamma)^\ell$ such that $\beta h = (\gamma_1, \ldots, \gamma_1)$. Clearly, $G^h \leqslant \mathsf{Aut}(\mathfrak{G}h)$; moreover $G^h \leqslant (L \wr (G\pi))^h = L \wr (G\pi)$ since $h \in L^\ell$. □

12.6 Direct product graphs and normal cartesian decompositions

In Section 12.2, we have already defined the cartesian product of graphs. Some results in this section are best expressed in terms of another graph product, namely, the direct product. If $\mathfrak{G}_1$ and $\mathfrak{G}_2$ are graphs, then the *direct product* of $\mathfrak{G}_1 \times \mathfrak{G}_2$ is defined as the graph whose vertex set is the cartesian product $\mathsf{V}(\mathfrak{G}_1) \times \mathsf{V}(\mathfrak{G}_2)$ and adjacency is defined by the rule that (γ_1, γ_2), (γ_1', γ_2') are adjacent in $\mathfrak{G}_1 \times \mathfrak{G}_2$ if and only if γ_1 is adjacent to γ_1' in $\mathfrak{G}_1$ and γ_2 is adjacent to γ_2' in $\mathfrak{G}_2$. The definition of direct product can be extended to an arbitrary finite number of factors, and in particular, if $\mathfrak{G}_1$ is a graph and $l \geqslant 2$, then one can define $(\mathfrak{G}_1)^{\times l}$ as the graph with vertex set $\mathsf{V}(\mathfrak{G}_1)^l$ such that $(\gamma_1, \ldots, \gamma_l)$ is adjacent to $(\gamma_1', \ldots, \gamma_l')$ in $(\mathfrak{G}_1)^{\times l}$ if and only if γ_i is adjacent to γ_i' in $\mathfrak{G}_1$ for all i. Note that if $\alpha_1 \in \mathfrak{G}_1$ and $\alpha = (\alpha_1, \ldots, \alpha_1)$ then the neighbourhood $(\mathfrak{G}_1)^{\times l}(\alpha)$ of α in $(\mathfrak{G}_1)^{\times l}$ is equal to the cartesian power $\mathfrak{G}_1(\alpha_1)^l$ of the neighbourhood of α_1 in $\mathfrak{G}_1$. See (Imrich & Klavžar 2000, Chapter 5) or (Hammack *et al.* 2011, Chapter 8) for more background on the direct products of graphs and their properties.

Proposition 12.17 *Suppose that Hypothesis 12.14 holds, and also that $\mathfrak{G}$ is M-arc-transitive and that the cartesian decomposition $\mathcal{E}$ is M-normal. Then $\mathfrak{G} \cong (\mathfrak{G}_1)^{\times \ell}$ where $\mathfrak{G}_1$ is the graph whose vertex set is Γ and edge set is the $M^{(1)}$-orbit of $\{\gamma, \alpha_1\}$ where $\alpha_1 \in \Gamma$ is chosen so that $(\gamma, \ldots, \gamma)$ and $(\alpha_1, \alpha_2, \ldots, \alpha_\ell)$ are adjacent in $\mathfrak{G}$. Moreover, $G\pi$ is transitive, $|\Gamma| \geqslant 3$, and $\mathfrak{G}$ is not $(G, 2)$-arc-transitive.*

Proof Since $\mathcal{E}$ is an M-normal cartesian decomposition, $G\pi$ is transitive (Theorem 4.22), and hence we may assume without loss of generality by Theorem 5.14 that $G \leqslant G^{(1)} \wr (G\pi)$ and, by Lemma 12.16, that there is $\alpha_1 \in \Gamma$ such that $\omega = (\gamma, \ldots, \gamma)$ and $\alpha = (\alpha_1, \ldots, \alpha_1)$ are adjacent in $\mathfrak{G}$. Since M is arc-transitive, M_ω is transitive in $\mathfrak{G}(\omega)$. Further, by Lemma 4.20, $M_\omega = \prod_j (M^{(j)})_\gamma$. As $\mathfrak{G}$ is M-arc-transitive, M is not regular on Ω. Thus $M^{(j)}$ is not regular on Γ, and so $M^{(j)}$ is the unique minimal normal subgroup of $G^{(j)}$, by Theorem 3.6. Since $G^{(j)} = G^{(1)}$, we have $M^{(j)} = M^{(1)}$ for all j. Thus

$$\mathfrak{G}(\omega) = \left(\alpha_1(M^{(1)})_\gamma\right) \times \cdots \times \left(\alpha_1(M^{(1)})_\gamma\right)$$

where the factor $\alpha_1(M^{(1)})_\gamma$ in the product above is taken ℓ times. If $\mathfrak{G}_1$ denotes the graph whose vertex set is Γ and whose edge set is the $M^{(1)}$-orbit containing $\{\gamma, \alpha_1\}$ then, by the remark before the proposition, the neighbourhoods of ω in $\mathfrak{G}$ and in $(\mathfrak{G}_1)^{\times\ell}$ are equal. Since M acts vertex transitively on both $\mathfrak{G}$ and $(\mathfrak{G}_1)^{\times\ell}$, we find that $\mathfrak{G} = (\mathfrak{G}_1)^{\times\ell}$. As $\mathfrak{G}$ is connected, the valency of $\mathfrak{G}$ is at least 2, and so $\mathfrak{G}_1$ also has valency at least 2, and in particular $|\mathsf{V}(\mathfrak{G})_1| = |\Gamma| \geqslant 3$. Hence if $\alpha_2 \in \alpha_1(M^{(1)})_\gamma \setminus \{\alpha_1\}$, then both $(\alpha_1, \ldots, \alpha_1, \alpha_2)$ and $(\alpha_1, \ldots, \alpha_1, \alpha_2, \alpha_2)$ are in $\mathfrak{G}(\omega) \setminus \{\alpha\}$, but no element of $L_{\gamma,\alpha_1} \wr \mathsf{S}_\ell$ can map one of these points to the other. Therefore G_ω is not 2-transitive on $\mathfrak{G}(\omega)$, and so $\mathfrak{G}$ is not $(G,2)$-arc-transitive, by Lemma 2.37 □

Next we study the case when $\mathfrak{G}$ is M-arc-transitive.

Corollary 12.18 *Assume Hypothesis 12.14, and also that $\mathfrak{G}$ is M-arc-transitive, that $G\pi$ is a transitive subgroup of S_ℓ and that $G \leqslant G^{(1)} \wr (G\pi)$. Then $I(\alpha, \beta) = \underline{\ell}$ and $\mathcal{S}(\mathfrak{G})$, defined in (12.5), contains a single graph $\mathfrak{G}_1$ with multiplicity ℓ. Moreover, if $\mathcal{E}$ is M-normal, then $\mathfrak{G} = (\mathfrak{G}_1)^{\times\ell}$.*

Proof Set $L = G^{(1)}$. By Lemma 12.16, we may assume that $\{\omega, \alpha\}$ is an edge in $\mathfrak{G}$ where $\alpha = (\alpha_1, \ldots, \alpha_1)$. Hence $|I(\alpha, \beta)| = \ell$ and $\mathcal{S}(\mathfrak{G})$ contains the graph $\mathfrak{G}_1$ with edge set $\{\gamma, \alpha_1\}L$ with multiplicity ℓ.

Now assume that $\mathcal{E}$ is M-normal and $M = M_1 \times \cdots \times M_\ell$ is the decomposition of M as in the proof of Proposition 12.17. Since $M_\omega = ((M_1)_\gamma)^\ell$ and M_ω is transitive on $\mathfrak{G}(\omega)$, we deduce that $(M_1)_\gamma$ is transitive on $\mathfrak{G}_1(\gamma)$. Then the structure of $\mathfrak{G}$ as a direct power follows as in the proof of Proposition 12.17. □

Lemma 12.19 *Suppose that Hypothesis 12.14 holds, and also that M is non-regular and that G_ω is quasiprimitive on $\mathfrak{G}(\omega)$. Then $\mathfrak{G}$ is M-arc-transitive, and so M_ω is transitive on $\mathfrak{G}(\omega)$.*

Proof By Hypothesis 12.14, M is vertex-transitive on $\mathfrak{G}$. As $M_\omega \trianglelefteq G_\omega$, either M_ω is transitive on $\mathfrak{G}(\omega)$ or M_ω is trivial on $\mathfrak{G}(\omega)$. Suppose that the latter holds. (We show that in this case M_ω fixes all vertices of $\mathfrak{G}$, which is impossible as M is not regular.) By assumption, M_ω fixes all vertices in $\{\omega\} \cup \mathfrak{G}(\omega)$; that is, it fixes all vertices of distance less than or equal to 1 from ω. Thus if $\beta \in \mathfrak{G}(\omega)$, then $M_\beta \subseteq M_\omega$ and the same argument shows that $M_\omega \subseteq M_\beta$. Thus $M_\omega = M_\beta$ and M_ω fixes pointwise all vertices of $\mathfrak{G}(\beta)$. Since this holds for all $\beta \in \mathfrak{G}(\omega)$, it follows that M_ω fixes all vertices with distance at most 2 from ω. Continuing by induction on the distance from ω and using the connectivity of $\mathfrak{G}$, M_ω fixes all vertices of $\mathfrak{G}$. This is a contradiction, and hence M_ω is transitive on $\mathfrak{G}(\omega)$. Since M is vertex-transitive, it follows from Lemma 2.37(i) that M is arc-transitive on $\mathfrak{G}$. □

Corollary 12.20 *Suppose that Hypothesis 12.14 holds, that M is not regular, and that $\mathcal{E}$ is M-normal. Then $\mathfrak{G}$ is not $(G,2)$-arc-transitive.*

Proof If $\mathfrak{G}$ is $(G,2)$-arc-transitive, then G_ω is 2-transitive on $\mathfrak{G}(\omega)$, by Lemma 2.37, and in particular G_ω is quasiprimitive on $\mathfrak{G}(\omega)$. Hence $\mathfrak{G}$ is M-arc-transitive, by Lemma 12.19, and then, by Proposition 12.17, we have a contradiction. Thus $\mathfrak{G}$ is not $(G,2)$-arc-transitive. □

We note that if G is a quasiprimitive group of type CD, then its unique minimal normal subgroup M is not regular and, by Theorem 11.12, $\mathcal{E}$ must be a blow-up decomposition. In particular $\mathcal{E}$ must be M-normal. Hence by Corollary 12.20, for such a group G, there exist no $(G,2)$-arc-transitive graphs. This argument gives an alternative proof of (Praeger 1993, Lemma 5.3(a)).

Let $\mathfrak{G}$ be a connected graph and $G \leqslant \mathsf{Aut}(\mathfrak{G})$. The *diameter* of $\mathfrak{G}$ is defined as $\mathrm{diam}(\mathfrak{G}) = \max\{d(\alpha,\beta) \mid \alpha,\ \beta \in \mathsf{V}(\mathfrak{G})\}$. The graph $\mathfrak{G}$ is called G-*distance-transitive* if G is transitive on $\mathfrak{G}_i = \{(\alpha,\beta) \mid d(\alpha,\beta) = i\}$ for all $i \leqslant \mathrm{diam}(\mathfrak{G})$. For a positive integer $j \leqslant \mathrm{diam}(\mathfrak{G})$, we say that $\mathfrak{G}$ is (G,j)-*distance-transitive* if G is transitive on $\mathfrak{G}_i$ for all $i \leqslant j$. Assuming that $\mathfrak{G}$ is G-vertex transitive and $\alpha \in \mathsf{V}(\mathfrak{G})$, we have that $\mathfrak{G}$ is (G,j)-distance-transitive if and only if G_α is transitive on $\mathfrak{G}_i(\alpha) = \{\beta \mid d(\alpha,\beta) = i\}$ for all $i \leqslant j$. A proof of this fact is analogous to the proof of Lemma 2.37(i) and we leave the details to the reader.

Theorem 12.21 *Suppose Hypothesis 12.14 is valid, and also that $\mathfrak{G}$*

is $(G,2)$-distance-transitive and M-arc-transitive and that $\mathcal{E}$ is M-normal. Then the component $\mathfrak{G}_1$ in Corollary 12.18 is the complete graph on Γ. Further, $\ell = 2$, and $\mathfrak{G}$ is the complement of the Hamming graph $H(2,\Gamma)$. Moreover, $\mathfrak{G}$ is $\mathsf{Aut}(\mathfrak{G})$-distance-transitive, but is not 2-arc-transitive.

Proof Proposition 12.17 implies that $\mathfrak{G} \cong (\mathfrak{G}_1)^{\times \ell}$, $\mathfrak{G}_1$ is not an empty graph, and $|\Gamma| \geqslant 3$. In fact, by possibly replacing the group G and the graph $\mathfrak{G}$ with isomorphic copies, we may assume by Corollary 12.18 that $\mathfrak{G} = (\mathfrak{G}_1)^{\times \ell}$. Recall that $\omega = (\gamma, \ldots, \gamma)$. Assume for seeking a contradiction that $\mathfrak{G}_1$ is not the complete graph on the vertex set Γ and choose γ', $\gamma_1 \in \Gamma$ such that $\{\gamma, \gamma'\}$ and $\{\gamma', \gamma_1\}$ are edges of $\mathfrak{G}_1$ and $d(\gamma, \gamma_1) = 2$. Set $\omega' = (\gamma', \ldots, \gamma')$, $\alpha = (\gamma, \ldots, \gamma, \gamma_1)$, and $\alpha' = (\gamma, \ldots, \gamma, \gamma_1, \gamma_1)$. Observe that $\{\omega, \omega'\}$, $\{\omega', \alpha\}$, $\{\omega', \alpha'\}$ are edges of $\mathfrak{G}$ and that $d(\omega, \alpha) = d(\omega, \alpha') = 2$. Since $G_\omega \leqslant (G^{(1)})_\gamma \wr \mathsf{S}_\ell$, no element of G_ω can map α to α', and so $\mathfrak{G}$ is not $(G,2)$-distance-transitive. Thus we obtain that $\mathfrak{G}_1 = K_\Gamma$.

Suppose now that $\ell \geqslant 3$. Then let $\gamma_1 \in \Gamma \setminus \{\gamma\}$, and consider the same α and α' as in the previous paragraph. Then α and α' both lie at distance two from ω in $\mathfrak{G}$, and no element of G_ω maps one to the other. Thus $\mathfrak{G}$ is not $(G,2)$-distance-transitive in this case either. Hence $\ell = 2$ and $\mathfrak{G} = K_\Gamma \times K_\Gamma$, as claimed. Therefore $\mathfrak{G}$ is isomorphic to the complement of $H(2,\Gamma)$. Thus, in this case, $\mathfrak{G}$ has diameter 2 with

$$
\begin{aligned}
\mathfrak{G}_1(\omega) &= \{(\gamma_1, \gamma_2) \mid \gamma_1 \neq \gamma \text{ and } \gamma_2 \neq \gamma\}; \\
\mathfrak{G}_2(\omega) &= \{(\gamma, \gamma_1) \mid \gamma_1 \neq \gamma\} \cup \{(\gamma_1, \gamma) \mid \gamma_1 \neq \gamma\}.
\end{aligned}
$$

The group $\mathsf{Aut}(\mathfrak{G}) = \mathsf{Aut}(H(2,\Gamma)) = \mathsf{Sym}\,\Gamma \wr \mathsf{S}_\ell$ is transitive on both $\mathfrak{G}_1(\omega)$ and $\mathfrak{G}_2(\omega)$, and hence $\mathfrak{G}$ is $\mathsf{Aut}(\mathfrak{G})$-distance-transitive. Also $\mathfrak{G}$ cannot be 2-arc-transitive by Proposition 12.17. □

12.7 2-arc-transitive graphs and permutation groups

In this final section we investigate pairs $(G, \mathfrak{G})$, where G is a finite group and $\mathfrak{G}$ is a finite graph, that satisfy Hypothesis 12.14 under the stronger condition that $\mathfrak{G}$ is $(G,2)$-arc-transitive. Our main theorem is the following.

Theorem 12.22 *Suppose that Γ is a finite set with $|\Gamma| \geqslant 2$ and let $\ell \geqslant 2$. Consider $W = \mathsf{Sym}\,\Gamma \wr \mathsf{S}_\ell$ as a permutation group acting in* *product action on $\Omega = \Gamma^\ell$. Let $\mathfrak{G}$ be a connected $(G,2)$-arc-transitive graph with vertex set Ω for some innately transitive group $G \leqslant W$ with plinth M such that M is either abelian or non-regular. Let $\mathcal{E}$ be the natural cartesian decomposition of Γ^ℓ. Then M must be non-abelian and one of the following is valid.*

(i) *G is quasiprimitive of type* As *and either*

   (a) *$M \cong \mathsf{A}_6$, $|\mathsf{Aut}(\mathsf{A}_6)\colon G| \in \{1,2\}$, $G \neq \mathsf{PGL}_2(9)$, $|\Omega| = 6^2$, and $\mathfrak{G}$ has valency 5; or*

   (b) *$M \cong \mathsf{Sp}_4(4)$, $G = \mathsf{Aut}(\mathsf{Sp}_4(4))$, $|\Omega| = 120^2$ and $\mathfrak{G}$ is a graph with valency 17.*

(ii) *M is not simple, G is transitive on $\mathcal{E}$ and $\mathcal{E} \in \mathsf{CD}_{2\not\sim}(G)$.*

Remark 12.23 (i) Computation with the computational algebra systems GAP (The GAP Group 2013) or Magma (Bosma *et al.* 1997) can verify that the 2-arc-transitive graphs with $T = \mathsf{A}_6$ and $T = \mathsf{Sp}_4(4)$ do in fact exist; see Remark 12.27. These two graphs are, to our knowledge, the only known connected $(G,2)$-arc-transitive graphs such that G is quasiprimitive with non-regular plinth and G preserves a cartesian decomposition of the vertex set. The graph with $T = \mathsf{A}_6$ is also known as Sylvester's Double Six Graph, it has 36 vertices and valency 5 (see (Brouwer, Cohen & Neumaier 1989, 13.1.2 Theorem)). Its full automorphism group is $\mathsf{Aut}(\mathsf{A}_6) \cong \mathsf{P\Gamma L}_2(9)$. The graph with $T = \mathsf{Sp}_4(4)$ has $120^2 = 14,400$ vertices and valency 17. Both graphs that appear in this context can be described using the generalised quadrangle associated with the non-degenerate alternating form stabilised by $\mathsf{Sp}_4(q)$, but giving such a description is beyond the scope of the book. On the other hand, the existence of examples in part (ii) is unresolved. Candidates appeared in (Li & Seress 2006), but we showed in (Li, Praeger & Schneider 2016) that the examples presented there do not admit cartesian decompositions.

(ii) If M is non-abelian and regular, then the graph $\mathfrak{G}$ is a Cayley graph and a theretical analysis with examples is given by R. W. Baddeley in (Baddeley 1993b). Thus if M is a non-abelian plinth of G, then we assume that M is non-regular.

The proof of Theorem 12.22 is completed in the next subsections which treat separately the cases where the plinth M is abelian, non-abelian simple, and non-abelian composite.

$\mathfrak{G}$	K_{n+1}	$\square_n$	$P_m(a)$	$\Gamma(C_{23})$	$\Gamma(C_{22})$
k	$\log_2(n+1)$	$n-1$	m^a	11	10
G_ω	$\mathsf{GL}_k(2)$	S_n	$\mathsf{P\Gamma L}_m(2^a)$	M_{23}	$\mathsf{M}_{22}\cdot 2$

Table 12.1. *The table for Theorem 12.24*

12.7.1 Abelian plinth

In this section we prove that, if all the hypotheses of Theorem 12.22 hold, then the group M cannot be abelian. The following theorem was proved in (Ivanov & Praeger 1993).

Theorem 12.24 *Let G be a finite primitive permutation group of type HA and let $\mathfrak{G}$ be a finite $(G,2)$-arc-transitive graph. Then $G = (\mathbb{F}_2)^k \rtimes G_\omega$ where G_ω is the stabiliser of a vertex ω and is also a subgroup of $\mathsf{GL}_k(2)$. Further, $\mathfrak{G}$, k, G_ω are contained in one of the lines of Table 12.1.*

See (Ivanov & Praeger 1993) for the notation concerning the graphs in Table 12.1.

Proposition 12.25 *Under Hypothesis 12.14, if $\mathfrak{G}$ is a finite $(G,2)$-arc-transitive graph, then M is non-abelian.*

Proof Suppose that M is abelian. From Hypothesis 12.14 we have $M = \mathbb{F}_p^k$ for some prime p, and $G \leqslant \mathsf{Sym}\,\Gamma \wr \mathsf{S}_\ell$. Thus G is an affine primitive permutation group and, for $\omega \in \Omega$, $G = M \rtimes G_\omega$ and M can be viewed as a k-dimensional vector space over $\mathbb{F}_p$ (see Section 7.1). Hence in this proof we use additive notation in M. Further, M, viewed as a G_ω-module, is irreducible (see Secton 7.1). By Theorem 8.3, $M \leqslant (\mathsf{Sym}\,\Gamma)^\ell$, and Theorem 8.4 implies that G_ω preserves a direct sum decomposition of M into ℓ components; that is, G_ω is imprimitive as a linear group acting on M (see the definitions before Theorem 8.4).

By Theorem 12.24, $p = 2$ and k and G_ω are contained in one of the columns of Table 12.1. In Columns 1–2, G_ω is $\mathsf{GL}_k(2)$ or S_{k+1} and these groups do not preserve non-trivial direct sum decompositions of their vertex sets. In Column 3, $G_\omega = \mathsf{P\Gamma L}_m(2^a)$ acting in dimension $k = m^a$ with $m \geqslant 3$ and $a \geqslant 1$. By the discussion in (Ivanov & Praeger 1993, (1.3)), M is an irreducible module for $\mathsf{PSL}_m(2^a)$. We claim that M is a primitive $\mathsf{PSL}_m(2^a)$-module. If M were imprimitive, then $\mathsf{PSL}_m(2^a)$ would permute non-trivially a direct decomposition $M = V_1 \oplus \cdots \oplus V_\ell$ with $\ell \leqslant m^a$. On the other hand, if $(m, 2^a) \neq (4, 2)$, the minimal degree permutation representation of $\mathsf{PSL}_m(2^a)$ is on the 1-spaces of the natural module and has degree $(2^{ma} - 1)/(2^a - 1) > 2^{(m-1)a}$, while the minimal degree of a permutation representation of $\mathsf{SL}_4(2)$ is 8; see (Cooperstein 1978, Table 1). As $m \geqslant 3$, we have $m < 2^{m-1}$, and so $m^a < 2^{(m-1)a}$. Further, $4^1 < 8$, and so $\mathsf{PSL}_m(2^a)$ cannot be imprimitive in these cases.

We have, in the remaining two columns, that $G_\omega \cong M_{23}$ or $G_\omega \cong M_{22}.2$ in dimension 10 or 11, respectively; these columns can be handled similarly. If the underlying module of M_{23} were imprimitive, then M_{23} would be acting on 11 points transitively, which is impossible. If the underlying module for $M_{22}.2$ were imprimitive, then either M_{22} would be acting in dimension less than 10 or it would be acting transitively on at most 10 points; both of these possibilities are clearly impossible by the information presented in the Atlas (Conway, Curtis, Norton, Parker & Wilson 1985). □

12.7.2 Finite simple plinth

In this section we will consider the situation described in Hypothesis 12.14 under the additional restriction that $\mathfrak{G}$ is $(G, 2)$-arc-transitive and $M = T$ is a finite non-abelian simple group. Unlike in the other cases, here we find two examples. Further, we show that these examples are the only possibilities. This proves Theorem 12.22 for groups with non-abelian simple plinth.

Theorem 12.26 *Suppose that Hypothesis 12.14 holds. Assume further that $M = T$ is a finite non-abelian simple group and that $\mathfrak{G}$ is $(G, 2)$-arc-transitive. Then one of the following is valid.*

(i) *$T = \mathsf{A}_6$, $|\mathsf{Aut}(\mathsf{A}_6) \colon G| \in \{1, 2\}$, $G \neq \mathsf{PGL}_2(9)$, and $|\mathsf{V}(\mathfrak{G})| = 36$, $\mathfrak{G}$ is Sylvester's Double Six Graph of valency 5.*

(ii) *$T = \mathsf{Sp}_4(4)$, $G = \mathsf{Aut}(\mathsf{Sp}_4(4))$, $|\mathsf{V}(\mathfrak{G})| = 14{,}400 = 120^2$, and $\mathfrak{G}$ is a graph of valency 17.*

In both cases, G is quasiprimitive.

Proof Let $\omega \in \mathsf{V}(\mathfrak{G})$ as in Hypothesis 12.14. By Theorem 8.18, T_ω is non-trivial. By Lemma 2.37, G_ω is 2-transitive, and in particular quasiprimitive on $\mathfrak{G}(\omega)$. Lemma 12.19 gives that $\mathfrak{G}$ is T-arc-transitive.

In particular $\mathfrak{G}(\omega)$ is a T_ω-orbit. As T is simple, the possibilities for T and T_ω are given in Theorem 8.17. In particular, T is isomorphic to one of the groups A_6, M_{12}, $\mathsf{Sp}_4(2^a)$, with $a \geqslant 2$, or $\mathsf{P\Omega}_8^+(q)$. We analyse each of these cases below.

$T = \mathsf{A}_6$, $|\mathsf{V}(\mathfrak{G})| = 36$: Calculation with GAP (The GAP Group 2013) or Magma (Bosma *et al.* 1997) shows that G has a suborbit of size 5 on which G_ω acts 2-transitively if and only if G has index at most 2 in $\mathsf{Aut}(T) = \mathsf{P\Gamma L}_2(9)$ and $G \neq \mathsf{PGL}_2(9)$. The corresponding 2-arc-transitive graph is undirected, connected and is of valency 5 as described in item (i).

$T = \mathsf{M}_{12}$ and $|\mathsf{V}(\mathfrak{G})| = 144$: Calculation with GAP (The GAP Group 2013) or Magma (Bosma *et al.* 1997) shows that T has no suborbit on which G_ω is 2-transitive and the corresponding graph is connected. Hence no graph arises in this case.

$T = \mathsf{Sp}_4(q)$ with $q = 2^a$ and $a \geqslant 2$: The group G is contained in $\mathsf{Aut}(T) \cong \mathsf{Sp}_4(q) \rtimes C_{2a}$ where C_{2a} is the group of field automorphisms extended by a graph automorphism. Further, $T_\omega = Z\langle\sigma\rangle$ where Z is a subgroup of order q^2+1 of a Singer cycle and σ is an element of order 4; see (Baddeley *et al.* 2004b, the proof of Lemma 5.2). Let $\alpha \in \mathfrak{G}(\omega)$ and let $x \in T$ such that $\alpha = \omega x$.

We claim that $Z \cap Z^x = 1$. Suppose that $1 \neq Z_1 = Z \cap Z^x$. Then Z_1 is cyclic of odd order. Let R be a cyclic subgroup of Z_1 with odd prime order r. Then $r \mid q^2+1$, and so $r \nmid q^2-1$, and hence $r \nmid q-1$ and $r \nmid q^3-1 = (q-1)(q^2+q+1)$. Thus r is a primitive prime divisor of q^4-1 and hence the centraliser of R in $\mathsf{GL}_4(q)$ is a full Singer cycle C_{q^4-1} (Huppert 1967, II.7.3 Satz). Hence Z and Z^x are subgroups of the same C_{q^4-1}, which implies that $Z = Z^x$. Since $T_\omega = \mathsf{N}_T(Z)$ and $T_\alpha = \mathsf{N}_T(Z^x)$ we find that $T_\omega = T_\alpha$; that is α is a fixed point of T_ω. By Lemma 2.19, $\mathsf{N}_T(T_\omega)$ is transitive on the set of fixed points of T_ω. Thus, as T_ω is self-normalising in T (see Lemma 8.14), we obtain that $\omega = \alpha$ which is a contradiction as $\alpha \in \mathfrak{G}(\omega)$. This shows that $Z \cap Z^x = 1$ as claimed.

As $Z \cap Z^x = 1$, Z is faithful on $\mathfrak{G}(\omega)$. As Z is normal in G_ω and G_ω is 2-transitive on $\mathfrak{G}(\omega)$, Z is transitive, and hence regular, on $\mathfrak{G}(\omega)$. Thus $|\mathfrak{G}(\omega)| = q^2+1$. As $|G_\omega| \mid 8a(q^2+1)$, and G_ω is 2-transitive on $\mathfrak{G}(\omega)$, it follows that $2^{2a} = q^2$ divides $8a$, and hence $a = 2$. If $a = 2$, then computation with GAP (The GAP Group 2013) or Magma (Bosma *et al.* 1997) shows that there exists a unique corresponding 2-arc-transitive graph $\mathfrak{G}$, which is given in item (ii).

$T = \mathsf{P\Omega}_8^+(q)$: In this case $T_\omega = G_2(q)$. If $q \geqslant 3$, then $G_2(q)$

does not have a 2-transitive permutation representation (Cameron 1999, Table 7.4), and so $q = 2$. On the other hand $G_2(2) \cong \mathsf{PSU}_3(3).2$ (see (Conway *et al.* 1985)) and has 2-transitive representations of degree $q^3 + 1 = 28$ and of degree 2. However, computation with GAP (The GAP Group 2013) or Magma (Bosma *et al.* 1997) shows that T_ω has no suborbit of size 28 or 2, and hence this case does not arise. □

Remark 12.27 The graphs in parts (i) and (ii) exist and they can easily be constructed using the following code in Magma (Bosma *et al.* 1997). The code as it stands below constructs the graph for the group $T = \mathsf{A}_6$ in Theorem 12.26(i). To obtain the graph in Theorem 12.26(ii), the reader needs to comment out the second line, by inserting '//' at the beginning, and remove '//' from the beginning of the third line.

```
// construct the group T
T := Alt( 6 ); a := 60; b := 36; c := 5;
// T := Sp(4,4); a := 8160; b := 14400; c := 17;

// find isomorphic maximal subgroups A and B
// such that AB=T and let C be their intersection
max := [ x`subgroup : x in MaximalSubgroups( T ) |
#x`subgroup eq a ];
A := max[1]; B := max[2];
assert #A*#B/#(A meet B ) eq #T;
C := A meet B;

// Consider T as a permutation group acting
// on the right cosets of C and find Aut( T )
// as the normaliser of T in the symmetric group
T := T@CosetAction( T, C );
G := Normalizer( Sym( b ), T );

// Find the G-suborbit corresponding to the
// required graph and construct the graph
suborb := [ x : x in Orbits( Stabilizer( G, 1 )) |
#x eq c ][1];
y := suborb[1];
edges := {1,y}^G;
Gr := Graph< b |  edges >;
```

```
// check that it is connected
assert IsConnected( Gr );

// check that G_1
// is 2-transitive on the neighbours of 1
assert Transitivity( Stabilizer( G, 1 ), suborb ) eq 2
```

The assertion in Theorem 12.26(i) concerning the possible groups G such that $\mathfrak{G}$ is $(G, 2)$-arc-transitive, can be verified with the following lines in Magma. The first two lines show that $T = \mathsf{A}_6$ is not 2-transitive on the neighbourhood of a vertex, and hence $\mathfrak{G}$ is not $(T, 2)$-arc-transitive. The last lines show that among the three maximal subgroups of $\mathsf{Aut}(T) = \mathsf{P\Gamma L}_2(9)$ that contain T, the ones with Magma identification $< 720, 765 >$ and $< 720, 763 >$ act 2-arc-transitively on $\mathfrak{G}$. These identification numbers correspond to the groups S_6 and M_{10}.

```
> Transitivity( Stabilizer( T, 1 ), suborb );
1
[ IdentifyGroup( x`subgroup ) :
      x in MaximalSubgroups( G ) |
      Transitivity( Stabilizer( x`subgroup, 1 ), suborb )
      eq 2 ];
[ <720, 765>, <720, 763> ]
```

12.7.3 Finite composite plinth

We continue working under Hypothesis 12.14. We know from Proposition 12.25 that if $\mathfrak{G}$ is $(G, 2)$-arc-transitive and finite, then M is non-abelian. Further, the situation when $\mathfrak{G}$ is finite and $(G, 2)$-arc-transitive and M is simple is fully described in Section 12.7.2. Hence from now on we will focus on the case when M is finite non-abelian, non-regular, and not simple. This is the last case required to complete the proof of Theorem 12.22, and for that we require also that M is not regular. (See the discussion in Remark 12.23(ii) of the case where M is regular.)

Theorem 12.28 *Suppose that Hypothesis 12.14 holds, that $\mathfrak{G}$ is a finite $(G, 2)$-arc-transitive graph and that M is non-regular and not simple. Then G is transitive on $\mathcal{E}$ and $\mathcal{E} \in \mathsf{CD}_{2\not\sim}(G)$.*

Proof We need to show that the cases when G is intransitive on $\mathcal{E}$ or $\mathcal{E} \in$

type of $\mathcal{E}$	n	q	k_1
intransitive	4	$2^a \geqslant 4$	k
$\mathsf{CD}_{2\sim}(G)$	4	$2^a \geqslant 4$	k
$\mathsf{CD}_3(G)$	6	2	$k \geqslant 3$
$\mathsf{CD}_{1S}(G)$	4	$2^a \geqslant 4$	$k/2 \geqslant 2$

Table 12.2.

$\mathsf{CD}_{1S}(G) \cup \mathsf{CD}_{2\sim}(G) \cup \mathsf{CD}_3(G)$ are impossible. Assume to the contrary that one of these cases is valid. By Corollary 12.20, G cannot preserve a non-trivial G-invariant normal cartesian decomposition of $\mathsf{V}(\mathfrak{G})$. Hence it follows from Theorems 9.15 and 10.12 (applying Theorem 9.15 in the G-transitive cases, while applying Theorem 10.12 when $\mathcal{E}$ is not G-transitive) that $T \cong \mathsf{Sp}_n(q)$ and there exists k_1 such that

$$(D_{q+1})^{k_1} \leqslant M_\omega \leqslant (D_{q+1} \cdot 2)^{k_1} \tag{12.6}$$

where each factor in the direct powers in (12.6) is a strip covering k/k_1 factors of $M \cong T^k$. The possibilities for the type of $\mathcal{E}$, n, q, and k_1 are summarised in Table 12.2.

Let $C = (C_{q+1})^{k_1}$, $A = (D_{q+1})^{k_1}$ and $B = (D_{q+1} \cdot 2)^{k_1}$, such that $C < A \leqslant M_\omega \leqslant B$. For $i \in \underline{k_1}$, let σ_i denote the i-th coordinate projection $B \to D_{q+1} \cdot 2$.

Claim 1. If N is a normal subgroup of G_ω contained in C, then $N = (C_s)^{k_1}$ where s is a divisor of $q+1$. Conversely, each such subgroup of C is normal in G_ω.

Proof of Claim 1. We have $N\sigma_i \leqslant C_{q+1}$. The subgroup C, being characteristic in M_ω, is normalised by G_ω, and G_ω preserves the direct decomposition $C = (C_{q+1})^{k_1}$. Thus $N\sigma_i$ is independent of i. Therefore $N\sigma_i$ is a cyclic subgroup of order s, say, of C_{q+1}, and hence N is a subdirect subgroup of $(C_s)^{k_1}$. Suppose that $x_1 \in C_s$. Then there is an element of the form $(x_1, \ldots, x_{k_1})$ in N with $x_i \in C_s$. Suppose that $x \in D_{q+1}$ is an involution that conjugates each element of C_{q+1} to its inverse. Since M_ω contains $(D_{q+1})^{k_1}$, the element $(1, x, \ldots, x) \in M_\omega$. Hence

$$(x_1, x_2, \ldots, x_{k_1})^{(1,x,\ldots,x)} = (x_1, x_2^{-1}, \ldots, x_{k_1}^{-1}) \in N,$$

which implies that

$$(x_1, x_2, \dots, x_{k_1})(x_1, x_2^{-1}, \dots, x_{k_1}^{-1}) = (x_1^2, 1, \dots, 1) \in N.$$

Since s is odd, the order $|x_1|$ of x_1 is coprime to 2, and it follows that $(x_1, 1, \dots, 1) \in N$. This shows that N contains $N\sigma_1$. The same argument shows that N contains $N\sigma_i$ for all i, and hence $N = (C_s)^{k_1}$ as claimed.

A subgroup of C of the form $(C_s)^{k_1}$ is a characteristic subgroup of C and, since $C \trianglelefteq G_\omega$ (as noted above), it follows that $(C_s)^{k_1}$ is normal in G_ω. □

Suppose that $\mathfrak{G}$ is a $(G,2)$-arc-transitive graph, as assumed in the theorem. By Lemma 2.37, $Q = G_\omega^{\mathfrak{G}(\omega)}$ is 2-transitive. In particular G_ω is primitive on $\mathfrak{G}(\omega)$ and so M_ω is either transitive on $\mathfrak{G}(\omega)$ or trivial. Since M is non-regular, M must be transitive (Lemma 12.19). Since $M_\omega \trianglelefteq G_\omega$ and M_ω is solvable, we find that Q is an affine 2-transitive group. In particular the socle V of Q is $V = \mathbb{F}_p^d$ for some prime p and $Q = V \rtimes Q_\beta = \mathbb{F}_p^d \rtimes G_{\omega,\beta}^{\mathfrak{G}(\omega)}$.

A linear group $H \leqslant \mathsf{GL}_d(p)$ with $d \geqslant 2$ is said to be *monomial* if H preserves setwise a direct sum decomposition $V_1 \oplus \cdots \oplus V_d$ of $V = (\mathbb{F}_p)^d$ with subspaces of dimension 1. If H is monomial, then H cannot be transitive on $V \setminus \{0\}$, and hence $V \rtimes H$ cannot be 2-transitive on V. In particular, the linear group Q_β above is not monomial.

Let $G_\omega^{[1]}$ be the kernel of G_ω in its action on $\mathfrak{G}(\omega)$. Define $M_\omega^{[1]}$ accordingly and note that $M_\omega^{[1]}$ is normal in G_ω. Recall that $C = (C_{q+1})^\ell$ as defined above and that C is normal in G_ω.

Claim 2. $C \leqslant M_\omega^{[1]}$.

Proof of Claim 2. Let $C_1 = C \cap M_\omega^{[1]} = C \cap G_\omega^{[1]}$ (since $C \leqslant M_\omega$). Then C_1 is normal in G_ω. Hence $C_1 = (C_s)^\ell$ for some $s \mid q+1$ by Claim 1. Thus $C/C_1 = C/(C \cap G_\omega^{[1]}) \cong CG_\omega^{[1]}/G_\omega^{[1]}$ can be considered as a normal subgroup of $G_\omega^{\mathfrak{G}(\omega)}$. If $C_1 \neq C$, then this is a non-trivial normal subgroup of $G_\omega^{\mathfrak{G}(\omega)}$ and hence it contains a minimal normal subgroup of the form $V = (C_r)^\ell$ for some prime divisor r of $q+1$. On the other hand, since G_ω permutes the ℓ factors of C and $\ell \geqslant 2$, the subgroup of $\mathsf{GL}(V) = \mathsf{GL}_\ell(r)$ induced by the conjugation action of G_ω on V is a subgroup of the group of monomial matrices of $\mathsf{GL}_\ell(r)$. In particular, G_ω cannot be transitive on the set of non-trivial elements of V (see the remark above). This is a contradiction, and hence $C_1 = C$, which means that $C \leqslant M_\omega^{[1]}$. □

Let $\beta \in \mathfrak{G}(\omega)$. By Claim 2, $C \leqslant M_\beta$. Since $M_\beta^{[1]}$ is G-conjugate to $M_\omega^{[1]}$, the subgroup $M_\beta^{[1]}$ contains the unique Hall $2'$-subgroup of M_β (see (Robinson 1996, Section 9.1)). On the other hand, this Hall $2'$-subgroup is equal to C, and hence $C \leqslant M_\beta^{[1]}$, and so $C \leqslant M_\beta$ for all $\beta \in \mathfrak{G}(\omega)$. The same argument implies that if $\delta \in \mathfrak{G}(\beta)$, then $C \leqslant M_\delta^{[1]}$. Since $\mathfrak{G}$ is assumed to be connected, recursively applying the argument a finite number of times shows that C acts trivially on the vertex set Ω of $\mathfrak{G}$, which is impossible, as G is assumed to be a subgroup of $\mathsf{Sym}\,\Omega$.

This shows that the assumption that either G is intransitive on $\mathcal{E}$ or that $\mathcal{E} \in \mathsf{CD}_{1S}(G) \cup \mathsf{CD}_{2\sim}(G) \cup \mathsf{CD}_3(G)$ always leads to a contradiction. Hence we must have $\mathcal{E} \in \mathsf{CD}_{2\not\sim}(G)$. □

Now Theorem 12.22 is a consequence of Proposition 12.25, Corollary 12.20, Theorem 12.26 and Theorem 12.28.

Appendix
Factorisations of simple and characteristically simple groups

In this appendix we review some results about factorisations of simple and characteristically simple groups that are necessary for some proofs in Chapters 9 and 10. These results do not form part of the main theme of the book, and are included here for the convenience of the reader. We keep the notation used in (Baddeley & Praeger 1998) and in (Praeger & Schneider 2002) to facilitate referencing the literature.

Some basic definitions and facts regarding group factorisations were already stated in Section 8.7. Formally, a *group factorisation* is a pair $(G, \{A, B\})$ where G is a group and A, B are subgroups of G such that $AB = G$. In this situation we also say that $\{A, B\}$ is a factorisation of G, and we often write that $G = AB$ is a factorisation. A factorisation is called *non-trivial* if both A and B are proper subgroups. In this appendix we only consider non-trivial factorisations.

A.1 Some factorisations of finite simple groups

As defined in Section 8.7, a factorisation $T = AB$ of a finite simple group T is said to be a full factorisation if whenever a prime p divides $|G|$, p also divides $|A|$ and $|B|$. A set $\{A_1, \ldots, A_k\}$ of proper subgroups of a possibly infinite simple group T is said to be a strong multiple factorisation if $T = A_i(A_j \cap A_l)$ whenever $|\{i, j, l\}| = 3$. Hence a set $\{A_1, \ldots, A_k\}$ of proper subgroups of T is a strong multiple factorisation if and only if $\{A_i, A_j, A_l\}$ is an abstract cartesian factorisation of T whenever $1 \leqslant i < j < l \leqslant k$. In particular, such a set $\{A, B, C\}$ is a strong multiple factorisation if and only if it is an abstract cartesian factorisation. We will extend these concepts for characteristically simple FCR-groups in Section A.2.

The proof of the following theorem can be found in (Baddeley & Praeger 1998).

Theorem A.1 *(A) Suppose that T is a finite simple group and let $T = AB$ be a full factorisation. Then T, A and B are as in one of the rows of Table A.1. Conversely, each row of Table A.1 yields a full factorisation of a simple group.*

	T	A	B
1	A_6	A_5	A_5
2	M_{12}	M_{11}	M_{11}, $\mathsf{PSL}_2(11)$
3	$\mathsf{P}\Omega_8^+(q)$, $q \geqslant 3$	$\Omega_7(q)$	$\Omega_7(q)$
4	$\mathsf{P}\Omega_8^+(2)$	$\mathsf{Sp}_6(2)$	A_7, A_8, S_7, S_8 $\mathsf{Sp}_6(2)$, $C_2^6 \rtimes \mathsf{A}_7$, $C_2^6 \rtimes \mathsf{A}_8$
		A_9	A_8, S_8, $\mathsf{Sp}_6(2)$, $C_2^6 \rtimes \mathsf{A}_7$, $C_2^6 \rtimes \mathsf{A}_8$
5	$\mathsf{Sp}_4(q)$, $q \geqslant 4$ even	$\mathsf{Sp}_2(q^2) \cdot 2$	$\mathsf{Sp}_2(q^2) \cdot 2$, $\mathsf{Sp}_2(q^2)$

Table A.1. *Full factorisations $\{A, B\}$ of finite simple groups T*

(B) If $\{A_1, \ldots, A_k\}$ is a strong multiple factorisation of a finite simple group T, then $k \leqslant 3$. Further if $\{A, B, C\}$ is a strong multiple factorisation of T, then T, A, B and C are as in one of the rows of Table A.2. Conversely, each row of Table A.2 yields a strong multiple factorisation of a simple group.

A.2 Some factorisations of finite characteristically simple groups

Recall that full factorisations were defined for finite simple groups in Section 8.7. Let $k \geqslant 2$ and let $M = T_1 \times \cdots \times T_k$ be a finite, non-abelian, characteristically simple group where $T_1, \ldots, T_k$ are pairwise isomorphic, simple normal subgroups. For each i, let $\sigma_i \colon M \to T_i$ denote the i-th projection map. Then a factorisation $M = K_1 K_2$ is said to be a *full factorisation* if, for each $i \in \{1, \ldots, k\}$,

(i) the subgroups $K_1\sigma_i$, $K_2\sigma_i$ are proper subgroups of T_i;

	T	A_1	A_2	A_3
1	$\mathsf{Sp}_{4a}(2)$, $a \geqslant 2$	$\mathsf{Sp}_{2a}(4) \cdot 2$	$\mathsf{O}^{-}_{4a}(2)$	$\mathsf{O}^{+}_{4a}(2)$
2	$\mathsf{P}\Omega^{+}_{8}(3)$	$\Omega_7(3)$	$C_3^6 \rtimes \mathsf{PSL}_4(3)$	$\mathsf{P}\Omega^{+}_{8}(2)$
3	$\mathsf{Sp}_6(2)$	$\mathsf{G}_2(2)$	$\mathsf{O}^{-}_{6}(2)$	$\mathsf{O}^{+}_{6}(2)$
		$\mathsf{G}_2(2)'$	$\mathsf{O}^{-}_{6}(2)$	$\mathsf{O}^{+}_{6}(2)$
		$\mathsf{G}_2(2)$	$\mathsf{O}^{-}_{6}(2)'$	$\mathsf{O}^{+}_{6}(2)$
		$\mathsf{G}_2(2)$	$\mathsf{O}^{-}_{6}(2)$	$\mathsf{O}^{+}_{6}(2)'$

Table A.2. *Strong multiple factorisations $\{A_1, A_2, A_3\}$ of finite simple groups T*

(ii) the factorisation $T_i = (K_1\sigma_i)(K_2\sigma_i)$ is a full factorisation of the finite simple group T_i (as defined after Lemma 8.13; see also Section A.1).

Full factorisations of finite characteristically simple groups were classified in (Praeger & Schneider 2002). The following result is a short summary of what we need to know about such factorisations to prove the results in this book.

Theorem A.2 *Suppose that $k \geqslant 2$ and $T_1, \ldots, T_k$ are pairwise isomorphic, finite, non-abelian simple groups, and set $M = T_1 \times \cdots \times T_k$. If $M = K_1K_2$ is a full factorisation then*

$$(K_j\sigma_1)' \times \cdots \times (K_j\sigma_k)' \leqslant K_j \quad \text{for} \quad j \in \{1,\ 2\}.$$

Further, for each $i \in \{1, \ldots, k\}$, the pair $(T_i, \{K_1\sigma_i, K_2\sigma_i\})$ occurs as $(T, \{A, B\})$ in one of the lines of Table A.1.

The following proposition describes the normalisers of the factors in a full factorisation of a characteristically simple group.

Proposition A.3 *Suppose, for $k \geqslant 2$, that $M = T_1 \times \cdots \times T_k \cong T^k$ is a finite characteristically simple group and $(M, \{K_1, K_2\})$ is a full factorisation such that, for all i, the pair $(T_i, \{K_1\sigma_i, K_2\sigma_i\})$ is as $(T, \{A, B\})$ in one of the rows of Table A.1.*

(i) *If T is as in one of rows 1–3 of Table A.1 then K_1, K_2, and $K_1 \cap K_2$ are self-normalising in M.*

(ii) *If row 4 of Table A.1 is valid then, for $j = 1,\ 2$, we have $\mathsf{N}_M(K_j) = \prod_i K_j\sigma_i$ and $\mathsf{N}_M(K_1 \cap K_2) = \mathsf{N}_M(K_1) \cap \mathsf{N}_M(K_2)$.*

Strong multiple factorisations for simple groups were defined in Section 8.7. Now we extend this concept for characteristically simple FCR-groups. For $k \geqslant 2$, let $M = T_1 \times \cdots \times T_k$ be a possibly infinite, non-abelian, characteristically simple FCR-group where the T_i are pairwise isomorphic non-abelian simple groups. As above, let $\sigma_i \colon M \to T_i$ denote the i-th coordinate projection. For subgroups $K_1, \ldots, K_\ell$ of M with $\ell \geqslant 3$, the pair $(M, \{K_1, \ldots, K_\ell\})$ is said to be a *strong multiple factorisation* if, for all $i \in \{1, \ldots, k\}$,

(i) $K_1\sigma_i, \ldots, K_\ell\sigma_i$ are proper subgroups of T_i; and
(ii) the set $\{K_1\sigma_i, \ldots, K_\ell\sigma_i\}$ is a strong multiple factorisation of the simple group T_i.

The following theorem, combining (Baddeley & Praeger 1998, Table V) and (Praeger & Schneider 2002, Theorem 1.7, Corollary 1.8), gives a characterisation of strong multiple factorisations of characteristically simple groups.

Theorem A.4 *A strong multiple factorisation of a finite characteristically simple group contains exactly three subgroups. If M is a non-abelian, characteristically simple group with simple normal subgroups $T_1, \ldots, T_k$, and $(M, \{K_1, K_2, K_3\})$ is a strong multiple factorisation, then $(K_i\sigma_1)' \times \cdots \times (K_i\sigma_k)' \leqslant K_i$ for $i = 1,\ 2,\ 3$, and, for $i = 1, \ldots, k$, the pair $(T_i, \{K_1\sigma_i, K_2\sigma_i, K_3\sigma_i\})$ occurs as $(T, \{A, B, C\})$ in one of the lines of Table A.2. Further, if one of the lines 1–2 of Table A.2 is valid then $K_i\sigma_1 \times \cdots \times K_i\sigma_k = K_i$ for $i = 1,\ 2,\ 3$.*

The concept of a full strip factorisation is defined for the purposes of this theory. Let $M = T_1 \times \cdots \times T_k$ be a non-abelian, characteristically simple FCR-group where the T_i are pairwise isomorphic simple groups and let $\sigma_i \colon M \to T_i$ be the i-th coordinate projection. Let D and K be proper subgroups of M. The triple (M, D, K) is said to be a *full strip factorisation* if

(i) $M = DK$;
(ii) D is a direct product of pairwise disjoint, non-trivial full strips;
(iii) for all $i,\ j \in \{1, \ldots, k\}$, $K\sigma_i$ is a proper subgroup of T_i and $K\sigma_i \cong K\sigma_j$.

Recall that if X is a strip involved in $T_1 \times \cdots \times T_k$, then the length of X is $|\{i \mid X\sigma_i \neq 1\}|$. The following lemma shows that in a full strip factorisation each full strip has length 2.

Lemma A.5 *If (M, D, K) is a full strip factorisation of a finite, characteristically simple group M, then each non-trivial, full strip involved in D has length 2.*

Proof Suppose without loss of generality that X is a non-trivial full strip involved in D, covering $T_1, \ldots, T_s$ for some $s \geqslant 2$. We let $I = \{T_1, \ldots, T_s\}$. Then $D\sigma_I = X$ and the factorisation $X(K\sigma_I) = T_1 \times \cdots \times T_s$ holds. Then Lemma 8.16 implies that $s \leqslant 3$, and if $s = 3$ then the simple direct factor T of M admits a strong multiple factorisation involving three subgroups isomorphic to the subgroups $K\sigma_i$, for $i = 1, 2, 3$. On the other hand, Table A.2 shows that finite simple groups do not admit strong multiple factorisations with isomorphic subgroups. This is a contradiction, and hence $s = 2$. □

The next result, which appeared as (Praeger & Schneider 2002, Theorem 1.5), provides a more detailed characterisation of full strip factorisations of characteristically simple groups.

Theorem A.6 *Let $M = T_1 \times \cdots \times T_{2k}$ be a finite characteristically simple group, where the T_i are non-abelian, simple groups, let $\varphi_i \colon T_i \to T_{i+k}$ be an isomorphism for $i = 1, \ldots, k$, and set*

$$D = \{(t_1, \ldots, t_k, t_1\varphi_1, \ldots, t_k\varphi_k \mid t_1 \in T_1, \ldots, t_k \in T_k\}.$$

If (M, D, K) is a full strip factorisation, then $(T_i, \{K\sigma_i, K\sigma_{i+k}\varphi_i^{-1}\})$ is a factorisation of T_i with isomorphic subgroups for all $i \in \{1, \ldots, k\}$, and $\prod_{i=1}^{2k}(K\sigma_i)' \leqslant K$. In particular, one of the lines of Table 8.2 holds for T_i, $K\sigma_i$, $K\sigma_{i+k}\varphi_i^{-1}$.

The last result of this appendix appeared as (Baddeley *et al.* 2008, Proposition 5.4).

Proposition A.7 *Let $M = T_1 \times \cdots \times T_{2k} = T^{2k}$, D and K be as in Theorem A.6, and suppose that $DK = M$.*

(i) *If T is as in one of the rows 1–3 of Table 8.2 then K and $K \cap D$ are self-normalising in M.*
(ii) *If T is as in row 4 of Table 8.2 then $\mathsf{N}_M(K) = \prod_i K\sigma_i$ and $\mathsf{N}_M(K \cap D) = D \cap \mathsf{N}_M(K)$.*

References

ARAÚJO, JOÃO, CAMERON, PETER J., & STEINBERG, BENJAMIN. Between primitive and 2-transitive: Synchronization and its friends. *EMS Surv. Math. Sci.*, 4(2):101–184 (2017). ISSN 2308-2151. doi:10.4171/EMSS/4-2-1.
URL: *http://www.ems-ph.org/doi/10.4171/EMSS/4-2-1* 41

ASCHBACHER, M. & SCOTT, L. Maximal subgroups of finite groups. *J. Algebra*, 92(1):44–80 (1985). ISSN 0021-8693. doi:10.1016/0021-8693(85)90145-0.
URL: *http://dx.doi.org/10.1016/0021-8693(85)90145-0* 7, 142, 146, 152, 166

ASCHBACHER, MICHAEL. Overgroups of primitive groups. *J. Aust. Math. Soc.*, 87(1):37–82 (2009a). ISSN 1446-7887. doi:10.1017/S1446788708000785.
URL: *http://dx.doi.org/10.1017/S1446788708000785* 8, 118, 277

ASCHBACHER, MICHAEL. Overgroups of primitive groups. II. *J. Algebra*, 322(5):1586–1626 (2009b). ISSN 0021-8693. doi:10.1016/j.jalgebra.2009.04.044.
URL: *http://dx.doi.org/10.1016/j.jalgebra.2009.04.044* 8, 118, 277

ASCHBACHER, MICHAEL & SHARESHIAN, JOHN. Restrictions on the structure of subgroup lattices of finite alternating and symmetric groups. *J. Algebra*, 322(7):2449–2463 (2009). ISSN 0021-8693. doi:10.1016/j.jalgebra.2009.05.042.
URL: *http://dx.doi.org/10.1016/j.jalgebra.2009.05.042* 118

BADDELEY, R. W. Primitive permutation groups with a regular nonabelian normal subgroup. *Proc. London Math. Soc. (3)*, 67(3):547–595 (1993a). ISSN 0024-6115. doi:10.1112/plms/s3-67.3.547.
URL: *http://dx.doi.org/10.1112/plms/s3-67.3.547* 132, 142

BADDELEY, ROBERT W. Two-arc transitive graphs and twisted wreath products. *J. Algebraic Combin.*, 2(3):215–237 (1993b). ISSN 0925-9899. doi:10.1023/A:1022447514654.
URL: *http://dx.doi.org/10.1023/A:1022447514654* 9, 291

BADDELEY, ROBERT W. & PRAEGER, CHERYL E. On classifying all full factorisations and multiple-factorisations of the finite almost simple groups. *J. Algebra*, 204(1):129–187 (1998). ISSN 0021-8693. 7, 12, 189, 222, 301, 303

BADDELEY, ROBERT W. & PRAEGER, CHERYL E. On primitive overgroups of quasiprimitive permutation groups. *J. Algebra*, 263(2):294–344 (2003). ISSN 0021-8693. 9, 10, 13, 94, 166, 262

BADDELEY, ROBERT W., PRAEGER, CHERYL E., & SCHNEIDER, CSABA. Identifying Cartesian decompositions preserved by transitive permutation groups. *Algebra Colloq.*, 11(1):1–10 (2004a). ISSN 1005-3867. 6

BADDELEY, ROBERT W., PRAEGER, CHERYL E., & SCHNEIDER, CSABA. Transitive simple subgroups of wreath products in product action. *J. Aust. Math. Soc.*, 77(1):55–72 (2004b). ISSN 1446-7887. 6, 191, 294

BADDELEY, ROBERT W., PRAEGER, CHERYL E., & SCHNEIDER, CSABA. Innately transitive subgroups of wreath products in product action. *Trans. Amer. Math. Soc.*, 358(4):1619–1641 (electronic) (2006). ISSN 0002-9947. 6

BADDELEY, ROBERT W., PRAEGER, CHERYL E., & SCHNEIDER, CSABA. Quasiprimitive groups and blow-up decompositions. *J. Algebra*, 311(1):337–351 (2007). ISSN 0021-8693. 6

BADDELEY, ROBERT W., PRAEGER, CHERYL E., & SCHNEIDER, CSABA. Intransitive Cartesian decompositions preserved by innately transitive permutation groups. *Trans. Amer. Math. Soc.*, 360(2):743–764 (electronic) (2008). ISSN 0002-9947. 6, 191, 304

BAMBERG, JOHN. *Innately transitive groups*. Ph.D. thesis, The University of Western Australia (2003). 9, 151, 171

BAMBERG, JOHN & PRAEGER, CHERYL E. Finite permutation groups with a transitive minimal normal subgroup. *Proc. London Math. Soc. (3)*, 89(1):71–103 (2004). ISSN 0024-6115. doi:10.1112/S0024611503014631.
URL: *http://dx.doi.org/10.1112/S0024611503014631* 9, 151, 153, 171, 253

BAUMEISTER, BARBARA. Factorizations of primitive permutation groups. *J. Algebra*, 194(2):631–653 (1997). ISSN 0021-8693. doi:10.1006/jabr.1997.7027.
URL: *http://dx.doi.org/10.1006/jabr.1997.7027* 10, 13, 189

BAUMEISTER, BARBARA. *Factorizations of groups and permutation groups* (2004). Habilitationsschrift, Martin-Luther-Universität Halle. 189

BERCOV, R. On groups without Abelian composition factors. *J. Algebra*, 5:106–109 (1967). ISSN 0021-8693. doi:10.1016/0021-8693(67)90029-4.
URL: *http://dx.doi.org/10.1016/0021-8693(67)90029-4* 138

BHATTACHARJEE, MEENAXI, MACPHERSON, DUGALD, MÖLLER, RÖGNVALDUR G., & NEUMANN, PETER M. *Notes on infinite permutation groups*, volume 12 of *Texts and Readings in Mathematics*. Hindustan Book Agency, New Delhi (1997). ISBN 81-85931-13-5; 3-540-64965-4. Lecture Notes in Mathematics, 1698. 121

BIGGS, NORMAN. *Algebraic graph theory*. Cambridge Mathematical Library. Cambridge University Press, Cambridge, second edition (1993). ISBN 0-521-45897-8. 267

BODNARCHUK, YU. V. The structure of the automorphism group of a nonstandard wreath product of a group. *Ukrain. Mat. Zh.*, 36(2):143–148 (1984). ISSN 0041-6053. 114

BOSMA, WIEB, CANNON, JOHN, & PLAYOUST, CATHERINE. The Magma algebra system. I. The user language. *J. Symbolic Comput.*, 24(3-4):235–265 (1997). ISSN 0747-7171. doi:10.1006/jsco.1996.0125. Computational algebra and number theory (London, 1993).
URL: *http://dx.doi.org/10.1006/jsco.1996.0125* 282, 284, 291,

294, 295

Brewster, Ben, Passman, D. S., & Wilcox, Elizabeth. The base group of a finite wreath product (2011). Unpublished manuscript, Last accessed on 7 September 2017.
URL: *http://www.math.wisc.edu/ passman/wreath.pdf* 113, 114

Brouwer, A. E., Cohen, A. M., & Neumaier, A. *Distance-regular graphs*, volume 18 of *Ergebnisse der Mathematik und ihrer Grenzgebiete (3) [Results in Mathematics and Related Areas (3)]*. Springer-Verlag, Berlin (1989). ISBN 3-540-50619-5. doi:10.1007/978-3-642-74341-2.
URL: *http://dx.doi.org/10.1007/978-3-642-74341-2* 291

Buekenhout, Francis. On a theorem of O'Nan and Scott. *Bull. Soc. Math. Belg. Sér. B*, 40(1):1–9 (1988). ISSN 0037-9476. 7, 166

Burnside, W. *Theory of Groups of Finite Order*. Cambridge University Press (1897).
URL: *http://books.google.com.br/books?id=fmoEAQAAIAAJ* 151

Cameron, P. J. The symmetric group, 12 (Peter Cameron's blog) (2011). Last accessed 28 March 2014.
URL: *http://cameroncounts.wordpress.com/2011/04/21/the-symmetric-group-12/* 152

Cameron, Peter J. Finite permutation groups and finite simple groups. *Bull. London Math. Soc.*, 13(1):1–22 (1981). ISSN 0024-6093. doi:10.1112/blms/13.1.1.
URL: *http://dx.doi.org/10.1112/blms/13.1.1* 152

Cameron, Peter J. *Permutation groups*, volume 45 of *London Mathematical Society Student Texts*. Cambridge University Press, Cambridge (1999). ISBN 0-521-65302-9; 0-521-65378-9. 7, 37, 62, 104, 118, 166, 281, 295

Cameron, Peter J. A note on Burnside's theorem (2001). Last accessed on 7 September 2017.
URL: *http://www.maths.qmul.ac.uk/~pjc/permgps/burnside.html* 62

Camina, Alan R. & Praeger, Cheryl E. Line-transitive, point quasiprimitive automorphism groups of finite linear spaces are affine or almost simple. *Aequationes Math.*, 61(3):221–232 (2001). ISSN 0001-9054. doi:10.1007/s000100050174.
URL: *http://dx.doi.org/10.1007/s000100050174* 9

Cara, Philippe, Devillers, Alice, Giudici, Michael, & Praeger, Cheryl E. Quotients of incidence geometries. *Des. Codes Cryptogr.*, 64(1-2):105–128 (2012). ISSN 0925-1022. doi:10.1007/s10623-011-9488-y.
URL: *http://dx.doi.org/10.1007/s10623-011-9488-y* 9

Conway, J. H., Curtis, R. T., Norton, S. P., Parker, R. A., & Wilson, R. A. *Atlas of finite groups*. Oxford University Press, Eynsham (1985). ISBN 0-19-853199-0. Maximal subgroups and ordinary characters for simple groups, With computational assistance from J. G. Thackray. 293, 295

Cooperstein, Bruce & Mason, Geoffrey (editors). *The Santa Cruz Conference on Finite Groups*, volume 37 of *Proceedings of Symposia in Pure Mathematics*. American Mathematical Society, Providence, R.I. (1980). ISBN 0-8218-1440-0. Held at the University of California, Santa Cruz, Calif., June 25–July 20, 1979. 152

Cooperstein, Bruce N. Minimal degree for a permutation representation of a classical group. *Israel J. Math.*, 30(3):213–235 (1978). ISSN 0021-2172. 293

DIXON, JOHN D. & MORTIMER, BRIAN. *Permutation groups*, volume 163 of *Graduate Texts in Mathematics*. Springer-Verlag, New York (1996). ISBN 0-387-94599-7. 7, 37, 43, 59, 62, 68, 104, 111, 115, 118, 145, 166, 261, 283

FANG, XIN-GUI. *Construction and classification of some families of almost simple 2-arc transitive graphs*. Ph.D. thesis, The University of Western Australia (1995). 9

FÖRSTER, P. & KOVÁCS, L. G. Finite primitive groups with a single non-abelian regular normal subgroup (1989). Technical Report 17, Australian National University, School of Mathematical Sciences. 147

GODSIL, CHRIS & ROYLE, GORDON. *Algebraic graph theory*, volume 207 of *Graduate Texts in Mathematics*. Springer-Verlag, New York (2001). ISBN 0-387-95241-1; 0-387-95220-9. doi:10.1007/978-1-4613-0163-9.
URL: *http://dx.doi.org/10.1007/978-1-4613-0163-9* 267

GORENSTEIN, DANIEL. *Finite simple groups*. University Series in Mathematics. Plenum Publishing Corp., New York (1982). ISBN 0-306-40779-5. An introduction to their classification. 94

GROSS, FLETCHER. On the uniqueness of wreath products. *J. Algebra*, 147(1):147–175 (1992). ISSN 0021-8693. doi:10.1016/0021-8693(92)90258-N.
URL: *http://dx.doi.org/10.1016/0021-8693(92)90258-N* 113, 114

GROSS, FLETCHER & KOVÁCS, L. G. On normal subgroups which are direct products. *J. Algebra*, 90(1):133–168 (1984). ISSN 0021-8693. doi:10.1016/0021-8693(84)90203-5.
URL: *http://dx.doi.org/10.1016/0021-8693(84)90203-5* 152, 156

HAMMACK, RICHARD, IMRICH, WILFRIED, & KLAVŽAR, SANDI. *Handbook of product graphs*. Discrete Mathematics and its Applications (Boca Raton). CRC Press, Boca Raton, FL, second edition (2011). ISBN 978-1-4398-1304-1. With a foreword by Peter Winkler. 270, 287

HASSANI, AKBAR, NOCHEFRANCA, LUZ R., & PRAEGER, CHERYL E. Two-arc transitive graphs admitting a two-dimensional projective linear group. *J. Group Theory*, 2(4):335–353 (1999). ISSN 1433-5883. doi:10.1515/jgth.1999.023.
URL: *http://dx.doi.org/10.1515/jgth.1999.023* 9

HIGMAN, D. G. Intersection matrices for finite permutation groups. *J. Algebra*, 6:22–42 (1967). ISSN 0021-8693. doi:10.1016/0021-8693(67)90011-7.
URL: *http://dx.doi.org/10.1016/0021-8693(67)90011-7* 41

HUPPERT, B. *Endliche Gruppen. I*. Die Grundlehren der Mathematischen Wissenschaften, Band 134. Springer-Verlag, Berlin-New York (1967). 294

HUPPERT, BERTRAM & BLACKBURN, NORMAN. *Finite groups. II*, volume 242 of *Grundlehren der Mathematischen Wissenschaften [Fundamental Principles of Mathematical Sciences]*. Springer-Verlag, Berlin (1982). ISBN 3-540-10632-4. AMD, 44. 234

IMRICH, WILFRIED & KLAVŽAR, SANDI. *Product graphs*. Wiley-Interscience Series in Discrete Mathematics and Optimization. Wiley-Interscience, New York (2000). ISBN 0-471-37039-8. Structure and recognition, with a foreword by Peter Winkler. 270, 287

IVANOV, A. A. & PRAEGER, CHERYL E. On finite affine 2-arc transitive graphs. *European J. Combin.*, 14(5):421–444 (1993). ISSN 0195-6698. doi:10.1006/eujc.1993.1047. Algebraic combinatorics (Vladimir, 1991).
URL: *http://dx.doi.org/10.1006/eujc.1993.1047* 9, 292

JONES, GARETH A. Classification and Galois conjugacy of Hamming maps. *Ars Math. Contemp.*, 4(2):313–328 (2011). ISSN 1855-3966. 284

JONES, GARETH A. & JONES, J. MARY. *Elementary number theory.* Springer Undergraduate Mathematics Series. Springer-Verlag London, Ltd., London (1998). ISBN 3-540-76197-7. doi:10.1007/978-1-4471-0613-5.
URL: *http://dx.doi.org/10.1007/978-1-4471-0613-5* 125

JONES, GARETH A. & KWON, YOUNG SOO. Classification of nonorientable regular embeddings of Hamming graphs. *European J. Combin.*, 33(8):1800–1807 (2012). ISSN 0195-6698.
URL: *http://dx.doi.org/10.1016/j.ejc.2012.04.001* 284

JORDAN, CAMILLE. *Traité des Substitutions et des Équationes Algébriques.* Gauthier-Villars, Paris (1870). 151

KANTOR, WILLIAM M. Automorphism groups of designs. *Math. Z.*, 109:246–252 (1969). ISSN 0025-5874. 37

KLEIDMAN, PETER B. The maximal subgroups of the finite 8-dimensional orthogonal groups $P\Omega_8^+(q)$ and of their automorphism groups. *J. Algebra*, 110(1):173–242 (1987). ISSN 0021-8693. 191, 198

KNAPP, WOLFGANG. On the point stabilizer in a primitive permutation group. *Math. Z.*, 133:137–168 (1973). ISSN 0025-5874. doi:10.1007/BF01237901.
URL: *http://dx.doi.org/10.1007/BF01237901* 9

KOVÁCS, L. G. Maximal subgroups in composite finite groups. *J. Algebra*, 99(1):114–131 (1986). ISSN 0021-8693. doi:10.1016/0021-8693(86) 90058-X.
URL: *http://dx.doi.org/10.1016/0021-8693(86)90058-X* 7, 146, 152

KOVÁCS, L. G. Primitive subgroups of wreath products in product action. *Proc. London Math. Soc. (3)*, 58(2):306–322 (1989a). ISSN 0024-6115. doi:10.1112/plms/s3-58.2.306.
URL: *http://dx.doi.org/10.1112/plms/s3-58.2.306* 8, 14, 121, 245, 247, 250, 251

KOVÁCS, L. G. Wreath decompositions of finite permutation groups. *Bull. Austral. Math. Soc.*, 40(2):255–279 (1989b). ISSN 0004-9727. doi:10.1017/S0004972700004366.
URL: *http://dx.doi.org/10.1017/S0004972700004366* 2

KURZWEIL, HANS & STELLMACHER, BERND. *The theory of finite groups.* Universitext. Springer-Verlag, New York (2004). ISBN 0-387-40510-0. doi:10.1007/b97433. An introduction, translated from the 1998 German original.
URL: *http://dx.doi.org/10.1007/b97433* 94

LAFUENTE, JULIO. On restricted twisted wreath products of groups. *Arch. Math. (Basel)*, 43(3):208–209 (1984). ISSN 0003-889X. doi:10.1007/BF01247564.
URL: *http://dx.doi.org/10.1007/BF01247564* 138

LI, CAI HENG. On finite s-transitive graphs of odd order. *J. Combin. Theory Ser. B*, 81(2):307–317 (2001). ISSN 0095-8956. doi:10.1006/jctb.2000.2012.
URL: *http://dx.doi.org/10.1006/jctb.2000.2012* 9

LI, CAI HENG. Finite s-arc transitive Cayley graphs and flag-transitive projective planes. *Proc. Amer. Math. Soc.*, 133(1):31–41 (electronic) (2005). ISSN 0002-9939. doi:10.1090/S0002-9939-04-07549-5.
URL: *http://dx.doi.org/10.1090/S0002-9939-04-07549-5* 9

LI, CAI-HENG, PRAEGER, CHERYL E., & SCHNEIDER, CSABA. Inclusions of innately transitive groups into wreath products in product action with applications to 2-arc-transitive graphs. *J. Pure Appl. Algebra*, 220(7):2683–2700 (2016). ISSN 0022-4049. doi:10.1016/j.jpaa.2015.12.005.
URL: *http://dx.doi.org/10.1016/j.jpaa.2015.12.005* 291

LI, CAI HENG, PRAEGER, CHERYL E., VENKATESH, AKSHAY, & ZHOU, SANMING. Finite locally-quasiprimitive graphs. *Discrete Math.*, 246(1-3):197–218 (2002). ISSN 0012-365X. doi:10.1016/S0012-365X(01)00258-8. Formal power series and algebraic combinatorics (Barcelona, 1999).
URL: *http://dx.doi.org/10.1016/S0012-365X(01)00258-8* 9

LI, CAI HENG & SERESS, ÁKOS. Constructions of quasiprimitive two-arc transitive graphs of product action type. In *Finite geometries, groups, and computation*, pages 115–123. Walter de Gruyter, Berlin (2006). 291

LIEBECK, MARTIN W., PRAEGER, CHERYL E., & SAXL, JAN. A classification of the maximal subgroups of the finite alternating and symmetric groups. *J. Algebra*, 111(2):365–383 (1987). ISSN 0021-8693. 117, 166, 169, 197, 198, 265

LIEBECK, MARTIN W., PRAEGER, CHERYL E., & SAXL, JAN. On the O'Nan-Scott theorem for finite primitive permutation groups. *J. Austral. Math. Soc. Ser. A*, 44(3):389–396 (1988). ISSN 0263-6115. 7, 153, 166

LIEBECK, MARTIN W., PRAEGER, CHERYL E., & SAXL, JAN. The maximal factorizations of the finite simple groups and their automorphism groups. *Mem. Amer. Math. Soc.*, 86(432):iv+151 (1990). ISSN 0065-9266. 7, 12, 188, 213, 233

MACPHERSON, DUGALD & PRAEGER, CHERYL E. Infinitary versions of the O'Nan-Scott theorem. *Proc. London Math. Soc. (3)*, 68(3):518–540 (1994). ISSN 0024-6115. doi:10.1112/plms/s3-68.3.518.
URL: *http://dx.doi.org/10.1112/plms/s3-68.3.518* 118, 151

MANNING, W. A. The order of primitive groups. III. *Trans. Amer. Math. Soc.*, 19(2):127–142 (1918). ISSN 0002-9947. doi:10.2307/1988916.
URL: *http://dx.doi.org/10.2307/1988916* 31

MELDRUM, J. D. P. *Wreath products of groups and semigroups*, volume 74 of *Pitman Monographs and Surveys in Pure and Applied Mathematics*. Longman, Harlow (1995). ISBN 0-582-02693-8. 114

MORRIS, JOY, PRAEGER, CHERYL E., & SPIGA, PABLO. Strongly regular edge-transitive graphs. *Ars Math. Contemp.*, 2(2):137–155 (2009). ISSN 1855-3966. 9

NEUMANN, B. H. Twisted wreath products of groups. *Arch. Math. (Basel)*, 14:1–6 (1963). ISSN 0003-889X. 132

NEUMANN, PETER M. On the structure of standard wreath products of groups. *Math. Z.*, 84:343–373 (1964). ISSN 0025-5874. doi:10.1007/BF01109904. 113, 114

NEUMANN, PETER M. The context of Burnside's contributions to group theory. In PETER M. NEUMANN, A. J. S. MANN & THOMPSON, JULIA C. (editors), *The collected papers of William Burnside*, volume 1, pages 15–38. Oxford University Press (2004). 151

NEUMANN, PETER M. The concept of primitivity in group theory and the second memoir of Galois. *Arch. Hist. Exact Sci.*, 60(4):379–429 (2006). ISSN 0003-9519. doi:10.1007/s00407-006-0111-y.
URL: *http://dx.doi.org/10.1007/s00407-006-0111-y* 9

NEUMANN, PETER M., PRAEGER, CHERYL E., & SMITH, SIMON M. Some infinite permutation groups and related finite linear groups. *J. Aust. Math. Soc.*, 102(1):136–149 (2017). ISSN 1446-7887. doi:10.1017/S1446788716000343.
URL: *http://dx.doi.org/10.1017/S1446788716000343* 60, 151

OBRAZTSOV, VIATCHESLAV N. Embedding into groups with well-described lattices of subgroups. *Bull. Austral. Math. Soc.*, 54(2):221–240 (1996). ISSN 0004-9727. doi:10.1017/S0004972700017688.
URL: *http://dx.doi.org/10.1017/S0004972700017688* 69

PÁLFY, P. P. & SAXL, J. Congruence lattices of finite algebras and factorizations of groups. *Comm. Algebra*, 18(9):2783–2790 (1990). ISSN 0092-7872. 13, 189

PÁLFY, PÉTER PÁL & PUDLÁK, PAVEL. Congruence lattices of finite algebras and intervals in subgroup lattices of finite groups. *Algebra Universalis*, 11(1):22–27 (1980). ISSN 0002-5240. doi:10.1007/BF02483080.
URL: *http://dx.doi.org/10.1007/BF02483080* 8

PAYNE, STANLEY E. & THAS, JOSEPH A. *Finite generalized quadrangles*. EMS Series of Lectures in Mathematics. European Mathematical Society (EMS), Zürich, second edition (2009). ISBN 978-3-03719-066-1. doi:10.4171/066.
URL: *http://dx.doi.org/10.4171/066* 269

PRAEGER, CHERYL E. The inclusion problem for finite primitive permutation groups. *Proc. London Math. Soc. (3)*, 60(1):68–88 (1990). ISSN 0024-6115. doi:10.1112/plms/s3-60.1.68.
URL: *http://dx.doi.org/10.1112/plms/s3-60.1.68* 8, 10, 245

PRAEGER, CHERYL E. An O'Nan-Scott theorem for finite quasiprimitive permutation groups and an application to 2-arc transitive graphs. *J. London Math. Soc. (2)*, 47(2):227–239 (1993). ISSN 0024-6107. doi:10.1112/jlms/s2-47.2.227.
URL: *http://dx.doi.org/10.1112/jlms/s2-47.2.227* 9, 10, 118, 151, 289

PRAEGER, CHERYL E., PYBER, LASZLÓ, SPIGA, PABLO, & SZABÓ, ENDRE. Graphs with automorphism groups admitting composition factors of bounded rank. *Proc. Amer. Math. Soc.*, 140(7):2307–2318 (2012). ISSN 0002-9939. doi:10.1090/S0002-9939-2011-11100-6.
URL: *http://dx.doi.org/10.1090/S0002-9939-2011-11100-6* 9

PRAEGER, CHERYL E., SAXL, JAN, & YOKOYAMA, KAZUHIRO. Distance transitive graphs and finite simple groups. *Proc. London Math. Soc. (3)*, 55(1):1–21 (1987). ISSN 0024-6115. doi:10.1112/plms/s3-55.1.1.
URL: *http://dx.doi.org/10.1112/plms/s3-55.1.1* 10

PRAEGER, CHERYL E. & SCHNEIDER, CSABA. Factorisations of characteristically simple groups. *J. Algebra*, 255(1):198–220 (2002). ISSN 0021-8693. 302, 303, 304

PRAEGER, CHERYL E. & SCHNEIDER, CSABA. Three types of inclusions of innately transitive permutation groups into wreath products in product action. *Israel J. Math.*, 158:65–104 (2007). ISSN 0021-2172. 6

PRAEGER, CHERYL E. & SCHNEIDER, CSABA. Embedding permutation groups into wreath products in product action. *J. Aust. Math. Soc.*, 92(1):127–136 (2012). ISSN 1446-7887. doi:10.1017/S1446788712000110.
URL: *http://dx.doi.org/10.1017/S1446788712000110* 6, 104, 118

PRAEGER, CHERYL E. & SCHNEIDER, CSABA. The contribution of L. G. Kovács

to the theory of permutation groups. *J. Aust. Math. Soc.*, 102(1):20–33 (2017a). ISSN 1446-7887. doi:10.1017/S1446788715000385.
URL: *http://dx.doi.org/10.1017/S1446788715000385* 146

PRAEGER, CHERYL E. & SCHNEIDER, CSABA. Group factorisations, uniform automorphisms, and permutation groups of simple diagonal type (2017b). To appear in *Israel J. Math.*
URL: *https://arxiv.org/abs/1611.01103* 94

ROBINSON, DEREK J. S. *A course in the theory of groups*, volume 80 of *Graduate Texts in Mathematics*. Springer-Verlag, New York, second edition (1996). ISBN 0-387-94461-3. 57, 58, 61, 103, 104, 148, 165, 188, 299

SABIDUSSI, GERT. Graph multiplication. *Math. Z.*, 72:446–457 (1959/1960). ISSN 0025-5874. doi:10.1007/BF01162967.
URL: *http://dx.doi.org/10.1007/BF01162967* 270

SCHARLAU, RUDOLF. Buildings. In *Handbook of incidence geometry*, pages 477–645. North-Holland, Amsterdam (1995). doi:10.1016/B978-044488355-1/50013-X.
URL: *http://dx.doi.org/10.1016/B978-044488355-1/50013-X* 277, 278

SCHMIDT, ROLAND. *Subgroup lattices of groups*, volume 14 of *de Gruyter Expositions in Mathematics*. Walter de Gruyter & Co., Berlin (1994). ISBN 3-11-011213-2. 77

SCOTT, L. L. Comments from Leonard Scott's home page on his various collaborators ($\sim$1997). Last accessed on 7 September 2017.
URL: *http://people.virginia.edu/ lls2l/collaborators.htm* 152

SCOTT, LEONARD L. Representations in characteristic p. In *The Santa Cruz Conference on Finite Groups (Univ. California, Santa Cruz, Calif., 1979)*, volume 37 of *Proc. Sympos. Pure Math.*, pages 319–331. Amer. Math. Soc., Providence, R.I. (1980). 7, 84, 152, 166

SEIFTER, NORBERT. Properties of graphs with polynomial growth. *J. Combin. Theory Ser. B*, 52(2):222–235 (1991). ISSN 0095-8956. doi:10.1016/0095-8956(91)90064-Q.
URL: *http://dx.doi.org/10.1016/0095-8956(91)90064-Q* 268

SERESS, ÁKOS. Toward the classification of s-arc transitive graphs. In *Groups St. Andrews 2005. Vol. 2*, volume 340 of *London Math. Soc. Lecture Note Ser.*, pages 401–414. Cambridge Univ. Press, Cambridge (2007). doi:10.1017/CBO9780511721205.003.
URL: *http://dx.doi.org/10.1017/CBO9780511721205.003* 268

SMITH, SIMON M. A classification of primitive permutation groups with finite stabilizers. *J. Algebra*, 432:12–21 (2015). ISSN 0021-8693. doi:10.1016/j.jalgebra.2015.01.023.
URL: *http://dx.doi.org/10.1016/j.jalgebra.2015.01.023* 60, 69, 118, 151, 165

SMITH, STEPHEN D. Applying the Classification of Finite Simple Groups: A User's Guide (2018). To be published by the American Mathematical Society. 166

SPRINGER, T. A. *Linear algebraic groups*. Modern Birkhäuser Classics. Birkhäuser Boston, Inc., Boston, MA, second edition (2009). ISBN 978-0-8176-4839-8. 94

SUZUKI, MICHIO. *Group theory. I*, volume 247 of *Grundlehren der Mathematischen Wissenschaften [Fundamental Principles of Mathematical Sciences]*. Springer-Verlag, Berlin (1982). ISBN 3-540-10915-3. Translated from the

Japanese by the author. 132

THE GAP GROUP. *GAP – Groups, Algorithms, and Programming, Version 4.6.5* (2013).
URL: *www.gap-system.org* 282, 284, 291, 294, 295

TITS, J. A local approach to buildings. In *The geometric vein*, pages 519–547. Springer, New York-Berlin (1981). 277

TUTTE, W. T. A family of cubical graphs. *Proc. Cambridge Philos. Soc.*, 43:459–474 (1947). 268

VIZING, V. G. The cartesian product of graphs. *Vyčisl. Sistemy No.*, 9:30–43 (1963). ISSN 0568-661X. 270

WEISS, RICHARD. The nonexistence of 8-transitive graphs. *Combinatorica*, 1(3):309–311 (1981). ISSN 0209-9683. doi:10.1007/BF02579337.
URL: *http://dx.doi.org/10.1007/BF02579337* 268

WILCOX, ELIZABETH. *Complete finite Frobenius groups and wreath products.* Ph.D. thesis, Binghamton University, State University of New York (2010). Last accessed on 29 March 2017.
URL: *https://tinyurl.com/y8n62x8w* 114

WILSON, ROBERT A. *The finite simple groups*, volume 251 of *Graduate Texts in Mathematics.* Springer-Verlag London, Ltd., London (2009). ISBN 978-1-84800-987-5. doi:10.1007/978-1-84800-988-2.
URL: *http://dx.doi.org/10.1007/978-1-84800-988-2* 166, 189

ZSIGMONDY, K. Zur Theorie der Potenzreste. *Monatsh. Math. Phys.*, 3(1):265–284 (1892). ISSN 0026-9255. doi:10.1007/BF01692444.
URL: *http://dx.doi.org/10.1007/BF01692444* 234

Glossary

$\mathsf{AGL}_d(p)$ affine general linear group. 154

$\mathsf{ASL}_d(p)$ affine special linear group. 154

AS primitive permutation group of almost simple type. 155

As quasiprimitive permutation group of almost simple type. 155

$\mathrm{AS}_{\mathrm{reg}}$ primitive permutation group of almost simple type with regular socle. 156

$\mathrm{As}_{\mathrm{reg}}$ quasiprimitive permutation group of almost simple type with regular socle. 156

$\mathsf{Aut}(\mathfrak{G})$ the group of automorphisms of the graph or digraph $\mathfrak{G}$. 38

$\mathsf{Aut}(G)$ the group of automorphisms of the group G. 20

$\mathsf{CD}(G)$ the set of cartesian decompositions preserved by G. 176

$\mathsf{CD}_{\mathrm{tr}}(G)$ the set of transitive cartesian decompositions preserved by G. 176

$\mathsf{CD}_1(G)$ a class of transitive cartesian decompositions preserved by G. 209

$\mathsf{CD}_{1\mathrm{S}}(G)$ a class of transitive cartesian decompositions preserved by G. 209

$\mathsf{CD}_{\mathrm{S}}(G)$ a class of transitive cartesian decompositions preserved by G. 209

$\mathsf{CD}_3(G)$ a class of transitive cartesian decompositions preserved by G. 209

$\mathsf{CD}_{2\not\sim}(G)$ a class of transitive cartesian decompositions preserved by G. 209

$\mathsf{CD}_{2\sim}(G)$ a class of transitive cartesian decompositions preserved by G. 209

$\mathcal{C}(\mathcal{E})$ the chamber system associated with the cartesian decomposition $\mathcal{E}$. 277

$\mathsf{C}_G(H)$ the centraliser of H in G. 45
CD primitive permutation group of compound diagonal type. 157
CD quasiprimitive permutation group of compound diagonal type. 157
$\mathsf{Core}_G\, H$ the core of H in G. 19

$\Delta(\alpha)$ $\{\beta \mid (\alpha, \beta) \in \Delta\}$. 35
Δ^* the paired orbital of Δ. 34
$\mathrm{diam}(\mathfrak{G})$ the diameter of the graph $\mathfrak{G}$. 289

$\mathcal{E} = \{\Gamma_1, \ldots, \Gamma_\ell\}$ a cartesian decomposition with partitions Γ_i. 4
$\mathsf{E}(\mathfrak{G})$ the edge set of the graph $\mathfrak{G}$. 37

$\mathcal{F}_i(\mathcal{E}, M, \omega)$ $\{K_j\sigma_i \mid j = 1, \ldots, \ell,\ K_j\sigma_i \neq T_i\}$. 204
$\mathsf{Fix}_\Omega(G)$ the fixed points of G in Ω. 29
$\mathbb{F}_p$ the field with p elements. 58
$\mathsf{Func}(\Delta, G)$ the set of functions from Δ to G. 102

$\mathfrak{G}_1 \Box \mathfrak{G}_2$ the cartesian product of the graphs $\mathfrak{G}_1$ and $\mathfrak{G}_2$. 269
$\mathfrak{G}(\alpha)$ the neighbourhood of α in the graph or digraph $\mathfrak{G}$. 40
γ_δ the γ-part in a cartesian decomposition. 115
$[G\colon H]$ the set of right cosets of H in G. 22
$\mathsf{GL}_k(p)$ the general linear group acting on $\mathbb{F}_p^k$. 58
$\mathsf{GL}(M)$ the general linear group acting on M. 58
G_ω the point stabiliser in G of ω. 18
G' the commutator subgroup of G. 57
$\mathfrak{Gr}(\mathcal{E})$ the graph of the cartesian decomposition $\mathcal{E}$. 269
$\mathfrak{G}_1 \times \mathfrak{G}_2$ the direct product of the graphs $\mathfrak{G}_1$ and $\mathfrak{G}_2$. 287
$(\mathfrak{G}_1)^{\times l}$ the l-fold direct power of the graph $\mathfrak{G}_1$. 287
$G \rtimes H$ the semidirect product of G and H. 104
$G \rtimes_\varphi H$ the semidirect product of G and H with respect to $\varphi\colon H \to \mathsf{Aut}(G)$. 104
$G_{(X)}$ the pointwise stabiliser of a set X. 18
G_X the setwise stabiliser of a set X. 18
G^X the permutation group induced on X by G_X. 18
$G \wr H$ the wreath product of G and H. 105

$H(\ell, q)$ the Hamming graph on the alphabet $\underline{q}$. 270
$H(\ell, X)$ the Hamming graph on the alphabet X. 270
$\mathsf{Hol}\, G$ the holomorph of G. 50
HA primitive permutation group of affine type. 154
HC primitive permutation group of holomorph of compound type. 156

HS primitive permutation group of holomorph of simple type. 156

id_X the identity map on X. 20
$\mathrm{Inn}(G)$ the group of inner automorphisms of the group G. 20

K_X the complete graph on the set X. 268
$K_{n,m}$ the complete bipartite graph of degree (n, m). 268
K_q the complete graph on the set $\underline{q}$. 270
$\mathcal{K}_\omega(\mathcal{E})$ the cartesian factorisation with respect to ω that corresponds to the cartesian decomposition $\mathcal{E}$. 181

λ_H the left coset action of $\mathsf{N}_G(H)$ on $[G\colon H]$. 46
λ_1 the left regular action of G on G. 47

$\underline{n}$ the set $\{1, \ldots, n\}$. 17
$\mathsf{N}_G(H)$ the normaliser of H in G. 29

ωG the G-orbit containing ω. 18
ωg the image of ω under g. 17
Ω^2 the set of ordered pairs of Ω. 33
$\Omega^{\{2\}}$ the set of unordered pairs of Ω. 33
$\Omega^{(2)}$ the set of ordered pairs of distinct elements of Ω. 33

PA primitive permutation group of product action type. 162
PA quasiprimitive permutation group of product action type. 162
$\mathsf{Prod}_i\, G_i$ the unrestricted direct product of the G_i. 57
$\prod_i G_i$ the restricted direct product of the G_i. 57
$\mathcal{P}^k(\Omega)$ the k-th power set of Ω. 18

ϱ the right regular action of a group G. 22
ϱ_H the right coset action of G on $[G\colon H]$ where G is the natural overgroup of H. 22
ϱ_H^G the right coset action of G on $[G\colon H]$. 22

σ_δ coordinate projection map. 102
σ_I coordinate projection. 72
σ_i coordinate projection. 72
SD primitive permutation group of simple diagonal type. 157
SD quasiprimitive permutation group of simple diagonal type. 156
$\mathsf{Soc}(G)$ the socle of G. 57
$\mathrm{Supp}_{\mathcal{G}}\, H$ the support of a strip. 81

$\mathsf{Sym}\,\Omega$ the symmetric group on a set Ω. 17

$T\,\mathrm{twr}_{\varphi}\,P$ the twisted wreath product of T and P with respect to $\varphi\colon Q \to \mathsf{Aut}(T)$ where $Q \leqslant P$. 136

TW primitive permutation group of twisted wreath type. 156

Tw quasiprimitive permutation group of twisted wreath type. 156

$\mathsf{V}(\mathfrak{G})$ the vertex set of the graph or digraph $\mathfrak{G}$. 37

X^{Γ_δ} the δ-component of X. 121

ξ_N^H the conjugation action of H on N. 51

X^Y the set of Y-conjugates of X. 187

$[x, y]$ the commutator $x^{-1}y^{-1}xy$ of x and y. 57

$\mathsf{Z}(G)$ the centre of a group G. 46

Index

Printed in the United States
by Baker & Taylor Publisher Services